W9-AHH-232

The Future Role of Pesticides in US Agriculture

WITHDRAWN
From Toronto Public Library

Committee on the Future Role of Pesticides in US Agriculture

Board on Agriculture and Natural Resources
and
Board on Environmental Studies and Toxicology

Commission on Life Sciences

National Research Council

NATIONAL ACADEMY PRESS
Washington, D.C.

NATIONAL ACADEMY PRESS 2101 Constitution Avenue, N.W. Washington, D.C. 20418

NOTICE: The project that is the subject of this report was approved by the Governing Board of the National Research Council, whose members are drawn from the councils of the National Academy of Sciences, the National Academy of Engineering, and the Institute of Medicine. The members of the committee responsible for the report were chosen for their special competences and with regard for appropriate balance.

This report has been prepared with funds provided by the US Department of Agriculture under grant number, 59-0700-5-119, Environmental Protection Agency under contract number 6W-1187-NANX, and the National Research Council. Any opinions, findings, conclusions, or recommendations expressed in this publication are those of the author(s) and do not necessarily reflect the views of the organizations or agencies that provided support for the project.

Library of Congress Cataloging-in-Publication Data

The future role of pesticides in US agriculture / Committee on the Future Role of Pesticides in US Agriculture, Board on Agriculture and Natural Resources and Board on Environmental Studies and Toxicology, Commission on Life Sciences.
p. cm.
Includes bibliographical references (p.).
ISBN 0-309-06526-7 (case bound)
1. Pesticides--United States. I. National Research Council (U.S.). Committee on the Future Role of Pesticides in US Agriculture. II. National Research Council (U.S.). Board on Environmental Studies and Toxicology. III. Title.
SB950.2.A1 F88 2000
632'.95'0973--dc21

00-011245

Additional copies of this report are available from National Academy Press, 2101 Constitution Avenue, N.W., Lockbox 285, Washington, D.C. 20055; (800) 624-6242 or (202) 334-3313 (in the Washington metropolitan area); Internet, http://www.nap.edu

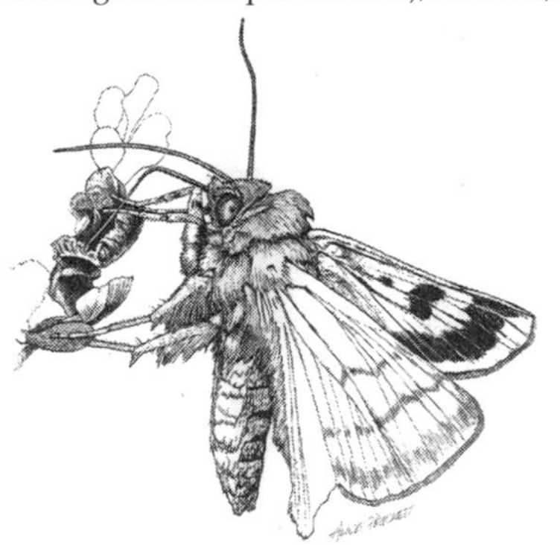

The illustration to the left, a corn earworm moth, *Heliothis zea*, was created by Alice Prickett of Urbana, Illinois, and was adapted for the image on the cover of this book.

Printed in the United States of America

Copyright 2000 by the National Academy of Sciences. All rights reserved.

THE NATIONAL ACADEMIES

National Academy of Sciences
National Academy of Engineering
Institute of Medicine
National Research Council

The **National Academy of Sciences** is a private, nonprofit, self-perpetuating society of distinguished scholars engaged in scientific and engineering research, dedicated to the furtherance of science and technology and to their use for the general welfare. Upon the authority of the charter granted to it by the Congress in 1863, the Academy has a mandate that requires it to advise the federal government on scientific and technical matters. Dr. Bruce M. Alberts is president of the National Academy of Sciences.

The **National Academy of Engineering** was established in 1964, under the charter of the National Academy of Sciences, as a parallel organization of outstanding engineers. It is autonomous in its administration and in the selection of its members, sharing with the National Academy of Sciences the responsibility for advising the federal government. The National Academy of Engineering also sponsors engineering programs aimed at meeting national needs, encourages education and research, and recognizes the superior achievements of engineers. Dr. William A. Wulf is president of the National Academy of Engineering.

The **Institute of Medicine** was established in 1970 by the National Academy of Sciences to secure the services of eminent members of appropriate professions in the examination of policy matters pertaining to the health of the public. The Institute acts under the responsibility given to the National Academy of Sciences by its congressional charter to be an adviser to the federal government and, upon its own initiative, to identify issues of medical care, research, and education. Dr. Kenneth I. Shine is president of the Institute of Medicine.

The **National Research Council** was organized by the National Academy of Sciences in 1916 to associate the broad community of science and technology with the Academy's purposes of furthering knowledge and advising the federal government. Functioning in accordance with general policies determined by the Academy, the Council has become the principal operating agency of both the National Academy of Sciences and the National Academy of Engineering in providing services to the government, the public, and the scientific and engineering communities. The Council is administered jointly by both Academies and the Institute of Medicine. Dr. Bruce M. Alberts and Dr. William A. Wulf are chairman and vice chairman, respectively, of the National Research Council.

COMMITTEE ON THE FUTURE ROLE OF PESTICIDES IN US AGRICULTURE

MAY BERENBAUM, *Chair*, Department of Entomology, University of Illinois, Urbana, Illinois
MARK BRUSSEAU, Department Soil, Water, and Environmental Science, University of Arizona, Tucson, Arizona
JOSEPH DIPIETRO, College of Veterinary Medicine, University of Florida, Gainesville, Florida
ROBERT GOODMAN, Department of Plant Pathology, University of Wisconsin, Madison, Wisconsin
FRED GOULD, Department of Entomology, North Carolina State University, Raleigh, North Carolina
JEFFREY GUNSOLUS, Department of Agronomy and Plant Genetics, University of Minnesota, St. Paul, Minnesota
BRUCE HAMMOCK, Department of Entomology, University of California, Davis, California
ROLF HARTUNG, Environmental Toxicology, University of Michigan, Ann Arbor, Michigan (Retired)
PAMELA MARRONE, AgraQuest, Inc., Davis, California
BRUCE MAXWELL, Department of Plant, Soil, and Environmental Sciences, Montana State University, Bozeman, Montana
KENNETH RAFFA, Department of Entomology, University of Wisconsin, Madison, Wisconsin
JOHN RYALS, Paradigm Genetics, Inc., Cary, North Carolina
DALE SHANER, American Cyanamid, Princeton, New Jersey
*JAMES SEIBER, Center for Environmental Sciences and Engineering and Department of Environmental Resource Sciences, University of Nevada, Reno, Nevada
DAVID ZILBERMAN, Department of Agricultural and Resource Economics, University of California, Berkeley, California

Consultant

ERIK LICHTENBERG, Department of Agricultural and Resource Economics, University of Maryland, College Park, Maryland

Staff

Kim Waddell, *Project Director*
Mary Jane Letaw, *Project Director* (through September 11, 1999)
Heather Christiansen, *Research Associate*
Karen Imhof, *Project Assistant*

*Resigned December, 1998 after changing affiliation to USDA-ARS

BOARD ON AGRICULTURE AND NATURAL RESOURCES

Harley W. Moon, *Chair*, Iowa State University
David H. Baker, University of Illinois
*Sandra S. Batie Department of Agricultural Economics Michigan State University East Lansing, Michigan
May R. Berenbaum, University of Illinois
*Anthony S. Earl Quarles & Brady Law Firm, Madison, Wisconsin
*Essex E. Finney, Jr. Agricultural Research Service, US Department of Agriculture, Mitchellville, Maryland (retired)
Cornelia B. Flora, Iowa State University
Robert T. Fraley, Monsanto Company, St. Louis, Missouri
Robert B. Fridley, University of California
W. R. (Reg) Gomes, University of California
Perry R. Hagenstein, Institute for Forest Analysis, Planning, and Policy, Wayland, Massachusetts
George R. Hallberg, The Cadmus Group, Inc., Waltham, Massachusetts
*Richard R. Harwood Crop and Soil Sciences Department, Michigan State University, East Lansing, Michigan
*T. Kent Kirk *Chair*, Department of Bacteriology, University of Wisconsin, Madison, Wisconsin
Calestous Juma, Harvard University
Gilbert A. Leveille, McNeil Consumer Healthcare, Denville, New Jersey
Whitney MacMillan, Cargill, Inc., Minneapolis, Minnesota (retired)
William L. Ogren, US Department of Agriculture (retired)
Nancy J. Rachman, International Life Science Institute, Washington, D.C.
G. Edward Schuh, University of Minnesota
John W. Suttie, University of Wisconsin
Thomas N. Urban, Pioneer Hi-Bred International, Inc., Des Moines, Iowa
Robert P. Wilson, Mississippi State University
James J. Zuiches, Washington State University

Staff

Warren Muir, *Executive Director*
Myron F. Uman, *Acting Executive Director* (through May 1999)
David L. Meeker, *Director* (since March 2000)
Charlotte Kirk Baer, *Associate Director*
Shirley Thatcher, *Administrative Assistant*

*Through December 1999

BOARD ON ENVIRONMENTAL STUDIES AND TOXICOLOGY

Gordon Orians, *Chair*, University of Washington, Seattle, Washington
Donald Mattison, *Vice Chair*, March of Dimes, White Plains, New York
David Allen, University of Texas, Austin, Texas
Ingrid C. Burke, Colorado State University, Fort Collins, Colorado
William L. Chameides, Georgia Institute of Technology, Atlanta, Georgia
John Doull, University of Kansas Medical Center, Kansas City, Kansas
Christopher B. Field, Carnegie Institute of Washington, Stanford, California
John Gerhart, University of California, Berkeley, California
J. Paul Gilman, Celera Genomics, Rockville, Maryland
Bruce D. Hammock, University of California, Davis, California
Mark Harwell, University of Miami, Miami, Florida
Rogene Henderson, Lovelace Respiratory Research Institute, Albuquerque, New Mexico
Carol Henry, Chemical Manufacturers Association, Arlington, Virginia
Barbara Hulka, University of North Carolina, Chapel Hill, North Carolina
James F. Kitchell, University of Wisconsin, Madison, Wisconsin
Daniel Krewski, University of Ottawa, Ottawa, Ontario
James A. MacMahon, Utah State University, Logan, Utah
Mario J. Molina, Massachusetts Institute of Technology, Cambridge, Massachusetts
Charles O'Melia, Johns Hopkins University, Baltimore, Maryland
Willem F. Passchier, Health Council of the Netherlands
Kirk Smith, University of California, Berkeley, California
Margaret Strand, Oppenheimer Wolff Donnelly & Bayh, LLP, Washington, DC
Terry F. Yosie, Chemical Manufacturers Association, Arlington, Virginia

Senior Staff

James J. Reisa, *Director*
David J. Policansky, *Associate Director and Senior Program Director for Applied Ecology*
Carol A. Maczka, *Senior Program Director for Toxicology and Risk Assessment*
Raymond A. Wassel, *Senior Program Director for Environmental Sciences and Engineering*
Kulbir Bakshi, *Program Director for the Committee on Toxicology*
Lee R. Paulson, *Program Director for Resource Management*
Roberta M. Wedge, *Program Director for Risk Analysis*

COMMISSION ON LIFE SCIENCES

Michael T. Clegg, *Chair*, University of California, Riverside, California
Paul Berg, *Vice Chair*, Stanford University, Stanford, California
Frederick R. Anderson, Cadwalader, Wickersham and Taft, Washington, D.C.
Joanna Burger, Rutgers University, Piscataway, New Jersey
James E. Cleaver, University of California, San Francisco, California
David S. Eisenberg, University of California, Los Angeles, California
John L. Emmerson, Fishers, Indiana
Neal L. First, University of Wisconsin, Madison, Wisconsin
David J. Galas, Keck Graduate Institute of Applied Life Science, Claremont, California
David V. Goeddel, Tularik, Inc., South San Francisco, California
Arturo Gomez-Pompa, University of California, Riverside, California
Corey S. Goodman, University of California, Berkeley, California
Jon W. Gordon, Mount Sinai School of Medicine, New York, New York
David G. Hoel, Medical University of South Carolina, Charleston, South Carolina
Barbara S. Hulka, University of North Carolina, Chapel Hill, North Carolina
Cynthia J. Kenyon, University of California, San Francisco, California
Bruce R. Levin, Emory University, Atlanta, Georgia
David M. Livingston, Dana-Farber Cancer Institute, Boston, Massachusetts
Donald R. Mattison, March of Dimes, White Plains, New York
Elliot M. Meyerowitz, California Institute of Technology, Pasadena, California
Robert T. Paine, University of Washington, Seattle, Washington
Ronald R. Sederoff, North Carolina State University, Raleigh, North Carolina
Robert R. Sokal, State University of New York, Stony Brook, New York
Charles F. Stevens, The Salk Institute for Biological Studies, La Jolla, California
Shirley M. Tilghman, Princeton University, Princeton, New Jersey
Raymond L. White, University of Utah, Salt Lake City, Utah

Staff

Warren R. Muir, *Executive Director*
Jacqueline K. Prince, *Financial Officer*
Barbara B. Smith, *Administrative Associate*
Laura T. Holliday, *Senior Program Assistant*

Preface

Predicting the future would appear to be an inherently unscientific process; there are few opportunities for constructing falsifiable hypotheses and testing them, short of waiting until the future actually becomes the present. Nonetheless, scientists are often called on to engage in the exercise; the ability to construct and project different futures can allow scientists to guide society in achieving a particular fate that it deems desirable. Relevant information gathered about the past and the present can be used as a basis for choosing intelligently among possible futures.

In January 1998, the National Research Council convened a committee of experts representing a broad range of disciplines to make an effort to predict the future of pesticide use in American agriculture. The effort was far from the first made by the National Research Council to evaluate aspects of pesticide use; the subject has been the focus of National Research Council attention for close to 5 decades, dating back to the earliest days of widespread adoption of synthetic organic pesticides (Appendix A). The format of evaluation has varied—some publications resulted from conferences or symposia and others, were the product of committee deliberations after extended study. Conclusions have also varied, sometimes dramatically. That they have varied is not surprising, given that, over the course of 5 decades technologies have changed, society's goals and values have changed, and even the biology of pest species has changed. In most cases, the publications have had a measurable impact and have influenced attitudes toward pesticides or pesticide use.

It is our hope that this report will influence attitudes and policies.

One prediction that can be made with some confidence, however, is that, irrespective of the impact of our committee's study and report, it will likely not be the last study commissioned by the National Research Council on the subject of pesticides. Whether there will be 5 more decades of debate or far fewer, depends on dimensions of society, technology, and biology that are impossible to predict even with the best analytical tools available today.

May R. Berenbaum
Chair
Committee on the Future Role of
Pesticides in US Agriculture

Acknowledgments

The committee was greatly assisted by many individuals and groups that generously shared facts and expertise during the information-gathering phase of this study. In particular, we thank the people who participated in the three public workshops held across the country to provide the committee with input on a wide variety of subjects:

MICHAEL ALAVANJA, Agricultural Health Study, National Cancer Institute
KATE AULTMAN, National Institute of Allergy and Infectious Diseases
JOE BODDIFORD, Georgia Peanut Commission, Sylvania, Georgia
BARRY BRENNAN, US Department of Agriculture, Cooperative State Research, Education, and Extension Service, Pesticide Applicator Training
JENNY BROOME, University of California Davis, Sustainable Agriculture Research and Education Program and Biological Integrated Farming Systems
MARGRIET CASWELL, US Department of Agriculture, Economic Research Service
RAY CARRUTHERS, US Department of Agriculture, Agricultural Research Service
PETER CAULKINS, US Environmental Protection Agency, Office of Pesticide Policy
FORREST CHUMLEY, Dupont Agricultural Products
RON CISNEY, Olocco Ag Services, Santa Maria, California

HAROLD COBLE, US Department of Agriculture, Cooperative State Research, Education, and Extension Service
JIM CRANNY, US Apple Association
RUPA DAS, California Department of Health Services, Berkeley, California
ERNIE DELFOSSE, US Department of Agriculture, Agricultural Research Service
HELENE DILLARD, New York State Agricultural Experiment Station
LARRY ELWORTH, Gettysburg, Pennsylvania
ROBERT EPSTEIN, US Department of Agriculture, Agricultural Marketing Service, Science and Technology Pesticide Data Program
LEONARD GIANESSI, National Center for Food and Agriculture Policy
BOB GILLIOM, US Geological Survey, Sacramento, California
DICK GUEST, US Department of Agriculture, Cooperative State Research, Education, and Extension Service Interregional-4 Program
JOHN G. HUFTALIN, Grower, Rochelle, Illinois
TOBI JONES, California Department of Pesticide Regulation
WOLFRAM KOELLER, Cornell University, New York State Agricultural Experiment Station, Geneva, New York
SAM LANG, Fairway Green, Raleigh, North Carolina
YOUNG LEE, US Food and Drug Administration, Center for Food Safety and Applied Nutrition
RAY MCALLISTER, American Crop Protection Association
MIKE MCKENRY, University of California Davis, Kearney Agricultural Experiment Station
CHARLES MELLINGER, Glades Crop Care, Jupiter, Florida
MIKE OWEN, Iowa State University, Ames
KATHLEEN MERRIGAN, Henry A. Wallace Institute, Washington, DC
MICHAEL O'MALLEY, University of California Davis, Employee Health Services
ELDON ORTMAN, Purdue University, West Lafayette, Indiana
STEVE PAVICH, Pavich Family Farms, Porterville, California
BOB PETERSON, Dow AgroSciences, Indianapolis, Indiana
DAVID PIMENTEL, Cornell University, Department of Entomology, Ithaca, New York
GEORGE PONDER, Curtice Burns Foods, Montezuma, Georgia
MARY PURCELL, US Department of Agriculture, Cooperative State Research, Education, and Extension Service, National Research Initiative Competitive Grants Program
BOB QUINN, Millenial Farm and Ranch, Big Sandy, Montana
NANCY RAGSDALE, US Department of Agriculture, Agricultural Research Service

SAM RIVES, US Department of Agriculture, National Agricultural Statistics Service
WAYNE SANDERSON, National Institute of Safety and Health, Cincinnati, Ohio
ANN SORENSEN, Center for Agriculture in the Environment, American Farmland Trust, Dekalb, Illinois
TOM SPARKS, Dow AgroSciences, Discovery Research, Indianapolis, Indiana
TONY THOMPSON, Willow Lake Farm, Windom, Minnesota
KEL WIEDER, US Department of Agriculture, Agriculture Cooperative State Research, Education, and Extension Service, National Research Initiative Competitive Grants
MARK WHALON, Pesticide Research Center, Michigan State University, East Lansing, Michigan
DEAN ZULEGER, Hartland Farms, Hancock, Wisconsin

The following are also acknowledged for assisting the National Research Council staff during preparation of the report by providing updated information and statistics: Arnold Aspelin, senior economist, US Environmental Protection Agency Office of Pesticides Programs; Eldon Ball, economist, US Department of Agriculture, Economic Research Service, Resource Economics Division; Jennifer Eppes, author of *The Future of Biopesticides,* Business Communications Company Inc.; Merritt Padgitt, agricultural conomist, US Department of Agriculture, Economic Research Service; James Parochetti, US Department of Agriculture, Cooperative State Research, Education, and Extension Service; and Patrick Stewart, Department of Political Science, Arkansas State University.

The committee and staff wish to acknowledge with special recognition the contributions of Nancy Ragsdale of the Agricultural Research Service, US Department of Agriculture. Her efforts catalyzed the interest in and subsequent support for this study from both the US Department of Agriculture and the National Research Council. The committee thanks her for unflagging support and patience with this report though its development.

The committee is grateful for the extraordinary efforts of the staff of the National Research Council Board on Agriculture and Natural Resources (BANR) in facilitating all stages of this study. Warren Muir, executive director of the Commission on Life Sciences and of BANR provided oversight and Charlotte Kirk Baer, acting as interim director and associate director of BANR, demonstrated an unremitting commitment to seeing this report through to completion. Mary Jane Letaw provided invaluable assistance as project director in early stages of the study, particularly in organizing the project and in arranging the workshops. Karen L.

Imhof, project assistant, ably provided enthusiastic support through all stages of report preparation and completion, despite many competing demands on her time. Kim Waddell, project director during later stages of the study, is particularly worthy of recognition for his exceptional efforts at shepherding the report through the review process in a timely and consummately professional way. Heather Christiansen, research associate, provided timely updates and improvements to many of the figures and tables throughout the report. The committee is also appreciative of Norman Grossblatt for his editorial refinement of the report.

This report has been reviewed in draft form by individuals chosen for their diverse perspectives and technical expertise, in accordance with procedures approved by the National Research Council Report Review Committee. The purpose of this independent review is to provide candid and critical comments that will assist the institution in making the published report as sound as possible and to ensure that the report meets institutional standards of objectivity, evidence, and responsiveness to the study charge. The review comments and draft manuscript remain confidential to protect the integrity of the deliberative process.

We wish to thank the following individuals for their participation in the review of this report: Sandra O. Archibald, Humphrey Institute of Public Affairs; Daniel Barolo, Jellinek, Schwartz & Connolly, Inc.; Ellis B. V. Cowling, North Carolina State University; William Fry, Cornell University; Maureen Hinkle, National Audubon Society (retired); Robert Hollingworth, Michigan State University; George G. Kennedy, North Carolina State University; William L. Ogren, US Department of Agriculture (retired); Steven Radosevich, Oregon State University; Mark Robson, Environmental and Occupational Health Sciences Institute; Len Saari, DuPont Agricultural Products; Thomas Sparks, DowAgro Sciences; and John Stark, Washington State University.

Although the individuals listed above have provided constructive comments and suggestions, it must be emphasized that responsibility for the final content of this report rests entirely with the author committee and the National Research Council.

Contents

TABLES, FIGURES, AND BOXES

Tables

Figures

Boxes

The Future Role of Pesticides in US Agriculture

Executive Summary

The National Research Council charged the committee on the Future Role of Pesticides in US Agriculture with providing insight and information on the future of chemical-pesticide use in US agriculture. The committee was given four specific charges:

- Identify circumstances under which chemical pesticides may be required in future pest management.
- Determine what types of chemical products are most appropriate tools for ecologically based pest management.
- Explore the most promising opportunities to increase benefits, and reduce health and environmental risks of pesticide use.
- Recommend an appropriate role for the public sector in research, product development, product testing and registration, implementation of pesticide use strategies, and public education about pesticides.

The scope of the study was to encompass pesticide use in production systems—processing, storage, and transportation of field crops, fruits, vegetables, ornamentals, fiber (including forest products), livestock, and the products of aquaculture. Pests to be considered included weeds, pathogens, and vertebrate and invertebrate organisms that must normally be managed to protect crops, livestock, and urban ecosystems. All aspects of pesticide research were to be considered—identification of pest behavior in the ecosystem, pest biochemistry and physiology, resistance management, impacts of pesticides on economic systems, and so on.

Because its task was so broad and it had a relatively short period for its study, the committee met five times over 11 months in 1998 and held three workshops to seek input from the public. A critical early challenge was to refine the charge. The committee defined the future of agriculture to be the next 10–20 years. Beyond 20 years, predicting technological innovations and their effects is extremely difficult. The committee also believed that the term pesticide required a precise definition for the purpose of this report. The legal definition set forth in the Federal Insecticide, Fungicide, and Rodenticide Act is in part inconsistent with biological definitions of pesticides. The definition has social aspects as well; public perceptions inevitably color policy discussions and decisions. Accordingly, in this report the committee took a broad view of the concept of pesticide to include both the strict legal definition and microbial pesticides, plant metabolites, and agents used in veterinary medicine to control insect and nematode pests. Recent innovations in pest-control technologies (notably, genetic engineering) might necessitate a reevaluation of the legal definition in the near future. The committee also embraced the phrase ecologically based pest management (EBPM) as representing a pest-management approach that depends primarily on knowledge of pest biology and secondarily on physical, chemical, and biological supplements. The foundation for this management approach is a working knowledge of the managed ecosystem, including natural processes that suppress pest populations. Practitioners of EBPM augment the natural processes with such tools as biological control organisms and products, resistant plants, and narrow-spectrum pesticides.

With respect to the first charge—to identify circumstances in which chemical pesticides will continue to be needed in pest management—the committee decided early during its deliberations that an assessment of the full range of agricultural pests and of the composition and deployment of chemical pesticides to control pests in various environments would be an impossible task because of the large volume of data and the number of analyses required to generate a credible evaluation. The committee reviewed the literature and received expert testimony on the potential effects of pesticides on productivity, environment, and human health and on the potential to reduce overall risks by improving approaches that use chemicals under diverse conditions—soils, crops, climates, and farm-management practices. The committee concluded that uses and potential effects of chemical pesticides and alternatives to improve pest management vary considerably among ecosystems. That conclusion was reinforced by expanded solicitation of expert opinion. Overall, the committee concluded that chemical pesticides will continue to play a role in pest management for the foreseeable future, in part because environmental compatibility of products is increasing—particularly with the growing proportion of reduced-risk pesticides being registered with the Environmental Protection Agency (EPA), and in part because competitive alternatives are not universally available. In many situations, the benefits of pesticide use are high relative to risks or there are no practical alternatives.

With respect to the second charge—to determine the types of chemical products that are most appropriate for ecologically based pest management—the committee concluded that societal concerns, scientific advances, and regulatory pressures have driven and continue to drive some of the more hazardous products from the marketplace. Synthetic organic insecticides traditionally associated with broad nontarget effects, with potentially hazardous residues, and with exposure risks to applicators are expected to occupy a decreasing market share. This trend has been promoted by regulatory changes that restricting use of older chemicals and by technological changes that lead to competitive alternative products. Many products registered in the last decade have safer properties and smaller environmental impacts than older synthetic organic pesticides. The novel chemical products that will dominate in the near future will most likely have a very different genesis from traditional synthetic organic insecticides; the number and diversity of biological sources will increase, and products that originate in chemistry laboratories will be designed with particular target sites or modes of action in mind. Innovations in pesticide-delivery systems (notably, in plants) promise to reduce adverse environmental impacts even further but are not expected to eliminate them.

The committee recognized, however, that the new products share many of the problems that have been presented by traditional synthetic organic insecticides. For example, there is no evidence that any of the new chemical and biotechnology products are completely free of the classic problems of resistance acquisition, nontarget effects, and residue exposure. Genetically engineered organisms that reduce pest pressure constitute a "new generation" of pest-management tools, but genetically engineered crops that express a control chemical can exert strong selection for resistance in pests. Similarly, genetically engineered crops that depend upon the concomitant use of a single chemical pesticide with a mode of action similar to that of the transgenically expressed trait could increase the development of pest resistance to the chemical. Moreover, adverse environmental impacts (e.g., against nontarget organisms) are still considerations, albeit at a scale smaller than those presented by traditional chemical products. Expression of control chemicals in plant parts to be consumed by humans or livestock raise concerns about health impacts. Thus, the use of transgenic crops will probably maintain, or even increase, the need for effective resistance-management programs, novel genes that protect crops, chemicals with new modes of action and nonpesticide management techniques. There remains a need for new chemicals that are compatible with ecologically based pest management and applicator and worker safety.

Recommendation 1. There is no justification for completely abandoning chemicals per se as components in the defensive toolbox used for managing pests. The committee recommends maintaining a diversity of tools for maximizing flexibility, precision, and stability of pest management.

No single pest-management strategy will work reliably in all managed or natural ecosystems. Indeed, such "magic bullet" fantasies have historically contributed to overuse and resistance problems. Chemical pesticides should not automatically be given the highest priority. Whether they should be considered tools of last resort depends on features of the particular system in which pest management is being used (for example, agriculture, forest, or household) and on the degree of exposure of humans and nontarget organisms. Pesticides should be evaluated in conjunction with all other alternative management practices not only with respect to efficacy, cost, and ease of implementation but also with respect to long-term sustainability, environmental impact, and health.

With regard to the second charge—identifying what types of chemical products will be required in specific managed ecosystems or localities in which particular chemical products will continue to be required—the committee decided that there is too much variability within and among systems to provide coherent and consistent recommendations. Differences among managed and natural ecosystems in biological factors, such as pest pressure, and in economic factors, such as profitability, make generalizations about particular products of little value. Indeed, generalizing across systems as to the necessity of pesticides is responsible in part for many concerns and conflicts of opinion.

As for the third charge—to identify the most promising opportunities for increasing benefits of and reducing risks posed by pesticide use—the committee identified these:

- Make research investments and policy changes that emphasize development of pesticides and application technologies that pose reduced health risks and are compatible with ecologically based pest management.
- Promote scientific and social initiatives to make development and use of alternatives to pesticides more competitive in a wide variety of managed and natural ecosystems.
- Increase the ability and motivation of agricultural workers to lessen

their exposure to potentially harmful chemicals and refine worker-protection regulations and enforce compliance with them.

- Reduce adverse off-target effects by judicious choice of chemical agents, implementation of precision application technology and determination of economic- and environmental-impact thresholds for pesticide use in more agricultural systems.
- Reduce the overall environmental impact of the agricultural enterprise.

The most promising opportunity for increasing benefits and reducing risks is to invest time, money, and effort into developing a diverse toolbox of pest-management strategies that include safe products and practices that integrate chemical approaches into an overall, ecologically based framework to optimize sustainable production, environmental quality, and human health.

With respect to the fourth charge—recommending an appropriate role for the public sector in research, product development, product testing and registration, implementation of pesticide use strategies, and public education about pesticides—the committee agreed that specific policy actions in research, education, regulation, and management can enhance the likelihood that the opportunity for public-sector contributions is not missed.

Research topics that should be targeted by the public sector include

- Minor use crops.
- Pest biology and ecology.
- Integration of several pest-management tools in managed and natural ecosystems.
- Targeted applications of pesticides.
- Risk perception and risk assessment of pesticides and their alternatives.
- Economic and social impacts of pesticide use.

PUBLIC-SECTOR ROLE IN RESEARCH

Pesticides provide economic benefits to producers and by extension to consumers. One of the major benefits of pesticides is protection of crop quality and yield. Pesticides can under some circumstances prevent large crop losses, thus raising agricultural output and farm income. Many farmers are responsible land stewards and are concerned with potential environmental impacts of pesticides, but it is unrealistic to expect most farmers to adopt alternative pest-management strategies that would decrease their profits without the use of some policy incentives and disincentives.

Organic foods and ecolabeling markets are creating new opportunities for growers who are willing to reduce or exclude synthetic chemicals in their production practices. Environmentally friendly products appeal to consumers, too; organic food sales are growing at a rate of 20%/year in the United States. Yet policy analysts report that only 0.1% of agricultural research is devoted to organic farming practices. Availability of alternative pest-management tools will be critical to meet the production standards and stiff competition expected in these niche markets.

Globalization policies and practices are affecting pest management on and off the farm. Reduction in trade barriers increases competitive pressures and provides extra incentives for United States farmers to reduce costs and increase crop yields. It is likely that trade will increase the spread of invasive pest species and pose risks to domestic plants and animals, as well as populations of native flora and fauna. To meet those emerging global pest problems, researchers will need to develop effective, environmentally compatible, and efficient pest controls as a complement to a suite of prevention strategies.

Recommendation 2. A concerted effort in research and policy should be made to increase the competitiveness of alternatives to chemical pesticides; this effort is a necessary prerequisite for diversifying the pest-management "toolbox" in an era of rapid economic and ecological change.

Nontarget effects of exposure of humans and the environment to pesticide residues are a continuing concern. The application of pesticides results in indirect effects on ecosystems by reducing local biodiversity and by changing the flow of energy and nutrients through the system as the biomass attributable to individual species is altered. Pesticide policies should be based on sound science; where there is uncertainty, expert judgment will become more important in decision-making. Across-the-board pesticide policies that do not account for biological and ecological factors and for socioeconomic influences are likely to be less effective.

Pesticide resistance now is universal across taxa. Pests will adapt to counter any control strategy that results in the death or reduced fitness of a substantial portion of their population. Cultural and biological controls are not immune to evolution of resistance. Pesticide resistance is conspicuous because of the intensity of selection by high-efficacy chemicals. By spreading the burden of crop protection over multiple tactics, rather than relying on a single tool, farmers will face less risk of crop loss and lower rates of pest adaptation to control measures. Because pests will continue to evolve in response to pest controls, research needs to support development of pest-management tools that reduce selection pressure, delay selection for resistance, and thus increase the life of chemical and other products.

Pests will continue to thrive and a strong science and technological base will be needed to support management decisions. We need to continue to use the best science to resolve these questions. Policy makers need to use the best science in their decision making. As new technologies develop, theoretical frameworks for resolving these questions continue to be developed.

Recommendation 3. Investments in research by the public sector should emphasize those areas of pest management that are not now being (and historically have never been) undertaken by private industry.

Federal funding of pesticide research has historically had a very narrow base. To diversify the range of tools available for managing pests, a diversity of approaches would be beneficial. The chief desirable policy changes to diversify the research enterprise are highlighted below.

Recommendation 3a. Investment in pest management research at USDA should be increased and restructured in particular to steadily increase the proportion and absolute amounts directed toward competitive grants in the National Research Initiative Competitive Grants Program (NRI), as opposed to earmarked projects.

A greater emphasis on research—not only on chemicals themselves, but also on the ecological consequences of pesticide use—can increase the probability that new products will be readily integrated into ecologically based pest-management systems.

Recommendation 3b. Total investment in pest management and the rate of new discoveries should be increased by broadening missions at funding agencies other than USDA—specifically, the National Institutes of Health (NIH), the NSF, EPA, Department of Energy (DOE), and the Food and Drug Administration (FDA) to address biological, biochemical, and chemical research that can be applied to ecologically based pest management.

Investment in basic research applicable to ecologically based pest management is consistent with the missions of the funding agencies. Such initiatives could include

- Obtaining the ecological and evolutionary biological information necessary for design and implementation of specific pest-management systems.

• Identifying ways to enhance the competitiveness of alternatives or adjuncts by investing in studies of cultural and biological control.

• Elucidating fundamental pest biochemistry, physiology, ecology, genomics, and genetics to generate information that can lead to novel pest-control approaches.

• Examining residue management, environmental fate (biological, physical, and chemical), and application technology to monitor and reduce environmental damage and adverse health effects of both pesticides and pesticide alternatives.

The lack of basic information on pest population spatial and temporal dynamics is a major impediment to implementation of ecologically based pest management. NSF and EPA could make an important contribution by funding research associated with understanding of pest-population and community dynamics. This type of research is funded by these agencies, but it focuses mostly on natural, as opposed to managed, ecosystems. In addition, all agencies could improve the basic understanding of pests and their impacts by funding longer-term projects that would adequately capture the variability in pest dynamics, including pesticide-resistance evolution, under alternative management systems.

Recommendation 3c. On-farm studies, in addition to laboratory and test-plot studies, are a necessary component of the research enterprise. Investment in implementation research, which helps to resolve the practical difficulties that hinder progression from basic findings to operational utility, is needed.

The idiosyncratic nature of individual agroecosystems limits the utility of both laboratory and test-plot studies in predicting the efficacy of pest-management strategies. An increased emphasis on large-scale and long-term on-farm studies through the use of the global positioning system (GPS) and global information system (GIS) technologies could contribute substantially to diversifying management tools and approaches. Such research programs should remove the gap between "basic" and "demonstration" research for all managed and natural ecosystems. USDA needs to fund applied research because there are limits to models that serve basic science as well. Such models advance fundamental knowledge, but often the major economic problems involve organisms that are hardly ideal from a fundamental scientific viewpoint. These problems can be best addressed by research on the organisms in question. The information generated by applied on-farm research is crucial to extension scientists, crop consultants, and producers.

Recommendation 3d. Basic research on public perceptions and on risk assessment and analysis would be useful in promoting wide-

spread acceptance and adoption of ecologically based management approaches.

The body of literature evaluating public responses to agrochemicals in particular and pest management in general is not extensive. Surveys that have been done indicate that communication with the public about pesticides and their alternatives has been ineffectual. Media coverage of integrated pest management in widely circulated urban newspapers is sketchy and tends to focus on urban issues, thus providing little useful information to readers relevant to integrated pest management in agricultural settings. Although there have been substantial advances in research on risk perception in recent years, risk communication is a relatively new discipline. Research priorities include elucidating impacts of increasing benefit perceptions in risk communication, developing empirical methods for more accurate characterization of public perceptions, identifying reasons for differing qualitative and quantitative perceptions about pesticide technology and agrobiotechnology, and determining whether risk communication can reduce the gap that exists between public perceptions and scientific risk assessments.

PUBLIC-SECTOR ROLE IN IMPLEMENTATION

The public sector consists of various layers of government (local, state, federal, and international). In theory, each level of government addresses problems that affect its constituency. The justifications of government intervention in the management of pest control include the need to address the externality problems associated with the human and environmental health effects of pesticides and the information uncertainties regarding pesticides and their impacts. The performance and value of pest-control technologies depend on their specific properties and the manner of their application. The regulatory process has been designed to screen out the riskier materials. However, few incentives exist for efficient and environmentally sound pest-control management strategies. Introduction of incentives that would reduce the reliance on riskier pest-control strategies and encourage the use of environmentally friendly strategies is likely to lead to increased efficiency in pesticide use.

Worker-safety concerns have emerged as a major problem associated with pesticide use. There have been some important improvements, but the search for more-efficient policies should continue. Development of these policies might entail investment in research to improve monitoring on the farm to allow more precise responses to changes in environmental conditions.

Sometimes objections to pesticides are an issue of subjective preference even when scientific evidence cannot support the objections. In this case, a government role in banning a pesticide might not be appropriate; an appropriate role might be to establish a legal framework that enables organic and pesticide-free markets to emerge and prosper so that consumers can be given an informed choice between lines of products that vary with pest management.

Although agricultural biotechnology is more successful now than it was 2-5 years ago, raising money for agricultural ventures is still difficult for various reasons. First and most important, few investors have expertise in agricultural biotechnology. Second, although investors who exited early from their investments in 1980s agricultural-biotechnology companies made good returns, there are no blockbuster successes among the agricultural biotechnology success stories. Third, agricultural biotechnology must compete with telecommunication, software, Internet, and health-care biotechnology for venture capital. Fourth, few new agricultural-biotechnology companies have continued to generate investor interest (there is no "critical mass"). For United States agriculture to stay in the forefront with safer, environmentally friendly pest-management tools, there needs to be a continuing cycle of innovative new companies that research risky cutting-edge technologies.

Recommendation 4. Government policies should be adapted to foster innovation and reward risk reduction in private industry and agriculture. The public sector has a unique role to play in supporting research on minor use cropping systems, where the inadequate availability of appropriate chemicals and the lack of environmentally and economically acceptable alternatives to synthetic chemicals contribute disproportionately to concerns about chemical impacts.

The public sector can foster innovation in product development and pest-management practices by continuing to reduce barriers to investment by the private sector and by increasing implementation of regulatory processes that encourage product and practice development. This points to several recommendations relevant to innovation and risk reduction:

Recommendation 4a. The Department of Commerce Advanced Technology Program should be encouraged to fund high-risk R&D for IPM, EBPM and alternatives that have commercial potential for early stage companies.

The Advanced Technology Program (ATP) funded by the Department of Commerce awards grants that average $1–5 million, making it a valuable source for an early stage company. Typically ATP awards companies grants for risky, cutting edge R&D that has commercial potential. New tools for ecologically based pest management could get a boost if new companies could successfully compete for ATP funding for developing new IPM tools and alternatives.

Recommendation 4b. Incentives should be increased for private companies to develop products and pest-management practices in crops with small acreages, including access to compete for Interregional Research Project 4 (IR4) funds used to obtain product registrations for minor-use crops.

The Interregional Research Project 4 (IR4) program exists to assist in getting products registered on minor crops. IR4 awards grants to university researchers for biopesticide research. It has a long history of success in getting registrations for products for minor crops when there are no incentives for large companies to do so. We expect over the next few years to see as much success with biopesticides as IR4 has had with chemicals. Private companies are not allowed to obtain grants, but they are most capable of moving new products to market. The IR4 program should broaden its scope and allow private companies to obtain grants. IR4 should also better measure the outcomes (such as impact on farmers) of its current biopesticide grant program for academic researchers.

Recommendation 4c. Redundancy in registration requirements should be reduced to expedite adoption of safer alternative products (such as biopesticides and reduced-risk conventional pesticides)

Incentives can also be put into place to foster the development of products and pest-management practices in minor-use crops. Currently, the EPA's Biopesticide and Pollution Prevention Division (BPPD) has responsibility for registering new microbial and biochemical pesticides under subdivision M of the FIFRA. It takes 12-24 months to complete a biopesticide registration in BPPD. If that time could be reduced to less than 12 months for minor crops without compromising human and environmental-safety screening for minor-use crops, there would be an even greater favorable financial impact on small companies, and farmers would benefit by having earlier access to products. Also EPA's BPPD and California's Department of Pesticide Regulation conduct duplicative reviews and could increase the sharing of the review work.

Recommendation 4d. Evaluation of the effectiveness of biocontrol agents should involve consideration of long-term impacts rather than only short-term yield, as is typically done for conventional practices.

Many biocontrol agents are not considered acceptable by growers, because they are evaluated for their immediate impact on pests (that is, they are expected to perform like pesticides). Some biocontrol pathogens used against weeds might cause as little as a 10% reduction in fecundity, which might not be a visible result but has a major long-term effect causing population decline. Low-efficacy biocontrol agents alone might not be acceptable for pest management but, in combination with other low-efficacy tactics, they could be preferable because they avoid the selection for resistance for that is associated with high-efficacy tactics.

Recommendation 4e. At the farm level, incentives for adopting efficient and environmentally sound integrated pest-management and ecologically based pest-management systems can come from

- Expanding crop insurance for adoption of integrated pest-management and ecologically based pest-management systems practices.
- Implementing taxes and fees on environmentally higher-risk practices.
- Setting up tradable permit systems to reduce overall pollution emissions.
- Ensuring availability of funds in support of resource conservation (such as the Conservation Reserve Program).
- Conditioning entitlement to government payment on environmental stewardship.
- Assessing and more stringently enforcing regulations designed to protect worker health and safety.

Innovative crop-insurance policies can be developed to promote the adoption of pesticide alternatives and to increase their economic competitiveness. USDA is developing and piloting some innovative crop-insurance programs to increase incentives to farmers to use alternative products and IPM systems that reduce the number of pesticide sprays. Another potential approach includes US Senator Richard Lugar's Farmers' Risk Management Act of 1999 (S. 1666), which would change the way crop insurance has traditionally been used to a risk-management approach that would involve landowners in helping them to financial viability. The Act, if it passes, would provide eligibility for crop insurance if IPM or crop advisers are used.

Recommendation 4f. Funds should be assigned to assess compliance with Worker Protection Standards and to improve worker health and safety in specialty crops.

Worker safety in specialty crops is a serious concern. There are two interacting problems: a more intensive interface between worker and crop and inadequate effort in developing safer products and practices. Those problems are exacerbated by the collective importance of such crops in diversifying and enriching the United States diet. Funds should be assigned to study worker health and safety in specialty crops and to assess compliance with WPS. Without more detailed objective information on compliance, there is a reasonable doubt that the 1992 WPS is accomplishing its goal. Conducting an objective study of compliance with WPS will be difficult but important. It is imperative that the organization and individuals conducting such a study be unbiased and have no conflict of interest. Farm workers typically do not know when or what pesticides have been applied to fields, so they must rely on their employers to protect them from hazardous exposure. Because some employers might not follow WPS regulations, funds should be assigned to develop pesticide formulations that contain specific odors or dyes that would provide farm workers with direct information on the presence of hazardous pesticide residues.

PUBLIC-SECTOR ROLE IN EDUCATION

It is clear from the committee's study that the general public has a critical function in determining the future role of pesticides in US agriculture. Consumer interest in food and other goods perceived as safe and healthy fuels the rapid growth of the organic-food market; at the same time, consumer use of pesticides in the home and on the lawn continues to grow. Many of the paradoxical decisions made by the voting and consuming public arise from a relatively poor grasp of the science behind crop protection.

The public sector has a responsibility to provide education and information. Because knowledge also has public-good properties, a major responsibility of the public sector is to provide basic knowledge and information for decision-makers, in both the public and private sectors Education in scientific and technical fields is designed to meet anticipated demands in the private and public sectors. As long as there is a demand for pesticide-based solutions to pest-control problems, the education system has to train people to work in this field and to provide independent pesticide expertise in the public sector. Because we agree that pest-control choices have to be determined in the context of a perspective that incorporates biophysical, ecological, and economic considerations, education should emphasis basic principles and knowledge that will lead to informed decisions.

The broad set of considerations associated with pest-control decisions requires more interdisciplinary education in land-grant universities. People trained in life sciences and agriculture should also have a strong background in decision theory, risk evaluation, ethics, and economics to be able to handle pest-control problems in the commercial world. Many of the decisions associated with pest control are subject to public choice and public debate. To obtain rational and efficient outcomes, it is essential that scientists be able to communicate with the public in a clear and non-technical manner about the tradeoffs associated with alternative pest-control issues.

All citizens should be familiar with the basic principles of applied biology and risk evaluation, which can be provided as part of basic education. The general public, including children from kindergarten to 12th grade, should be educated about basic principles of environmental risk and of pest and disease control.

Recommendation 5. The public sector must act on its responsibility to provide quality education to ensure well-informed decision-making in both the private and public sectors.

This effort encompasses efforts in the agricultural sector, in the academic sector, and in the public sector at large.

Recommendation 5a. In the agricultural sector, a transition should be made toward principle-based (as opposed to product-based) decision-making. The transition should be encouraged throughout the continuum from basic to implementation research in universities, in extension, in the USDA Agricultural Research Service, and among producers.

Formulaic approaches to pest problems that are aimed at yield maximization rather than at sustainability approaches (product-based decision-making) have contributed to many of the problems plaguing agriculture. A sound grasp of fundamental principles should provide decision-makers with the flexibility needed to select from a menu of alternatives and to tailor practices to particular production systems

Recommendation 5b. Land-grant universities should emphasize systems-based interdisciplinary research and teaching and foster instruction in applied biology and risk evaluation for nonscientists.

There is a need to educate legislators and the general public about ecologically based pest management in research and in practice. Investment in increasing K–12 exposure to concepts of risk evaluation, food agriculture, and general biology can also have enormous benefits in creating a more knowledgeable and educated electorate.

5c. An effort should be made, in the government and in the land-grant system, to educate and train scientists about the value of public outreach.

The public sector should provide incentives and training for scientists to communicate effectively to the public about principles and practices of ecologically based pest management. Such incentives are almost nonexistent in many institutions, particularly outside the agriculture colleges. Outreach efforts, because of their overall value in providing popular support for the research enterprise, should be accorded some commensurate value in decisions related to career advancement and professional stature.

CODA

Our goal in agriculture should be the production of high-quality food and fiber at low cost and with minimal deleterious effects on humans or the environment. To make agriculture more productive and profitable in the face of rising costs and rising standards of human and environmental health, we will have to use the best combination of available technologies. These technologies should include chemical, as well as biological and recombinant, methods of pest control integrated into ecologically balanced programs. The effort to reach the goal must be based on sound fundamental and applied research, and decisions must be based on science. Accomplishing the goal requires expansion of the research effort in government, industry, and university laboratories.

1

History and Context

DEFINITION OF *PESTICIDE*

The word *pesticide* and its more specific variants *insecticide*, *fungicide*, and *herbicide* have been in common parlance for over a century. Although there is general agreement as to their essential meaning, there is by no means widespread consistency in their use by the general public, by scientists, or by legislators and regulators. The literal meaning of *pesticide* is "pest-killer", but common usage has restricted its meaning considerably. Although physical agents—such as heat, cold, and even shoe leather—have proved to be effective pest-killers over the years, they are rarely if ever considered pesticides. The entry of the terms into the language coincided with and was occasioned by the introduction and adoption of chemical agents and today these terms are used almost exclusively to refer to chemical agents that kill pests. It is important to define the terms precisely and to use them consistently; the definitions will affect the changing status of pesticides in regulatory contexts and their economic impact in the broader context of US agriculture.

The original definition of *pesticide* in the Federal Insecticide, Fungicide, and Rodenticide Act (FIFRA) came from earlier California law. The legal definition has since been modified at the federal and state levels. According to FIFRA, "a pesticide is any substance (or mixture of substances) intended for a pesticidal purpose, i.e., use for the purpose of preventing, destroying, repelling, or mitigating any pest or use as a plant regulator, defoliant, or desiccant, except that the term 'pesticides' shall not include any article that is a 'new animal drug' within the meaning of section 201(w) of the Federal Food, Drug, and Cosmetic Act (31 USD 321 (w)), that has been determined by the Secretary of Health and Human Services not to be a new animal drug by a regulation establishing conditions of use for the article, or that is an animal feed." What can be considered a pesticide is determined in part by what is specified as a pest. According to FIFRA, "an organism is declared to be a pest under circumstances that make it deleterious to man or the environment." FIFRA also lays out the foundation for the regulation of pesticides by the federal government: according to the July 1, 1997, edition of FIFRA, as amended, "no person may distribute or sell any pesticide product that is not registered under the act except as provided in (152.20, 152.25 and 152.30)." Pesticides are regulated by three federal agencies: the US Department of Agriculture (USDA), the Environmental Protection Agency (EPA), and the Food and Drug Administration (FDA).

Although the legal definition of a pesticide might appear straightforward, its biological underpinnings are tenuous at best. Among the most important complications is the broad definition of *pest* itself, a term with no biological validity (Newsome, 1967). Simply put, a pest is any organism that appears in a place where it is unwelcome, and *unwelcome* is defined in strictly human terms. Pest status does not adhere to taxonomic lines; some families of insects, for example, have members that are pest species and others that are regarded as beneficial (the beetle family Coccinellidae, for example, includes economically damaging herbivorous crop pests and predaceous biological control agents). By the same token, some species are regarded as an economic boon in some localities and a bane in others. Even in the same locality, different constituencies might regard the same species from different perspectives. In some parts of the southern United States, for example, Johnsongrass is considered a noxious weed by crop producers and an important component of dove habitat by hunters. Clearly, the same species of plant can be a crop or a weed, depending on circumstances. The fact that little if any taxonomic consistency underlies pest status suggests that selectivity for pest species, generally considered a positive attribute of a pesticide, is unlikely to be achieved easily.

Confusion in the general public as to the nature and origin of pesticides is a reflection of inconsistencies in usage among various constituencies. According to FIFRA, substances that repel but do not kill pests and substances that mitigate pest problems are considered pesticides—a designation that is neither intuitive nor etymologically consistent. Moreover, in a legal context, origins serve as the basis of differentiation. According to EPA, for example, substances produced naturally by plants to defend against insects and pathogenic microbes—and the genetic material required for the production of such substances—are considered "plant-pesticides". The term *pesticide* without modification implies a material synthesized by humans. There is a regulatory program in which incentives are provided for reduced-risk pesticides (EPA 1992), that is, pesticides with minimal impacts on nontarget species (including humans) and ecosystems. Such pesticides might include natural metabolites, which often are highly biodegradable and can be selective in their mode of action. The implication is that these materials are produced by living organisms as opposed to being synthesized industrially.

Recent developments in the production of pesticides might cloud that distinction further in the near future. For example, a biological step (using an esterase for chiral resolution) instead of a chemical catalysis is used in the synthesis of Pydrin® insecticide in Japan; however, this material is considered a synthetic pesticide, not a natural metabolite. In the pharmaceutical arena, pregnenolone and progesterone have been synthesized in yeast via deletion of one gene and insertion of five genes (Duport et al. 1998). The question arises as to how extensively organisms can be manipulated and still be considered to produce a natural metabolite. The reduction in restrictions for pesticidal natural metabolites offers a needed boost to companies seeking to market materials generally considered to be more environmentally compatible. This influx of new products will likely increase the diversity of materials available to pest-management specialists and create new confusion in the minds of customers.

Other considerations in defining *pesticide* are the mode of application and the intent of the applicator. Applied biological control involves human input, but the agents involved—natural enemies of pest species—are not regarded as pesticides by FIFRA. Classical biological control involves the intentional release of a beneficial, generally nonindigenous species to control a deleterious species. On occasion, the population of the beneficial species must be augmented by human intervention, often in the form of inundative numbers to flood a problem area rather than inoculative numbers to establish long-term populations. FIFRA does not regulate macroorganisms, such as parasitic wasps and nonpathogenic nematodes, and microorganisms are exempt from oversight if their vector is a nematode (NRC 1996). In general, biological control agents offer a high degree of specificity, but there is a discernible trend toward a pesticide application mentality with inundative use of these technologies. The use of microbial pesticides is an example of how nonchemical control techniques can have greater similarities to chemical control techniques than to other, more natural measures of pest control.

For the purposes of this report, we use the FIFRA definition of pesticide with a clear extension to encompass biopesticides (which include microbial pesticides and plant metabolites (box 1-1), but we will specifi-

BOX 1-1
Biopesticide Categories

Product	Definition	Examples
Microbial pesticides	Microorganisms that operate as the active ingredient	Bacteria, fungi, viruses, virus coat proteins
Plant-pesticides	Substances that plants produce from genetic material that has been added to them	*Bacillus thuringiensis* pesticidal protein, potato leafroll virus (PLRV)-resistance gene produced in potato plants
Biochemical pesticides	Naturally occurring substances that control pests by nontoxic mechanisms	Pheromones, floral attractants and plant volatiles, natural insect-growth regulators, plant-growth regulators and herbicides, repellents

Source: EPA 1999

cally exclude agents that are normally considered to be biological control agents and plant varieties that are produced by traditional breeding programs. We will extend the term to agents used in veterinary medicine to control insect and nematode pests.

Pesticides used in companion animals and livestock include pesticides that fit the FIFRA definition and that have been used widely in the treatment of insect infestations of animals. With the advent of the avermectin class of endectocides in the late 1970s—ivermectin, doramectin, milbemycin, and moxidectin—compounds that are active against both external insects and internal parasites became available (figure 1-1). Thus, some pesticides used on animals undergo approval through the FDA New Animal Drug Application (NADA) policy and do not undergo review and approval through FIFRA policy.

HISTORY OF PEST CONTROL

From their earliest days agriculturists have been beset by pests. Carvings dating back to 2300 BC in tombs in Egypt, one of the centers of plant domestication, depict locusts eating grain. Biblical passages allude to locusts (Exodus 10:3-6, 12-17, 19) and agricultural pests of other kinds (Joel 1:4, 7, 10-12, 17, 18, 2:19, 25; Deuteronomy 28:38, 39, 42; Amos 4:9,

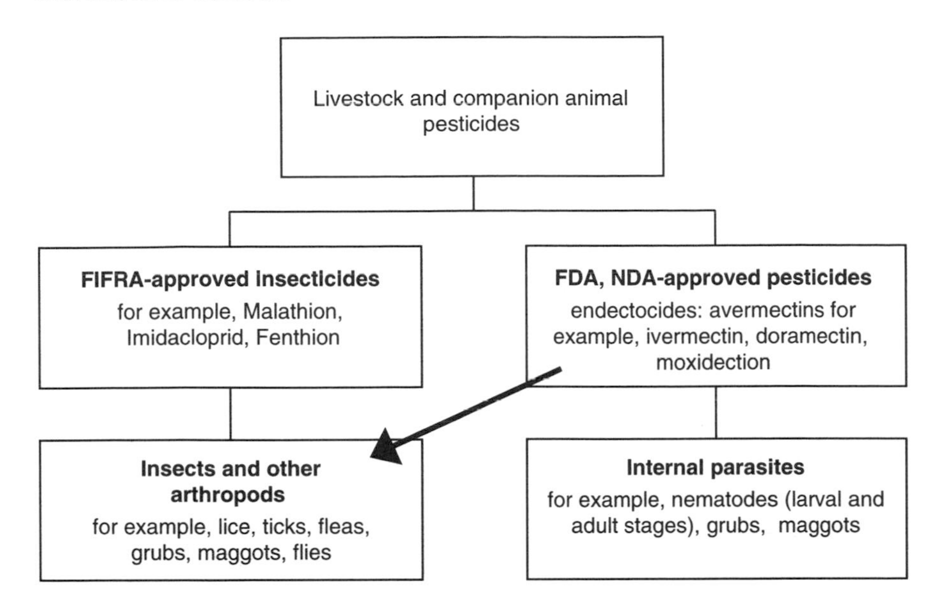

FIGURE 1-1 Relationship between FIFRA-approved insecticides and FDA NADA-approved pesticides used on companion animals and livestock.

19), including weeds (Jonah 2:5). Missing from the writings of the time, however, are descriptions of pest management. Little practical guidance was provided for doing anything other than surrendering to pests, possibly because infestations were often regarded as visitations by angry deities—condign retribution for impure thoughts or disobedience. The most common recourse in the case of infestation was to appeal to a deity for assistance.

The Greeks and Romans also had to deal with outbreaks of insects, weeds, and plant pathogens and similarly resorted to divine intervention for assistance. In the classical era, however, people began to take it upon themselves to rid themselves of pests, and pest-management practices are described in several of the more utilitarian writings that survive. From those texts, for example, Theophrastus' *Enquiry into Plants*, it is apparent that the ancients had a good grasp of many concepts of pest ecology—that herbivorous insects are host-specific, that population sizes depend on climate and geography, and that there is intraspecific variation in susceptibility to particular pests. Knowledge of life cycles was useful in designing pest-control programs; Aristotle's *Generation of Animals* was a seminal work in that regard. Pliny the Elder was the authoritative voice on pest control for centuries. Pliny was basically a politician with a penchant for writing—his *Natural History* encompassed 32 volumes and covered almost all the knowledge of the day. Because he recorded everything dutifully, however, without commentary or criticism, it is difficult to differentiate between what was regarded as sound science and what was discredited as folk wisdom. Rightly or wrongly, his work influenced almost all later writings about pest control until the 17th century.

According to available sources, the Greeks and Romans relied on several approaches to pest control. Mechanical methods were used extensively, particularly during locust plagues. Systematic manual collecting was even legislated; in Cyrene, according to Pliny, locusts were collected by law three times a year, and in Lemnos the law required that each man take a specific quantity of locusts to the local magistrates (Beavis 1988). Biological methods of control were less easily regulated, but many classical authors noted that birds (including jays and jackdaws) feast on insects. Cultural control, too, was remarkably sophisticated for the era. Many authors mention intercropping cabbages with vetch or with squill to control cabbageworm; given the host specificity of *Pieris brassicae* and *P. rapae*, the European cabbageworms, such a practice would likely interfere with host-finding and reduce infestation (Tahavainen and Root 1972).

Chemical control was, for the most part, even more notably like today's. Elemental sulfur, still in use today as an insecticide, was mixed with oil and used as an insect repellent and was boiled along with bitumen and olive oil leaves as a fumigant. As well, Greeks and Roman agriculturists relied heavily upon plant extracts for protecting cultivated plants from insects, weeds, and pathogens. Extracts could be made in water or oil, from fresh or burned plant material, sprinkled alone or in combination with other ingredients. A wide variety of plants were used, but the list is by no means a random one. Many of the plants specifically mentioned by classical authors as conferring protection against insect herbivores have been shown to contain highly insecticidal components. The recommendation of a solution of "dead larvae from another garden" (Beavis 1988) raises the possibility that alarm pheromones or insect pathogens were exploited for control purposes.

Europeans did little to advance the front of pest management after the collapse of the Roman Empire; classical authorities were dutifully copied and knowledge progressed slowly. What typified the era in general was not so much progress as regression; instead of relying on scientific observation, people tended to rely on divine intervention (of a sort) to deal with pest problems. There arose in the ninth century the curious and remarkably durable practice of prosecution, excommunication, or anathematization of insects and other animals in ecclesiastical courts of law. Historical records document at least 13 such cases in the 9th through 15th centuries. In the 100-year span of the 16th century, 18 trials were conducted. They then began to taper off, but the practice continued well into the 19th century (Evans 1906).

As early as the 17th century, more scientific measures were being adopted to combat insect pests. Francis Bacon, lord chancellor of England in 1561–1626, is widely credited with introducing the scientific method. Although classical authorities continued to be acknowledged as sources, their accounts no longer took precedence over direct observation and experimentation. Moreover, throughout the 17th and 18th centuries, the stepped-up pace of world travel and exploration led to the discovery of new cultures, approaches, and materials—tobacco, for example, a New World plant known as a fumitory, also proved to be an exceptional insecticide. Most of the magical practices fell into abeyance and were virtually forgotten.

By the beginning of the 19th century, the pest controller's armamentarium was restricted primarily to botanical preparations, elemental sulfur, oil soaps and kerosene emulsions to combat insects, and lime and sodium chloride for weed control. By the end of that century, however, a fundamental change had taken effect, owing in large part to two major innovations. Classical biological control—the importation of natural enemies from the original area of an introduced pest—was popularized by the dramatic success of the release of *Rodolia cardinalis*, the vedalia beetle, to stop an outbreak of *Icerya purchasi*, cottony cushion scale, in California in 1873. By 1902, biological control was extended to weed management when efforts were launched to identify natural enemies of the weedy shrub *Lantana camara* for importation into Hawaii, where it was choking out native vegetation (Clausen 1978).

Almost contemporaneous with pioneering developments in biological control was the fortuitous discovery that compounds containing heavy metals have pesticidal properties (Ordish, 1976). Although Paris green, a component of paints that consists of a mixture of copper and arsenic acid, was applied initially to grape foliage in France to discourage thievery, its efficacy in reducing herbivore damage did not escape notice. Its effectiveness, coupled with that of Bordeaux mixture as a fungicide, stimulated extensive testing and adoption of a variety of inorganic compounds as pesticides. Copper sulfate was introduced for use against charlock in wheat fields (and remains in use today as an algicide) (Timmons 1969). Hydrocyanic acid came into use as a fumigant in 1886 against insects, and carbon bisulfide as a fumigant against weeds in 1906. Arsenicals were used for both insect and weed control—lead arsenate for insects in 1892 and sodium arsenite for annual weeds in terrestrial and aquatic systems. Accompanying the proliferation of pesticides was an intensified effort to design and manufacture equipment to increase application efficiency.

The heavy-metal compounds, making up the so-called first generation of chemical pesticides, also had some drawbacks. Their broad-spectrum toxicity extended to human applicators, and under particular environmental conditions they could cause extensive damage to crop foliage. Moreover, without regulation, unscrupulous manufacturers proliferated, offering for sale products with little or no active ingredient. In California, the extensive fraud led to the Insecticide Law of 1901, the nation's first pesticide law, which standardized arsenic content in arsenicals. By 1901, California alone was spending over $250,000 on pesticides for food production (Stoll 1995).

Despite the problems associated with inorganic insecticides, they supplanted the use of biological control agents in most segments of the agricultural community. Thus, upon their discovery, the so-called second generation of insecticides, the synthetic organic compounds, found ready markets. Dichlorodiphenyl trichloroethane (DDT) had been synthesized by Otto Ziedler in 1874, but its insecticidal properties were not discovered until 1939, when Paul Muller tested it as part of an extensive screening study. DDT was cheap, persistent, and extraordinarily toxic to a wide variety of arthropods. Its utility in suppressing a typhus epidemic in Italy in 1943-1944 enhanced its appeal, as did its use in reducing wartime casualties in the Pacific Theater from malaria and other insectborne diseases. At the end of World War II, DDT became available for public use and was widely adopted (Dunlap 1981). Its astonishing efficacy led to the development of a variety of chlorinated hydrocarbons, such as gamma-lindane and toxaphene. Another product of wartime research was the organophosphates, which, despite their greater toxicity to mammals and other nontarget species, enjoyed considerable popularity because of their broad-spectrum efficacy and low cost (Casida and Quistad 1998).

Herbicides developed in parallel with insecticides during this era. Inorganic agents predominated in the early part of the 20th century. Sodium chlorate was used to control deep-rooted perennial weeds in noncrop areas and in small patches of field bindweed in cultivated fields; borates were found to control weeds without the flammability of the chlorates and are still used under asphalt to prevent weed growth (Timmons 1969). The compound that in 1942 dramatically illustrated the value of synthetic organic compounds for weed control was 2,4-dichlorophenoxyacetic acid (2,4-D); this compound was widely and enthusiastically embraced because of its selectivity for broadleaf weeds. Introduced soon after were substituted ureas and uracils, *S*-triazines and triazoles, and phenoxyethyl and acid sulfate compounds. From 1950 to 1969, the number of herbicides registered and released for use tripled (Timmons 1969).

The virtues of many of the synthetic organic insecticides proved in the long term to be environmental liabilities (Metcalf, 1980). Persistence reduced costs of application but also increased the probability of the evolution of resistance. Moreover, persistence and broad-spectrum toxicity led to widespread nontarget effects. Predators and parasites were especially vulnerable because of biomagnification, increases in concentration as material moves up a food chain. Unintended devastation of natural enemies led to the appearance of secondary pests, species that, before extensive insecticide use, were kept in check by their enemies. Biomagnification also led to concern about human health effects of repeated long-term exposure to residual pesticides as environmental and dietary contaminants.

Momentum for reassessments of pesticide safety and efficacy increased through the late 1950s. Public concern about pesticides crystallized with the publication of the book *Silent Spring* by Rachel Carson in 1962. The thesis of the book was that current patterns of pesticide use were harmful and that environmentally compatible alternatives to pesticides were available and should be used more extensively. Although Carson did not advocate a total ban on pesticide use (rather, she argued for their intelligent use), reaction in the chemical community was swift and critical. Nonetheless, her well-written arguments won her a substantial constituency. In the years following, environmental awareness grew among the public. By the end of the 1960s, a major effort to reevaluate the role of pesticides in US agriculture emerged; it culminated in the creation of the Environmental Protection Agency, a ban on DDT (and later other organochlorine insecticides) for all agricultural uses, and passage of legislation regulating the production and use of pesticides. Since 1971, many synthetic organic insecticides have been canceled or restricted because they posed health or environmental risks (table 1-1).

In the wake of *Silent Spring*, alternatives to synthetic organic insecticides have received more attention in the research community. In 1959, the first insect pheromone, bombykol, was characterized; by 1966, the sex pheromone of the cabbage looper *Trichoplusia ni*, a pest of several row crop species, was identified and ushered in an era in which pheromones have been exploited for trapping, mating disruption, and monitoring of pests. Advances in insect physiology allowed the development of control chemicals targeted at disrupting hormones that regulate insect growth, such as methoprene, a compound that works as a juvenile hormone analogue to interfere with maturation. The advantages of these so-called third-generation and fourth-generation pesticides were numerous; among them were that they offered specificity and environmental degradability. Nonchemical alternatives also proliferated, notably, control techniques based on microbial pathogens.

By the 1990s a new approach to pesticide development was made possible by advances in molecular biology and genetic engineering.

TABLE 1-1 EPA Regulatory Actions and Special Review Status of Selected Pesticides Used in Field-Crop Production, 1972-June 1995

Pesticides	Regulatory Action	Date
Alachlor	Uses restricted and label warning, under EPA review for groundwater contamination	1987
Aldicarb	Use canceled on bananas, posing dietary risk	1992
Aldrin	All uses canceled except termite control	1972
Captafol	All uses canceled	1987
Chlordimeform	All uses canceled, use of existing inventory until 1989.	1988
Cyanazine	Manufacturers voluntarily phasing out production by 2000, but stock can be used until 2003	1995
DDT*	All uses canceled (except control of vector diseases, health quarantine, and body lice)	1972
Diazinon	All uses on golf courses and sod farms canceled	1990
Dimethoate	Dust formulation denied and label changed	1981
Dinoseb	All uses canceled	1989
EBDC** (Mancozeb, Maneb, Metiram, Nabam, Zineb)	Protective-clothing and wildlife hazard warning	1982
Endrin	All uses canceled	1985
Ethalfluralin	Benefits exceeded risks; additional data required	1985
Heptachlor	All uses canceled except homeowner termite control	1988
Linuron	No regulatory action needed	1989
Methyl bromide	Annual production and use limited to 1991 levels; use to be terminated in 2001	1993
Mevinphos	Voluntary cancellation of all uses	1994
Monocrotophos	All uses canceled	1988
Parathion	Use on field crops only, under EPA review with toxicity data requested	1991
Propargite	Registered use for 10 crops canceled, use for other crops remains legal	1996
Toxaphene	Most uses canceled except emergency use for specific insect infestation of corn, cotton, and small grains	1982
Trifluralin	Restrictions on product formulation	1982
2,4-D*** (2,4-DB, 2,4-DP)	Industry agreed to reduce exposure through label change and user education	1992

*dichlorodiphenyltrichloroethane
**ethylene bisdithiocarbamic acid
***2,4-dichlorophenoxyacetic acid

Sources: EPA, 1998; Lin et al., 1995.

Casida and Quistad (1998) highlighted the actual and potential contributions of genetic engineering to pesticide productivity

- Enhancement of pesticide manufacturing capabilities (via cell-culture techniques).
- Improvement in rates of discovery of novel compounds (via novel screening techniques).
- Facilitation of development of transgenic crops that produce their own defenses.
- Facilitation of development of transgenic natural enemies that can withstand exposure to conventional pesticides.
- Evolution of new pesticides, in terms of mode of action, structure, specificity and origins.

GOALS OF AGRICULTURE

Pesticides have been an integral part of US agriculture since its earliest days. The goals of US agriculture have historically been to

- Ensure an adequate supply of high-quality safe food and other agricultural products to meet the nutritional needs of consumers.
- Sustain a competitive food and agricultural economy.
- Enhance quality of life and economic opportunity for rural citizens and society as a whole.
- Maintain a quality environment and natural-resource base (USDA 1999).

The use of pesticides in the past has been motivated and justified by their perceived effectiveness in allowing the agricultural sector to achieve its goals. Agriculture and related activities are evaluated by their contribution to the well-being of the nation (Just et al. 1982) which can be thought of as the sum of the well-being of members of a society. Historically, various agencies in the public sector (such as the US Department of Agriculture, the Food and Drug Administration, and land grant universities) have been extensively involved in

- Research, development, and testing of pest controls.
- Oversight of those controls (chemical and, increasingly, biological and genetic) for safety and efficacy.
- Transfer of the control technologies to producers.
- Education of the public about pest management.

Public support for research has traditionally been justified by the public-good argument for research that does not lead to commercial innovations. Research that does not generate marketable products that will repay the research is of high priority for public support. It can be research that leads to ecological and evolutionary biological information, to new cultural practices, or to biological control techniques for pest management. Studies in biochemistry, physiology, genomics, and genetics may also contribute to enrichment of these disciplines such that embodied pest control innovations that are marketable may eventually be developed.

Today public-sector researchers face concerns about their continued participation in this enterprise. Federal funding of pesticide research has historically had a very narrow base, evidenced by the pattern of Agricultural Research Service (ARS) funding (table 1-2). Table 1-2 provides a summary of the funding of 85 chemical-pesticide research projects in the USDA ARS in 1999. Most ARS chemical-pesticide research is directed to field studies and efficacy (including postharvest, laboratory efficacy research, and pest-management predictive models). Of the $20,764, 416 that funds the 85 projects, 57% is in that category. The three categories of field studies and efficacy (11), environmental fate (7), and residue analysis (15) are funded at $18.4 million, about 89% of the total for the 85 projects. The more basic chemical-pesticide research (categories 1-7) received about $5.3 million, about 25% of the total; categories 8-15 make up the remaining $15.5 million. Environmental-fate research receives 77% of the funding in the basic categories; field studies-efficacy receives 76% of the funds supporting the more applied research. It is important to note that little ARS funding is directed toward basic research (for example, toxicology or mode-of-action research) that supports development of new chemical pesticides.

Agencies other than USDA contribute to public sector support of pesticide research, albeit at a lower level. Table 1-3 shows National Science Foundation (NSF) awards for the divisions of environmental biology and chemistry for 1995–1999. The table reveals that only 11% of the funded studies involved agriculture, and less than 4 % involved agricultural pests or pesticides.

Advances in science that provide alternatives to chemical controls and that lead to greater understanding of how chemical controls work have changed the atmosphere in which public research decisions are made. Also, as a consequence of those advances, public agencies sensitive to how the scientific basis of criteria for determining pesticide use can change are redefining pesticide use and acceptable amounts of residues in food products. The increasing availability of alternatives to chemical pest controls appears to be emerging as a consideration in their decision-making.

TABLE 1-2 USDA Agricultural Research Service Funding of Chemical-Pesticide Research, 1999

Category	Research subject	Funding Dollars	Percentage of Total
1	Natural products	392,285	1.89
2	Pest biochemistry	75,468	0.36
3	Structure-activity relations, molecular modeling	62,952	0.30
4	Toxicology, mode of action	0	0.00
5	Metabolism: pest, host, and mammalian	62,952	0.30
6	Pest resistance (to pesticides)	693,827	3.34
7	Environmental fate (includes uptake and transport in plants and environmental degradation	4,051,198	19.51
8	Pesticide-screening bioassays	0	0.00
9	Application-delivery technology	850,640	4.10
10	Disposal technology	0	0.00
11	Field studies-efficacy (includes postharvest, laboratory efficacy research, and pest-management predictive models)	11,838,807	57.02
12	Resistance management	182,978	0.88
13	Farmworker protection	0	0.00
14	Human exposure	0	0.00
15	Residue analysis (includes chemical detection and methods for residue analysis)	2,553,309	12.30
	TOTAL[1]	20,764,416	100.00

[1]Includes salaries and operating expenses in addition to research.

Source: Pesticide funding data from Nancy Ragsdale, USDA-ARS, 1999

In light of shifting priorities in pesticide research and management policies, the USDA and EPA requested the National Research Council convene a committee of experts to address the future role of pesticides in agriculture and the nature of research that would be required to support development and use of new chemical pesticides.

In its charge, the committee was asked to:

- Identify the circumstances under which chemical pesticides may be required in future pest management.
- Determine what types of chemical products are the most appropriate tools for ecologically based pest management.

TABLE 1-3 National Science Foundation Award Data Relevant to Pesticide Research (1995–1999)

Key Words	No. of Projects[a]	% of Projects
CHEMISTRY		
Agriculture	12	0.62
Agriculture + chemical	5	0.26
Pesticide	4	0.21
Insecticide	2	0.10
Herbicide	3	0.15
Fungicide	0	0.00
Combinatorial chemistry	14	0.72
Combinatorial chemistry + pesticide	0	0.00
Structure activity	9	0.46
Mechanism of action	4	0.21
Novel target	0	0.00
DIVISION OF ENVIRONMENTAL BIOLOGY		
Agriculture	174	11.04
Pest	128	8.12
Agriculture + pest	49	3.11
Insect + pest	74	4.70
Pathogen	54	3.43
Weed	20	1.27
Resistance	139	8.82
Pest + resistance	27	1.71
Pesticide	13	0.82
Insecticide	7	0.44
Pesticide + resistance	6	0.38
Ecology + pest	95	6.03
Biological control	40	2.54
Ecosystem	529	33.57
Ecosystem + pest	79	5.01
LTER[b]	48	3.05
LTER[b] + agriculture	10	0.63
LTER[b] + agriculture + field crop	1	0.06
LTER[b] + forest	19	1.21

[a]Total number of projects awarded after in 1995–1999 is 42,850. Of them, less than 1%were related to agriculture and pesticides. Three keyword combinations were used to search for projects that were related to agriculture and pesticides: agriculture, pest + agriculture, and pesticide. Numbers of projects found were 325 (.076% of the total), 88 (.021%), and 99 (.023%) respectively. There are two subdivisions within NSF where such projects would most likely be categorized: Chemistry (CHE) and the Division of Environmental Biology (DEB). CHE accounts for 1,936 (4.52%) of total NSF projects related to agriculture and pesticides. DEB accounts for 1,576 (3.68%). Within the two entities, projects are further divided by keywords used to search for them, as the table shows.

[b]LTER=long term ecological research.

Source: NSF, 1999.

• Explore the most promising opportunities to increase benefits and reduce health and environmental risks of pesticide use.

• Recommend an appropriate role for the public sector in research, product development, product testing and registration, implementation of pesticide use strategies, and public education about pesticides.

REFERENCES

Beavis, I.C. 1988. Insects and Other Invertebrates in Classical Antiquity. Devon: University of Exeter.

Carson, R. 1962. Pp. 161–172 in Silent Spring. Boston: Houghton Mifflin.

Casida, J.E., and G.B. Quistad 1998. Golden age of insecticide research; past present or future? Annu. Rev. Entomol. 43:1–16.

Clausen, C.P., ed. 1978. Introduced Parasites and Predators of Arthropod Pests and Weeds: A World Review. Agriculture Handbook/USDA no. 480. Washington, DC: US Dept. of Agriculture, Agricultural Research Service.

Dunlap, T.R. 1981. DDT: Scientists, Citizens, and Public Policy. NJ: Princeton University Press.

Duport, C., R. Spagnoli, E. Degryse, and D. Pompon. 1998. Self-sufficient biosynthesis of pregnenolone and progesterone in engineered yeast. Nature Biotech. 16(2):186–189.

EPA (US Environmental Protection Agency). 1992. Incentives for Development and Registration of Reduced Risk Pesticides. Federal Register. 57(139):32140–32145.

EPA (US Environmental Protection Agency). 1998. Status of Pesticides in Registration, Reregistration, and Special Review (Rainbow Report). Washington, DC: EPA, Office of Pesticide Programs.

EPA (US Environmental Protection Agency). 1999. What are biopesticides. Office of Pesticide Program. [Online]. Available: http://www.epa.gov/oppbppd1/biopesticides/what_are_biopesticides.htm

Evans, E.P. 1906. The Criminal Prosecution and Capital Punishment of Animals. Reprint London: Faber and Faber 1986.

Just, R.E., D.L. Hueth, and A. Schmitz. 1982. Applied Welfare Economics and Public Policy. Englewood Cliffs, NJ: Prentice Hall.

Lin, B.-H., M. Padgitt, L. Bull, H. Delvo, D. Shank, and H. Taylor. 1995. Pesticide and Fertilzer Use and Trends in US Agriculture. Agricultural Economic Report Number 717. Washington, DC: US Department of Agriculture, Economic Research Service.

Metcalf, R.L. 1980. Changing role of insecticides in crop protection. Ann. Rev. Entomol. 25:219–256.

Newsome, L.D. 1967. Consequences of insecticide use on nontarget organisms. Am. Rev. Ent. 12:257–258.

NRC (National Research Council). 1996. Ecologically Based Pest Management: New Solutions for a New Century. Washington, DC: National Academy Press

NSF (National Science Foundation). 1999. Award search results. [Online]. Available: http://www.nsf.gov/verity/srchawdf.

Ordish, G. 1976. The Constant Pest: A Short History of Pests and their Control. New York: Charles Scribner and Sons.

Stoll, S., 1995. Insects and institutions: university science and the fruit business in California. Ag. Hist. 69:216–269.

Tahavainen, J., and R.B. Root. 1972. The influence of vegetational diversity on the population ecology of a specialized herbivore, *Phyllotreta cruciferae* (Coleoptera: Chrysomelidae). Oecologia 10:321–346.

Timmons, F.L. 1969. A history of weed control in the United States and Canada. Weed Sci. 17:294–307.

USDA (US Department of Agriculture). 1999. Agricultural Research Service mission. [Online]. Available: http://www.ars.usda.gov/afm/mr.html.

2

Benefits, Costs and Contemporary Use Patterns

BENEFITS OF PESTICIDES

Pesticides are an integral component of US agriculture and account for about 4.5% of total farm production costs (Aspelin and Grube, 1999). Pesticide use in the United States averaged over 1.2 billion pounds of active ingredient in 1997, and was associated with expenditures exceeding $11.9 billion; this use involved over 20,700 products and more than 890 active ingredients. Herbicides account for the greatest use in volume and expenditure; in 1997, 568 million pounds was used in agriculture, commerce, home, and garden. Insecticide applications constituted 168 million pounds, and fungicides 165 million pounds. Use patterns of pesticides vary with crop type, locality, climate, and user needs (Aspelin and Grube, 1999).

Pesticides are used so extensively because they provide many benefits to farmers and by extension to consumers. From the time when synthetic pesticides were developed after World War II, there have been major increases in agricultural productivity accompanied by an increase in efficiency, with fewer farmers on fewer farms producing more food for more people (Figure 2-1) (Rasmussen et al. 1998). A major factor in the changing productivity patterns, either directly or indirectly, has been the use of pesticides. In maize, for example, there has been 3-fold increase in yields since 1950. Although to a large extent this increase is attributable to the adoption of new hybrids with increased disease and insect resistance and with the ability to use more nitrogen fertilizer, another major factor

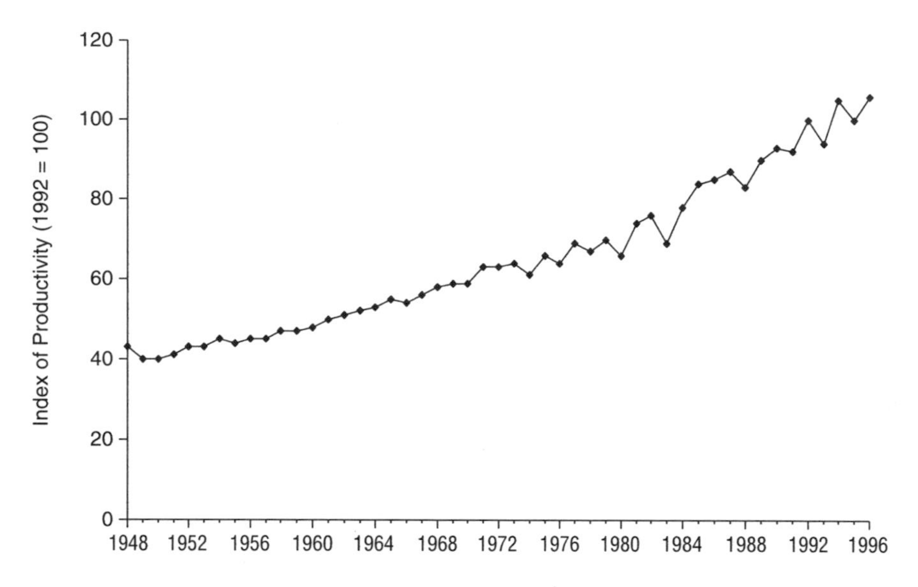

FIGURE 2-1 Index of farm productivity in the United States, 1948–1996.

[a]The index of productivity was determined by dividing all input values (such as feed, seed, livestock purchases, and pesticides) by all output values (such as feed crops, poultry, and eggs). Input and output values are unit-free quantity indexes that measure change over time as weighted by price. They were determined with Fisher's Ideal Index number procedure (also known as geometric mean of Laspayres and Paache indexes).

Source Data from Ball et al., 1997.

has been changes in planting practices facilitated by the availability of effective herbicides. Historically, for example, corn was planted in hills of three or more plants and, in many cases, in check rows, which allowed farmers to cultivate the corn in two directions for weed control. With the advent of effective herbicides, farmers switched from hill planting to drilled, narrow-row planting. The plant population increased from 10,000–12,000 plants per acre to 25,000–30,000 plants per acre. That led to the development of new high-yield hybrids that could tolerate the high population densities. Herbicides also allowed corn to be planted earlier in the growing season, and this resulted in a higher yield potential for the crop. Before herbicides, corn had to be planted later so that the first flushes of weeds could be killed with tillage. The development of soil-applied insecticides also allowed more farmers to grow maize for multiple years and increased productivity on an area-wide basis. Wheat production has also benefited from the use of herbicides; earlier, broadleaf and grass weeds caused great losses in yield. Cultivation is not practical in cereals, in contrast with maize. The introduction of 2,4-D and grass herbicides increased yields by controlling the weeds without damaging the crop (Warren 1998).

The beneficial impact of insecticides is illustrated by patterns of cotton cultivation in the southern United States. When *Anthonomus grandis*, the boll weevil, crossed the Rio Grande in 1892, it rapidly spread through the lower Southeast and drove major cotton production out of many states. With the advent of synthetic organic insecticides, farmers were able to return previously infested areas to cotton cultivation. Boll weevil eradication programs combining chemical control with other management practices have further expanded acreage in cotton (Carlson and Sugiyama 1985). That example also illustrates the complexity of pesticide issues. After the boll weevil outbreaks exerted their initial damage, southern farmers were forced to diversify their crops. The long-term result was of such strong economic benefit that the citizens of Enterprise, Alabama, erected a monument in their town square inscribed "in profound appreciation of the boll weevil and what it has done as the herald of prosperity" (Pfadt 1978).

Plant disease can be devastating for crop production, as was tragically illustrated in the Irish potato famine of 1845–1847; indeed, this disaster led to the development of the science of plant pathology (Agrios 1988). Disease is still a major problem in potato production, and over 90% of the acreage in the United States is treated with a fungicide to prevent yield loss. In the Columbia Basin, a late blight epidemic (caused by new aggressive strains of *Phytophthora infestans*) occurred in 1995, and affected 65,000 ha (Johnson et al. 1997). This area accounts for about 20% of US potato production. Left unchecked, the late blight epidemic could have decreased yield by 30–100%. With the use of several fungicides, the epidemic was controlled, and there was only a 4–6% loss in yield and no increase in storage loss compared with previous years (Johnson et al. 1997).

One of the major benefits of pesticides is the protection of yield. According to one estimate (Oerke et al. 1994), yields of many crops could decrease by as much as 50%, particularly because of insect and disease damage, without crop-protection. Knutson et al. (1990) estimated that removing pesticides from US agriculture would cause crop-production to decline, particularly in the southern states, and increase cultivated acreage by 10% to compensate for crop yield losses. Crop yield losses were estimated at 24–57%, depending on the crop species, if no pesticides or alternative crop protection measures were used. Moreover, exports in this scenario would decrease by 50%, and consumer expenditures for food would increase by about $228 per year and be accompanied by an increase in inflation as food prices increased. However, those estimates failed to take into account the possibility that other pest-control strategies could be used or that new technologies could be developed in the absence of chemical control (Jaenicke 1997).

A survey conducted by the Weed Science Society of America (Bridges and Anderson, 1992) estimated that the total US crop loss due to weeds is about $4 billion a year. In the absence of herbicides and best management practices, this loss could theoretically increase to $19.5 billion. The estimated loss in crops grown without herbicides ranged from 20% for corn and wheat up to 80% in peanuts (Bridges and Anderson, 1992).

Pesticide use also provides some benefits directly to consumers. Zilberman et al. (1991) estimated that every $1 increase in pesticide expenditure raises gross agricultural output by $3.00–6.50. Most of that benefit is passed on to consumers in the form of lower prices for food. Major losses prevented by pesticide use are those experienced during transport and storage. Oerke et al. (1994) estimated that about 50% of the harvested crop, particularly of such perishable crops as fruits and vegetables, could be lost in transport and storage because of insects and disease in the absence of pesticide use. Moreover, pesticide use can improve food quality in storage by reducing the incidence of such fungal contaminants as aflatoxins, known liver carcinogens, which are responsive to fungicides.

The use of herbicides has reduced the need for growers to cultivate to control weeds and that reduction has led to an increase in the practices associated with conservation tillage. These include no-till, ridge-till, strip-till, and mulch-till—practices that leave at least 30% cover after planting. Leaving cover after planting reduces soil loss due to wind and water erosion up to 90%, and it increases crop residue (organic matter) on the soil surfaces up to 40% (CTIC, 1998a). Conservation tillage in the United States has increased from 26.1% of the total acreage in 1990 to 37.2% of the total acreage in 1998 (1998b). Without herbicides, widespread adoption of conservation tillage would likely not have taken place.

Although agriculture accounts for two-thirds of all expenditures on pesticides and three-fourths of total volume used, nonagricultural uses of pesticides are also substantial. Pesticides are used on some 2 million US farms but they are also used in some 74 million households (albeit at much lower rates). Expenditures for home and garden use of pesticides in US households were approaching $2 billion a year in 1996, most of it on insecticides ($1.34 billion), fungicides and repellents ($185 million), and herbicides ($479 million) (Aspelin and Grube, 1999).

Estimating the economic benefit of household pesticides is difficult in that in most cases no tangible product is sold for a profit. Benefits are often aesthetic rather than economic (although aesthetic improvement can increase traffic at a place of business or increase the resale value of a residence). Even if difficult to measure, the aesthetic benefits of controlling pests in homes, gardens, and lawns, must be sufficient for homeowners to use pesticide products despite reservations about their safety (Potter and Bessin 1998); close to 85% of US households contain at least one pesticide product in storage (Whitmore et al., 1993).

Control of household pests can potentially provide health benefits because insect allergens (including those present in cockroach excrement and body parts) contribute to asthma, particularly in children. The presence of domiciliary cockroaches is strongly associated with sensitization to cockroach allergens, and sensitization has been associated with the incidence of bronchial asthma (Duffy et al., 1998). About 70% of urban residents with asthma are sensitive to cockroach allergens. The high mortality and morbidity of inner-city children due to asthma are linked to exposure to cockroach allergens (Petersen and Shurdut, 1999). That cockroach control could reduce the incidence of asthma is suggested by the positive correlation between the degree of cockroach sensitivity and the number of cockroaches seen in infested dwellings by residents. Helm et al. (1993), for example, established a quantitative relationship between cockroach density and the amount of cockroach aeroallergens. However, particularly in multifamily households, reducing cockroach numbers does not always lower the incidence of asthma (Gergen et al., 1999).

The decision of whether to treat for cockroaches at present is determined not by an economic injury level (EIL), but rather by an aesthetic level. EILs cannot be calculated, because an economic value of human life cannot be easily assessed. No observed-effect levels (NOEL) based on detectable levels of cholinesterase depression, however, can be established for the organophosphate agents used for cockroach control. Assessments of air and surface residues and biological monitoring have been used to evaluate multiple exposures of residents of homes undergoing crack and crevice treatment with organophosphates (summarized in Peterson and Shurdut, 1999). Maximum daily exposures were calculated at 2.4–8% of the reference dose (RfD) for children and less than 1% of the RfD for adults (RfD is the dose at or below which aggregate exposure every day over the course of a lifetime does not pose a significant risk). Use of chlorpyrifos, the agent of choice for crack and crevice treatment, was thought to result in minimal exposures and did not pose an appreciable risk to residents. Thus, even if aesthetic and health benefits are difficult to quantify, they still are expected to be offset by very low risk factors for chemical agents currently in use. On June 8, 2000, US EPA revised their risk assessment for this compound based on the mandates of the Food Quality Protection Act (FQPA) and eliminated chlorpyrifos for residential use. After December 31, 2001, retailers will not be able to sell any chlorpyrifos for home use except in baits with child-resistant packaging

(EPA 2000). Risks are not so easily quantified for nonchemical alternatives for cockroach abatement (such as baits, cleaning, and microbial agents), but they are expected to be low (Peterson and Shurdut, 1999).

Control of stinging hymenopterans, which kill about 40 people in the United States every year (Merck & Co., 2000), has considerable health benefits, which are difficult to quantify given the problems associated with assigning value to human life.

In summary, in the context of production agriculture and ancillary enterprises, pesticides are intended to

- Increase yields.
- Increase farming efficiency.
- Increase availability of fruits and vegetables.
- Supply low-cost food and fiber for consumers.
- Improve food quality.
- Decrease loss of food during transport and storage.
- Improve soil conservation.
- Ensure a stable and predictable food supply.

Contemporary Pesticide Use on US Crops

Broadly speaking, pesticides are used extensively in US agriculture; but they are used most intensively on fruits and vegetables. Intensity of pesticide use is measured by the amounts applied per acreage—which is much higher for fruits and vegetables than for other crops. For example, vegetables represent less than 2% of the crop acreage but received 17% of the total pesticides used (Lin et al. 1995). Current information on pesticide use is available from USDA surveys on corn, wheat, soybeans, cotton, potatoes, other vegetables, citrus, apples, and other fruits (ERS, 1997). Those crops account for about 80% of both planted crop acreage and sales of agricultural products and can thus be taken as broadly representative of US agriculture (USDA, 1996). Data on pesticide use include amounts of active ingredients applied and shares of acreage treated in toto and by major category (ERS, 1997). In 1996, corn, wheat, cotton, and soybeans together accounted for almost two-thirds of all pesticides applied to those crops (ERS, 1997). Corn herbicides alone accounted for about one-third of the total, and soybean herbicides for about one-eighth. Herbicides and insecticides applied to cotton each accounted for 5–6% of the total, and herbicides applied to wheat accounted for 4%. Other pesticides applied to potatoes, mainly soil fumigants, accounted for over one-eighth of the total.

The extent of pesticide use on any given crop can also be captured by the share of acreage treated. By that measure, herbicide use is widespread.

Herbicides are applied to 92–97% of acreage planted in corn, cotton, soybeans, and citrus; 87% of potato acreage; three-fourths of vegetable acreage; and two-thirds of the acreage planted in apples and other fruit (ERS, 1997; see Table 2-1). Herbicide use is least extensive on winter wheat (56%). In contrast, insecticide use is much less widespread. Among row crops, insecticides are used most extensively in cotton, tobacco, and potatoes. About 30% of corn acreage is treated annually with insecticides, and insecticides are applied to 12% or less of wheat and soybean acreage. Insecticide use is quite prevalent, however, on fruit and vegetable crops. Nearly all apples, citrus, and potatoes and about 90% of other vegetable and other fruit crop acreage are treated with insecticides (ERS, 1997). Fungicide use is similarly highly prevalent on potatoes and fruit crops. Among row crops, only in cotton, tobacco and potato are fungicides used regularly; less than 10% of cotton acreage is typically treated with fungicides. The category "other pesticides" includes defoliants, growth regulators, and soil fumigants, which are used widely on cotton and potatoes. Potatoes are particularly pesticide intensive—almost 90% of the acreage is treated, with fungicides and soil fumigants as the dominant types of treatment (ERS, 1997).

One measure of the intensity of pesticide use is reflected by calculat-

TABLE 2-1 Pesticide Use in US Row Crops, Fruits, and Vegetables

	Proportion of Area Treated, %		
Crop	Herbicide	Insecticide	Fungicide
Row crops[a]			
Maize	97	30	<1
Cotton	92	79	6
Soybean	97	1	<1
Winter wheat	56	12	1
Spring wheat	88	3	<1
Tobacco	75	96	49
Potato	87	83	89
Fruits and vegetables[b]			
Apple	63	98	93
Oranges	97	94	69
Peaches	66	97	97
Grapes	74	67	90
Tomato, fresh	52	94	91
Lettuce, head	60	100	77

[a]Data for 1996. Fungicide amounts do not include seed treatments. Source: Agricultural Chemical Usage 1996 Field Crop Summary USDA September 1997.

[b]Data for 1995. Source:Agricultural Statistics 1997, NASS Crop Branch (202) 720-2127.

ing the amount of active ingredient applied per treated acre. We do that by dividing the total amount of each material used by the number of treated acres, which we estimate by multiplying planted acreage by the fraction of acreage treated. If the fraction of acreage treated is not reported (for instance, for other pesticides used on fruits and vegetables), we use planted acreage instead (Table 2-2). Planted acreage is likely to be larger than treated acreage, because some acreage is not treated, so this procedure can result in an estimate of application rate that is lower than the actual rate.

Potatoes are the most pesticide-intensive US crop, because of their heavy use of soil fumigants (Table 2-3). Other vegetables and apples are the next most intensive, receiving a total of about 20 lb of pesticides per treated acre. Citrus (9.6 lb/acre) is also highly pesticide-intensive. (Lin et al., 1995). In contrast, cotton, the most pesticide-intensive of the major crops, received only about 5 lb/acre, about one-fourth to one-half as much as most fruit and vegetable crops. Corn received less than 3 lb/acre, and soybeans and wheat 1lb/acre or less. Only in the case of herbicides are application rates per treated acre comparable between major crops and fruits and vegetables. Corn and cotton receive roughly the same amounts of active ingredients per acre as potatoes, other vegetables, apples, and other fruits (Lin et al., 1995).

The total amount of pesticides applied to some major crops (Figure 2-2) increased over the last few years after declining for over a decade. Pesticide use in US agriculture increased steadily from the late 1940s until around 1980, because of the spread of herbicide use on corn and soybeans (ERS 1997, Osteen and Szmedra 1989). Pesticide use on major grain and oilseed crops has fallen consistently since the early 1980s. The adoption of pest-management programs that take advantage of the strengths of new pesticides has contributed to decreasing the amount of pesticides used. For example, a 1992 survey showed that pesticide use in Missouri grain crops had decreased by 6% since 1975 while the total quantity of herbicide and insecticide active ingredients had decreased by 38%; the decrease in herbicide use by Missouri corn and soybean farmers from 1984 to 1992 amounted to 3 million pounds. Those decreases were attributable to the availability of more effective herbicides with lower application rates (NAPIAP, 1997). Similarly, a survey in North Dakota in 1996 showed that many farmers had adopted new cultural and management practices that enhanced the effectiveness of pest management. For example, 75% of the farmers surveyed used field monitoring and crop rotation as part of their integrated program. In addition, several thousand wheat growers were trained in field monitoring, insect identification, and other practices, which resulted in a 75% decrease in the number of acres treated for orange wheat blossom midge.

Increases in pesticide use over the last 10 years are due for the most part to increases in the use of fungicides and other pesticides, mainly soil fumigants, on potatoes and other vegetables (Padgitt et al., 2000). Increases in pesticide use for a given crop can be the result of additional acreage being planted. For example, pesticide use in cotton has increased due largely to the resurgence of cotton production in the southeastern United States, which is itself attributable to the success of the boll weevil eradication program administered by USDA (Carlson et al., 1989).

Those trends suggest that differences in the intensity of pesticide use among crops appear to have become greater over time, mainly because of increases in the use of fungicides, such other pesticides as soil fumigants, and growth regulators. Over the last 2 decades, major crops (for example, grains, oilseeds, and cotton) have become less pesticide-intensive. Insecticide use and herbicide use on potatoes, other vegetables, citrus and apples have remained roughly constant since the early 1980s, whereas the use of fungicides and other pesticides has increased. The use of all types of pesticides on other fruits increased between 1980 and 1990 and has since remained roughly constant.

Pesticide-Related Productivity in US Agriculture

Gauging the productivity of pesticide use in agriculture is difficult. The aggregate concept "pesticides" has considerable currency in policy discussions but is hard to define precisely. Pesticides in the aggregate encompass a wide variety of chemicals with different properties and effects. As a result, there might be no consistent correlation between crop output and common measures of pesticide use, such as the amount of active ingredient applied or the acreage treated with pesticides. For example, reducing the application of a given compound might lead to reductions in output whereas a switch from a less-toxic to a more-toxic compound that results in the same reduction in weight of active ingredient applied might not. Despite the conceptual difficulties, it is important to have at least a rough sense of how pesticide use in the aggregate influences agricultural productivity.

Zilberman et al. (1991) pointed out that the productivity of pesticides—and thus the effects of reducing pesticide use—depends in large measure on substitution possibilities within the agricultural economy. Some substitutes are available only in the short term, when land allocations, cropping patterns, and consumption are fixed. Others are available in the long-term. Substitution between pesticides and other inputs can occur at the farm level or at the regional and national levels. Short-term substitutes for pesticides at the farm level include labor (such as, hand weeding), capital and energy (such as cultivation to control weeds),

TABLE 2-2 Pounds of Pesticide Active Ingredient per Planted Acre in Major US crops, 1990–1997

	1964	1966	1971	1976	1982	1990
Crop Herbicides						
Corn	0.387	0.693	1.362	2.454	2.974	2.932
Cotton	0.312	0.631	1.587	1.571	1.829	1.710
Wheat	0.165	0.152	0.216	0.273	0.226	0.215
Soybeans	0.133	0.279	0.840	1.614	1.880	12.870
Potatoes	0.989	1.482	1.521	1.254	1.256	1.687
Other vegetables	0.670	1.005	1.061	1.696	1.984	1.735
Citrus	0.265	0.397	0.457	3.970	5.556	6.635
Apples	0.617	0.924	0.389	1.427	1.548	0.815
Crop Insecticides						
Corn	0.238	0.356	0.344	0.379	0.368	0.313
Cotton	5.259	9.271	5.937	5.503	1.692	1.100
Wheat	0.016	0.016	0.032	0.090	0.033	0.013
Soybeans	0.158	0.086	0.129	0.157	0.164	0.000
Potatoes	1.111	1.984	1.934	2.318	2.898	2.566
Other vegetables	2.532	2.352	2.610	1.775	2.039	1.662
Citrus	1.825	3.213	2.554	3.843	4.687	4.678
Apples	23.993	20.185	12.011	8.960	7.898	8.220
Crop Fungicides						
Corn	0.000	0.000	0.000	0.000	0.000	0.000
Cotton	0.012	0.036	0.018	0.004	0.018	0.080
Wheat	0.000	0.000	0.000	0.011	0.013	0.002
Soybeans	0.000	0.000	0.000	0.004	0.001	0.000
Potatoes	2.463	2.357	2.880	2.962	3.094	2.006
Other vegetables	1.384	1.179	1.789	1.581	3.056	4.553
Citrus	6.314	4.559	7.754	4.922	4.312	3.950
Apples	17.173	20.190	17.919	16.093	13.512	9.778
Crop Other pesticides[a]						
Corn	0.001	0.008	0.006	0.006	0.002	0.000
Cotton	0.838	1.373	1.513	1.088	0.824	1.230
Wheat	0.000	0.001	0.005	0.000	0.000	0.000
Soybeans	0.000	0.001	0.001	0.040	0.034	0.000
Potatoes	0.069	0.006	4.467	6.095	11.658	25.055
Other vegetables	1.777	0.164	1.084	1.584	2.834	6.100
Citrus	1.971	0.766	1.072	0.179	0.006	0.016
Apples	2.298	2.564	1.363	1.424	1.003	0.104

1991	1992	1993	1994	1995	1996	1997
2.767	2.829	2.758	2.723	2.615	2.661	2.640
1.850	1.949	1.756	2.085	1.943	1.893	2.115
0.195	0.241	0.254	0.294	0.289	0.403	0.342
1.181	1.139	1.066	1.124	1.088	1.212	1.181
1.777	1.643	1.805	2.048	2.074	1.992	1.762
1.700	1.658	1.671	1.681	1.909	2.126	2.127
7.176	6.208	5.385	4.908	4.455	4.170	3.913
0.823	0.883	0.868	1.307	1.739	1.735	1.987
0.303	0.264	0.253	0.219	0.211	0.202	0.218
0.584	1.156	1.146	1.742	1.772	1.278	1.398
0.003	0.017	0.003	0.028	0.013	0.030	0.017
0.007	0.007	0.005	0.003	0.008	0.006	0.011
2.559	2.614	2.816	3.107	2.217	1.717	2.423
1.627	1.572	1.554	1.545	1.511	1.491	1.503
4.706	5.079	5.597	5.215	4.929	4.805	4.783
8.230	8.609	8.894	8.279	7.609	7.375	7.285
0.000	0.000	0.000	0.000	0.000	0.000	0.000
0.050	0.060	0.052	0.080	0.059	0.034	0.065
0.001	0.017	0.010	0.014	0.007	0.003	0.001
0.000	0.001	0.000	0.000	0.000	0.000	0.000
2.274	2.689	3.177	4.449	5.722	4.945	7.930
4.738	4.946	5.482	6.045	6.469	6.902	6.920
4.235	3.837	3.485	3.681	3.791	3.626	3.478
9.259	9.713	9.978	10.022	10.000	11.497	13.245
0.000	0.000	0.000	0.000	0.000	0.000	0.000
1.103	1.193	0.945	1.137	1.163	1.278	1.340
0.000	0.000	0.000	0.000	0.000	0.000	0.000
0.000	0.000	0.000	0.000	0.000	0.000	0.000
18.621	24.122	28.664	35.734	27.969	25.343	30.837
6.510	6.918	8.062	9.217	10.764	12.341	12.337
—[a]	—[a]	—[a]	0.102	0.200	0.600	1.100
0.206	0.221	0.217	0.218	0.217	0.217	0.221

Table continued on next page

TABLE 2-2 Continued

	1964	1966	1971	1976	1982	1990
Crop All pesticide types						
Corn	0.626	1.057	1.712	2.839	3.344	3.245
Cotton	6.421	11.311	9.055	8.166	4.363	4.120
Wheat	0.181	0.169	0.253	0.374	0.272	0.230
Soybeans	0.291	0.366	0.970	1.815	2.079	12.870
Potatoes	4.632	5.829	10.802	12.629	18.906	31.314
Other vegetables	6.363	4.700	6.544	6.636	9.913	14.050
Citrus	10.375	8.935	11.837	12.914	14.561	15.279
Apples	44.081	43.863	31.682	27.904	23.961	18.917

[a]none reported or too little reported to make an estimate.

Source: Lin et al.,1995; Padgitt et al., 2000.

changes in cultural methods (such as changing planting dates), and human capital (such as scouting). Long-term substitutes at the farm level include shifts in crop varieties (for example, to more-resistant but possibly lower-yielding cultivars), biological pest controls (such as released predators and pheromone traps), crop rotation, and land allocations (increases or shifts in cultivated acreage). Long-term substitutes at the regional and national levels include regional shifts in production, such changes in consumption as increases in imports and decreases in exports, changes in diet composition, and changes in feed mixes.

In general, pesticide productivity will tend to be low in situations where substitution possibilities are large. For example, the United States has a great deal of land suitable for growing grain and oilseed crops. Real prices of fuel and electricity have been falling since 1980 while the price of durable equipment has remained roughly constant (Ball et al. 1997). The prices of both types of inputs have fallen relative to agricultural chemical prices; this suggests that cultivation has become more attractive relative to herbicides as a form of weed control. Because US grain and oilseed production is highly mechanized, those pricing patterns suggest an abundance of cost-effective substitutes for pesticides and thus relatively low pesticide productivity on these crops. In contrast, the United States has a limited supply of land in areas with climates suitable for growing fresh fruits and vegetables year-round. The prices of hired and self-employed labor have risen steadily, both in real terms and relative to agricultural chemical prices, and this suggests that labor-intensive pest-control methods, such as hand weeding and scouting, have become less attractive

1991	1992	1993	1994	1995	1996	1997
3.070	3.092	3.011	2.943	2.825	2.864	2.858
3.580	4.350	3.892	5.036	4.943	4.483	4.925
0.197	0.273	0.265	0.338	0.311	0.435	0.362
1.190	1.146	1.071	1.127	1.098	1.216	1.193
25.302	31.068	36.462	45.339	37.983	33.997	42.952
14.575	15.094	16.799	18.487	20.679	22.860	22.887
16.118	15.237	14.467	13.906	13.270	13.146	13.043
18.724	19.426	20.174	19.826	19.565	21.041	22.737

relative to pesticide use. Fruit and vegetable production is relatively labor-intensive, so those facts suggest a greater paucity of cost-effective substitutes for pesticides and thus higher pesticide productivity on fruits and vegetables.

Pesticide productivity has been estimated in three general ways: with partial-budget models based on agronomic projections, with combinations of budget and market models, and with econometric models.

Partial-Budget Models

Partial-budget models estimate productivity effects of changes in pesticide use by constructing alternative production scenarios. Each scenario consists of a set of input usage rates and crop yields. Current production methods can be obtained from crop budgets when available. Otherwise, most studies rely on the opinions of experts familiar with crop-production conditions in different growing regions to construct scenarios characterizing current production methods and input usage sets likely to occur under various assumptions of pesticide-use reductions. Changes in yield are derived from pesticide field trials when available and from expert opinion otherwise. Current input and output prices are then used to translate changes in input use and crop yields into monetary terms.

The most widely cited studies on pesticide productivity, those of Pimentel and various coauthors, use this method (see, for example, Pimentel et al. 1978; 1991; 1992). Cramer's (1967) assessment of global crop losses also falls into this category, as does the Knutson et al. (1993)

TABLE 2-3 Acreage and Amounts of Pesticides Applied to Major US Crops, 1997

Crop	Acres[a]	Herbicides[b]	Insecticides[b]	Fungicides[b]	Other Pesticides[b]	All Pesticides[b]	Pounds per Acre
Corn	80,227	211,800	17,500	0	0	229,300	2.86
Cotton	13,808	29,200	19,300	900	18,500	67,900	4.92
Wheat	70,989	24,300	1,200	100	0	25,600	0.36
Soybeans	70,850	83,700	800	0	0	84,500	1.19
Potatoes	1,362	2,400	3,300	10,800	42,000	58,500	42.95
Other Vegetables	3,526	7,500	5,300	24,400	43,500	80,700	22.89
Citrus	1,150	450	5,500	4,000	1,100	11,050	9.61
Apples	453	900	3,300	6,000	100	10,300	22.74
Total	242,365	360,250	56,200	46,200	105,200	567,850	

[a]In thousands.
[b]In thousands of pounds.
Source: Padgitt et al. 2000.

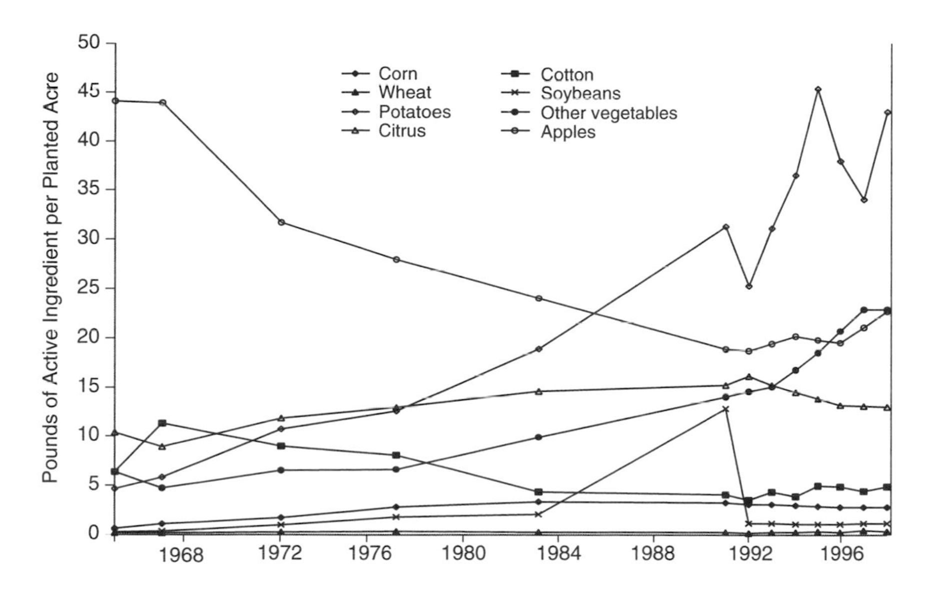

FIGURE 2-2 Total pesticides application on major US crops, 1964–1997.
Source: Padgitt et al., 2000.

study of pesticide use on fruit and vegetable crops. Those studies use data from field trials and expert opinion to estimate pest-induced losses crop by crop basis with current pesticide use, without pesticides, and with a 50% reduction in pesticide use. They construct alternative production scenarios for each crop to estimate changes in input use. Current prices are then used to value changes in per-acre production costs and per-acre yield losses, which are added to obtain an estimate of the costs of changes in pesticide use.

Pimentel and associates compile estimates of crop losses due to insects, diseases, and weeds crop by crop. They then add losses due to each class of pest on each crop to obtain estimates of aggregate crop losses in US agriculture. As they acknowledge, this procedure overestimates crop losses because of overlaps in damage caused by insects, diseases, and weeds. Crop losses are valued at current crop prices. One of the more recent of these studies (Pimentel et al. 1991) estimates that aggregate crop losses amounted to 37% of total output in 1986, up from 33% in 1974. Those estimates were compared with USDA assessments for the 1940s and 1950s, which estimated aggregate crop losses due to pests at 34% and 31% respectively, of total output. In comparison, Cramer (1967) estimated crop losses of around 28% due to all pests in all of North and Central America. Pimentel and associates interpret the temporal coincidence of rising pesticide use and rising crop losses as additional evidence that US agriculture has been on a "pesticide treadmill" in which disruption of agroecosystems by pesticides forces farmers to use ever-increasing amounts of pesticides. The study also estimates that pesticide use could be reduced by 50% by the substitution of nonchemical pest controls, such as crop rotation, use of resistant crop varieties, scouting, and field sanitation. It estimates that such a reduction in pesticide use would not result in any additional crop losses but would increase pest-control costs by $1 billion, or 25%.

Partial-budget models of this kind generally overstate pesticide productivity (and thus the economic effects of changes in pesticide use) because they consider only a small subset of substitution possibilities (Lichtenberg et al. 1988). They consider only short-run substitution, that is, changes in production methods while land allocations (acreage, crop rotations, and cropping patterns) and consumption patterns are kept fixed. They generally consider only a single substitution possibility for each crop. At best, they consider one alternative production scenario per region. Input usage rates and yields are assumed to be constant on all farms, so input-output ratios (input use per unit of output) are treated as fixed. The models thus ignore even short-run, farm-level substitution possibilities caused by differences in land quality, human capital, and other characteristics of farm operations.

The data used by the studies are problematic. Crop losses cannot generally be observed directly, because they involve comparisons of actual output with output that would have been obtained under conditions that do not typically occur. The empirical basis of the experts' projections is thus not very rigorous. Field trials can hold constant all production practices except pesticide use, deliberately ignoring substitution possibilities. Moreover, they are often conducted in areas with heavier than normal pest pressure, where pesticide productivity is probably higher (Pimentel et al. 1991). As a result, studies that rely on data like those of Pimentel and associates tend to overestimate crop losses in US agriculture.

In addition, estimates of crop losses at 37% are questionably high. The costs of pesticides and nonchemical pest-control methods alike are low relative to crop prices and total production costs (see, for example, the crop budgets in Economic Research Service, 1997). Crop losses of the magnitude estimated by Pimentel et al. (1991) should be sufficient to make it profitable to use chemical and nonchemical pest controls at much greater rates than observed today. In other words, if crop losses were as high as Pimentel et al. (1991) estimate, pesticide use would not now be as low as it is in most crops.

Combined Budget-Market Models

Other studies have attempted to estimate pesticide related and effects of large reductions in pesticide use by combining partial- budget models with models of output markets. These studies use the same approach as partial-budget models in estimating yield and cost effects of changes in pesticide use. Alternative production scenarios consisting of input usage rates and crop yields are constructed for each crop, possibly varying across production regions. Most studies use expert opinion to construct the scenarios. Crop budgets are used to estimate changes in per-acre production expenses. Projected changes in per-acre expenses and yields are then incorporated into models of agricultural-commodity markets and used to project changes in output prices and consumption in market equilibrium.

Zilberman et al. (1991) used this approach to estimate the likely effects of pesticide bans such as the failed 1990 California ballot initiative 128, popularly know as the "Big Green," on the production of five major fruit and vegetable crops. Zilberman et al. obtained alternative scenarios from crop scientists for production of five major California fruit and vegetable crops (almonds, grapes, lettuce, oranges, and strawberries) under current conditions and under restrictions on pesticide use that would be imposed by the Big Green initiative. Budgets were used to calculate changes in per-acre production expenses. Changes in per-acre production expenses and in crop yields were used to calculate shifts in the supply curves of each commodity (Lichtenberg et al. 1988). Estimates of supply and demand elasticities were then used to calculate changes in production and output prices for each crop. The authors found that Big Green would reduce output by 10–25%, increase crop prices by 13–57%, and cost consumers $883 million, about 25% of expenditures on the five commodities at the time (if prices go up more than quantities go down, then expenditures increase).

Models of this type incorporate some, but by no means all, substitution possibilities. They tend to capture changes in land use, such as changes in cropping patterns and regional shifts in production. Because they rely on a limited set of production scenarios, however, they tend not to capture all possible short-term substitutes at the farm level, such as shifts in labor, energy, and machinery; use of biological pest controls; and long-term substitutes, such as changes in crop varieties or crop rotation. The Zilberman et al. (1991) findings indicate that the potential for substitution in fruit and vegetable crops is more limited than found in other crop production systems, so pesticide productivity is higher for these crops.

Econometric Models

It is possible to estimate pesticide productivity directly with econometric models. Statistical methods can be used to estimate parameters of models that link output with input use. Varied substitution possibilities are implicit in the parameters of these models. Specification of models that are nonlinear in input use allows rates of substitution between inputs to vary as input usage changes. Alternatively, nonstatistical methods can be used to derive input-output relationships. Models of this kind are commonly used to estimate factor productivity and productivity growth in the US agricultural economy (see, for example, Griliches 1963, Ball 1985, Capalbo and Antle 1988, Chavas and Cox 1988, Chambers and Pope 1994).

Econometric models capture all forms of substitution in production, including short-run and long-run substitutes for pesticides on individual farms and at the regional and national levels. They do have some limitations. The data used to estimate these models limit them to reflect only forms of substitution actually experienced. In addition, the models consider only supply (production) and so do not capture substitution possibilities in consumption, such as changes in diet composition, in feed mix, and in imports and exports. Separate models of demand and market clearing are needed to capture equilibrium changes in consumption.

Headley (1968) estimated such a model by using state-level cross-sectional data for the year 1963. He used crop sales to measure output and expenditures on fertilizers, labor, land and buildings, machinery, pesticides, and other inputs as measures of input use and found that an additional dollar spent on pesticides increased the value of output by about $4—a high level of productivity for that period. Such a finding also indicates that increasing pesticide use would increase the profitability of the agricultural sector substantially. Headley attributed his result to the fact that use of herbicides was still in a relatively early stage of diffusion at that time.

Headley's model, like most economic models, generates estimates of the marginal productivity associated with pesticides, that is, the additional amount (value) of output obtained by using an additional unit of pesticides. Economists believe that in most circumstances (including agriculture) marginal productivity is falling at actual input usage levels; this implies that marginal productivity is less than average productivity. Multiplying the marginal productivity of pesticides by the quantity of pesticides used as, for example, Pimentel et al. (1992) do thus understates the total value added by pesticides. The total value added can be calculated more accurately by multiplying the value of the average product of pesticides (which exceeds the value of the marginal product) by the amount used.

Profit maximization implies using pesticides up to the amount where the value of their marginal product equals their unit price. Pesticides are for the most part relatively inexpensive in the United States, so one would expect the value of their marginal product to be correspondingly low. But a low value of marginal product does not imply a low total product. If the average product of an input is substantially higher than its marginal product, the total value added by the input will be quite high even if its value at the margin is low. Water is often cited as a case in point. The marginal value of water is low because so much is used, but its total value is large—perhaps even infinite—because it is essential for life.

There are several reasons to believe that Headley's estimate of marginal pesticide productivity could be too high. First, using sales as a measure of output tends to bias productivity estimates upward because output price tends to be positively correlated with input demand. Second, Headley's specification assumes that pesticides are an essential input, that is, that production is impossible without pesticides. This assumption, too, tends to bias the estimate of productivity upward. Finally, as Lichtenberg and Zilberman (1986) argue, the specification that Headley uses does not allow pesticide productivity to decline as fast as it should, again leading to upwardly biased estimates of pesticide productivity.

A useful approach to estimating pesticide productivity is that proposed by Lichtenberg and Zilberman (1986). They argue that pesticides do not generally affect potential output; rather, they prevent losses and are thus best conceptualized as producing damage abatement (avoidance of crop loss), an intermediate input. One appealing specification arising from this approach is treating realized output as the product of potential output and damage abatement. The former is produced by normal inputs, such as land, water, and fertilizer. The latter is produced by damage-control inputs, such as pesticides. Crop loss avoided (damage abated) must lie between 0 and 1. Crop losses can be estimated implicitly as the inverse of damage abatement (that is, as 1 minus the fraction of crop loss avoided).

Carrasco-Tauber and Moffitt (1992) applied this approach to state-level cross-sectional data on sales and input expenditures like those used by Headley (1968). Their use of sales as a dependent variable suggests that their estimate of pesticide productivity should be biased upward. Only one of the abatement specifications they used does not assume that pesticides are essential inputs. That specification generated an implicit estimate of aggregate US crop losses in 1987 of 7.3% at average pesticide use, far less than the Pimentel et al. (1991) estimate.

Chambers and Lichtenberg (1994) developed a dual form of this model based on the assumptions of profit maximization and separability between normal and damage-control inputs. They used this dual formulation to specify production relationships under two specifications of damage abatement, neither of which imposed the assumption that pesticides are essential inputs. They estimated the parameters of these models for the aggregate US agricultural economy, using time-series data or aggregate input, output, and price index for the period 1948-1989 developed by Ball (1988). Those data have been used widely to investigate factor productivity in US agriculture (see, for example, Ball et al. 1997. As in most factor-productivity studies in agriculture, Chambers and Lichtenberg (1994) assumed constant returns to scale, so that all inputs and outputs were measured on a per-acre basis. A later paper compared four alternative damage-abatement specifications (Chambers and Lichtenberg, forthcoming). Statistical tests confirmed the hypothesis that pesticides are not essential. Statistical tests also showed that crop losses declined as pesticide use increased.

Implicit crop losses in 1987 estimated from those models ranged from 9% to 11%, close to the Carrasco-Tauber and Moffit (1992) estimate but only about one-fourth to one-third that size estimated by Pimentel et al. (1991). The estimated parameters of the models imply that a 50% reduction in pesticide use from 1989 would increase crop losses by 3–4 percentage points to as much as 12–15 percentage points under average pest pressure. Assuming no change in crop prices, farm income would decrease by about $3 billion, or 6%, considerably more than estimated by Pimentel et al. (1991). Because those are aggregate estimates, they give no indication of how those reductions in pesticide use—with consequent increases in damage—would be apportioned among crops; the general depiction of the importance of pesticides in US agriculture presented here suggests that the bulk of the reductions would come from grains and oilseeds. Finally, estimated crop losses with zero pesticide use ranged from 17% to 20%.

Quality, Storage, and Risk

Reducing crop loss is the primary motivation for pesticide use, but pesticides also render other important services in agricultural production: enhancing product quality, prolonging storage life, and reducing production and income risk.

Product Quality

Pesticides can enhance product quality by reducing mold, scarring, blemishes, contamination with insect parts and weeds, and other unsightly and unsanitary features. The benefits of such enhanced product quality have several forms. Cleaner products can receive a higher price. For example, apple-growers are docked for impurities, and excessively blemished fruits might be salable only for low-value processing uses, such as juice. Processors set limits on impurities, such as insect parts or weeds in fruits and vegetables. They can impose price penalties on violative shipments or reject such shipments altogether. Federal food-purity regulations set limits on impurities in fresh produce; produce found to exceed the limits might not be salable. Federal and state grading standards for fresh fruits and vegetables prescribe lower grades—and thus lower prices—for produce that has more blemishes.

Some have argued that many forms of product quality are purely cosmetic and without a basis in health, nutritional value, or consumer demand (van den Bosch et al. 1975, Pimentel et al. 1991). Federal food-purity standards and grading standards have been especially criticized for that basis. It has been argued that those standards are stricter than necessary and that the stringency of these standards induces farmers to use pesticides more intensively than they otherwise would. Studies of purchases of peaches (Parker and Zilberman 1993), wheat (Ulrich et al. 1987), and other commodities show that consumers are willing to pay more for agricultural products that have fewer blemishes and impurities. Some consumer surveys also indicate unwillingness to purchase produce that has cosmetic defects or insect damage (see, for example, Ott 1990). Studies of pesticide productivity, such as those discussed previously, ignore quality considerations and thus understate benefits of pesticide use. In some cases, aesthetic quality is the primary consideration for pest control, however. Examples include such high-value ornamental crops as flowers, Christmas trees, and woody ornamentals.

Storage

Postharvest uses of pesticides include treatment with fungicides or growth regulators to prolong storage life and fumigation to prevent insect contamination. Prolongation of storage life of fresh fruits and vegetables increases consumer and producer welfare by increasing the availability of fresh produce between harvest seasons (Lichtenberg and Zilberman 1997). Prolongation of storage life of grains essentially lowers the cost of producing grain. As noted, insect parts in grain are considered impurities, and buyers might impose dockage or reject shipments; fumigation of grain thus enhances grain quality. Fumigation of fresh fruits and vegetables plays an important role in facilitating trade; many countries require fumigation to prevent inadvertent importation of exotic pest species (Yarkin et al. 1994). Fumigation has been required in interstate commerce in the

United States as well (for example, fumigation of citrus to prevent the spread of the Mediterranean fruit fly). Studies of pesticide productivity discussed previously ignore postharvest pesticide uses and thus also understate the benefits of pesticide use.

Risk

It has been argued (van den Bosch and Stern 1962, Carlson and Main 1976, Norgaard 1976) that aversion to risk is an important motivation for pesticide use, especially preventive insecticide treatment aimed at mobile insect pests. In this view, many preventive applications serve primarily to provide farmers with insurance; in other words, pesticide application provides little protection, because pest pressure in most years is too low to cause significant damage. Because pesticides are relatively inexpensive, farmers apply them as insurance against the (small) chance that pest pressure will be large enough to cause appreciable damage. That line of reasoning suggests that such applications could be eliminated at little or no cost in increased crop losses or reduced crop quality (van den Bosch and Stern 1962; Carlson and Main 1976; Norgaard 1976). Scouting and economic thresholds have been advocated as one means of eliminating such unnecessary preventive treatments. They have become quite prevalent in US agriculture. In 1994-1995, about 90% of cotton acreage, 85% of potato acreage, 80% of wheat acreage, 75% each of corn and soybean acreage, 65% of fruit acreage, and 21% of vegetable acreage were scouted for pests (ERS 1997). Public provision of all-peril crop insurance has been suggested as a means of eliminating incentives for uneconomical preventive treatment (Carlson and Main 1976, Norgaard 1976).

Preventive treatment might be economically efficient even if there is only a small probability that pest pressure will cause substantial damage and farmers are not risk-averse and thus have no demand for insurance. The average return on preventive treatment will be high when the value of the crop is high and when the material and application costs of the pesticide are low. The average return on reactive treatment will be low when scouting is expensive, when pest pressure is measured with low accuracy, and when the efficacy of rescue treatments is low. For example, by the time symptoms of disease in apples become observable, rescue treatment accomplishes little or nothing; preventive treatment with fungicides is thus standard in many growing regions (see, for example, Babcock et al. 1992). Under those circumstances (or combinations of them), growers will find preventive treatment more profitable (and hence less expenseive) on the average than reactive treatment, and the attractiveness of scouting will be limited; reducing or eliminating preventive treatment would reduce crop productivity and increase production cost.

Recent research suggests that in many circumstances pesticide use can increase income risk rather than reduce it. Theoretical analysis (Horowitz and Lichtenberg 1994) and simulation studies (Pannell 1991) show that pesticides tend to be risk-increasing when crop growth and pest pressure are positively correlated, as often occurs with weed and insect pressure in dryland farming. Econometric analyses of cotton in California (Farnsworth and Moffitt 1981), corn in the US corn belt (Horowitz and Lichtenberg 1993), and wheat in Kansas (Saha et al. 1997) and Switzerland (Gotsch and Regev 1996, Regev et al. 1997) confirm that greater pesticide use is associated with higher income variability for these crops.

Pesticides do play an important role in reducing risk when used to enforce quarantines and thereby prevent establishment of invasive pests. Such treatments can be applied either at the source or at the point of entry and can help to maintain interstate and international embargoes.

CURRENT PROBLEMS ASSOCIATED WITH PESTICIDES

Resistance to Pesticides

Resistance to pesticides (or, more specifically, synthetic organic chemicals) is almost universal among pest taxa (NRC 1986). Bacteria have developed resistance to antibiotics, protozoa to antimalarial drugs, green mold (*Penicillium digitatum*) to biphenyl fungicides, chickweed to the herbicide 2,4-D, schistosomiasis-carrying snails to the molluscicide sodium-pentachlorophenate, rats to warfarin, and pine voles to endrin. Resistance of nematodes to soil fumigants has not yet been observed but systemic nematocides are relatively new and it is probably only a matter of time until resistance appears. The evolution of resistance is more the rule than the exception in pest populations (Brattsten and Ahmad 1986).

Acquisition of resistance is an evolutionary phenomenon—that is, it is the result of changes in gene frequencies in a population over time. Normal unexposed populations are polymorphic in that some individuals are "preadapted" to cope with the selective agent. The application of selection pressure in the form of pesticide-induced mortality confers an advantage to those genotypes whose fitness is not affected by pesticide exposure; these genotypes reproduce preferentially, and as a consequence, population composition changes over time. It is important to note that the use of pesticides does not generally lead to resistance as a result of mutagenicity; although they can lower fitness, they do not, for the most part, act to change the genotype of an individual (indeed, genetically based resistance is not acquired during the lifetime of the individual).

It must also be noted that chemical pesticides are not the only mortality agents that can select for resistance. For over 2 decades, crop rotation was used as the principal management strategy for corn rootworm (*Diabrotica* spp.) in the midwestern states. Eggs of these species are deposited in the soil by gravid females; in the spring, hatching grubs attack roots of corn and other graminaceous hosts. By alternating corn with soybeans, growers greatly reduced rootworm problems; that grubs hatched in soybean fields encountered no corn roots, emerged, and starved. However, rootworm problems began to arise in corn planted after soybeans throughout the region where rotation was heavily used. These problems were attributable to two kinds of responses. Populations of *D. barberi* rootworms with extended egg-stage diapause—lasting 2 years—appeared (Levine and Oloumi-Sadeghi 1991), as did populations of western corn rootworm *(D. virgifera)* in which individuals, on reaching maturity, abandoned cornfields to lay eggs in adjacent soybean fields (Onstad et al. 1999). Thus, even cultural methods of control are prone to counteradaptation by pests. Any control strategy that results in the death or reduced fitness of a substantial portion of the pest population is prone to the evolution of resistance. Pesticide resistance is conspicuous, however, because of the intensity of the selection pressure exerted by the pesticide chemicals, which are designed to effect high mortality.

That insects could become resistant to the toxic effects of insecticides was first recognized in 1908 (Melander 1914), when a strain of San Jose scale (*Aspidiotus perniciosus*) was discovered to be resistant to lime sulfur. California red scale resistance to hydrogen cyanide was reported in 1916, and resistance to lead arsenate was found in codling moths in 1917. The number of resistant species has since increased almost exponentially. In 1976, resistance was recorded in 364 insect species (Georghiou and Mellon 1983). In 1986, the number was over 400 (Ku 1987); and it now exceeds 500. Over 150 microbe species and 270 weed species are resistant to at least one chemical pesticide (Jacobsen 1997). The majority of cases of insecticide resistance involve chlorinated hydrocarbons which have been in widespread continuous use for the longest time. Depending on the taxon, resistance to these chemicals is usually acquired within 10 years of first exposure, and acquisition of one form of resistance can facilitate the acquisition of other forms. (For example, the kdr gene in insects played a key role in the genetic evolution of DDT resistance and it confers protection against pyrethroids, another class of pesticides that has the sodium channel as its target site). The naive notion that insects would not develop resistance to the so-called third- and fourth-generation insecticides (which rely on interfering with hormone or pheromone function) was shattered first by Dyte (1972), who described *Tribolium castaneum* populations that were cross-resistant to juvenile hormones. Cross or induced resistance was later observed in many species (Sparks and Hammock 1983).

In the middle 1950s, Harper (1957) predicted that herbicide-tolerant weed species and resistant phenotypes would become common in crop systems that received annual applications of herbicides. In 1970, triazine resistance was discovered in *Senecio vulgaris*; and by 1977, there were 188 documented cases of weeds that were herbicide-resistant in 42 countries to over 15 families of chemicals in 42 countries (Heap 1997). Most cases involve a weed species with a single resistance mechanism, but there are exceptions; *Lolium rigidum* in Australia, for example, has evolved multiple resistance to all the herbicides used in the cropping system (seven chemicals in five families). It is likely that auto lactate synthase (ALS) and acetyl-CoA carboxylase (ACCase) inhibitor resistant weeds will present farmers with major problems in the next 5 years because of their proportional representation in the large-acreage crop markets and the rapid evolution of resistance to herbicides with similar structure or modes of action. Thirty-eight weed species have evolved resistance to ALS inhibitors in cereal, corn and soybean, and rice production systems (Heap 1997).

Herbicide resistance is still relatively restricted and has been slower to develop than insecticide or fungicide resistance. The relative rarity of herbicide resistance is likely due to the low persistence of many herbicides relative to the generation time of the pest, the lower fitness of some resistant genotypes than of the susceptible genotypes, the ability of susceptible weeds that escape death to produce large amounts of seed, and a large reserve of susceptible genotypes in the seed bank (Radosevich et al. 1997). Herbicide-resistant weeds have the greatest economic impact on crop production when there are no alternative herbicides to control the resistant genotypes or when the available alternatives are relatively expensive. Weeds that have multiple resistance are therefore the greatest concern of producers.

If food production is to become more sustainable, it will be imperative to preserve the efficacy of acceptable weed-control methods by defending against weed adaptation. Rapid and continuing weed adaptation might explain the persistent nature of yield losses to weeds despite technological advances (Jordon and Jannink 1997).

Biological factors can affect the speed of selection. One important factor is the generation time (under selection pressure): the larger the number of generations per year, the faster the selection of resistance. Root maggots, such as *Hylemya* spp., with three or four generations per year evolved resistance to cyclodienes in only 3–4 years, whereas *Diabrotica virgifera*, the corn rootworm, with only one generation per year developed resistance in 8–10 years. According to Georghiou and Taylor (1976), "it would be safe to predict that none of us will be present when the 17-year locust, *Magicicada septendecim*, will have developed resistance." The importance of generation time in evolution of resistance is underscored by the paucity of predaceous insects that are resistant to insecticides. Although there are insecticide-resistant insect predators such as *Coleomegilla maculata,* they are far outnumbered by the resistant herbivores. Predators that have acquired resistance, notably predaceous mites (Hoy et al., 1998), have intrinsic reproductive rates comparable with those of prey species. Rosenheim and Tabashnik (1990), however, dispute the importance of generation time, mainly because the probability of mutation is the same in every generation. They examined the literature and found no correlation between the rapidity with which resistance is acquired and the number of generations per year.

Dispersal and migration also influence the rate of evolution of resistance; if susceptible immigrants invade a local population treated with insecticide every year, evolution of resistance will take longer. Curtis et al. (1978) applied the idea to a theoretical control program: grid-application to ensure a supply of susceptible immigrants. That approach is being used to delay acquisition of resistance to *Bacillus thuringiensis* endotoxin expressed in transgenic plants; interplanting transgenic plants with susceptible plants provides a refuge for susceptible genotypes. Fortuitous survival might be another factor in delaying resistance acquisition. Fortuitous survival may be a factor in cases of polyphagous insects only one of whose hosts is treated or insects that avoid contact with insecticides by concealment in plant structures.

Finally, operational factors affect the speed of evolution of resistance. The selection regimen most likely to result in resistance acquisition is to reach and destroy a high percentage of the population, use a pesticide with prolonged environmental persistence, apply a pesticide thoroughly so as to leave no refugia, apply at low population densities to reduce the probability of survival of susceptible genotypes, apply to every generation, and apply over a large area to prevent immigration of susceptible individuals into the target population (Georghiou 1972).

The standard economic conceptualization of resistance treats susceptibility to a pesticide (or class of pesticides) as an exhaustible resource that is depleted gradually by pesticide application. In many cases, as resistance spreads, pesticide application rates rise while pesticide effectiveness falls, so that growers experience gradually increasing pest-control costs and gradually decreasing yields (Hueth and Regev 1974). Growers can find it profitable to switch to an alternative pest-control method or even an alternative crop (Regev et al. 1983). Using crop rotation or cycling of pesticides to retard the spread of resistance or renew the level of susceptibility in the pest population is implicit in this conceptualization, although not developed explicitly in the literature.

Despite the widespread occurrence of resistance, its effects on US agriculture as a whole are difficult to quantify. US agriculture has experienced neither rising pest-control costs nor falling yields in the aggregate. From 1980 (when the diffusion of herbicide use was largely completed) to 1993, real pesticide expenditures remained roughly constant (Figure 2-3), while total factor productivity (output corrected for changes in input use) grew at an average annual rate of 2.6–2.9% (Ball et al. 1997). But growers have been able to switch to substitute chemicals and use alternative methods. For example, resistance had already led cotton growers to switch from DDT to organophosphate insecticides by the time the registration of DDT was canceled (Carlson 1977). In other cases, producers consciously rotate pesticides to prevent or retard the spread of resistance. Thus, the most easily measured impact of resistance to date has been to shorten economic life of some pesticides. Detailed economic studies documenting the costs (or lack of costs) of pesticide resistance in specific agricultural systems are needed to assess the economic impact of pesticide resistance.

Pesticide manufacturers have incentives to respond to resistance both by developing resistance-management strategies for existing chemicals and by searching for new substitutes. Development of resistance-management strategies helps to maintain market share of existing products. Because fixed R&D costs make up a large share of the total costs of producing pesticides, longer product life increases a manufacturer's return on investment. At the same time, resistance functions like planned obso-

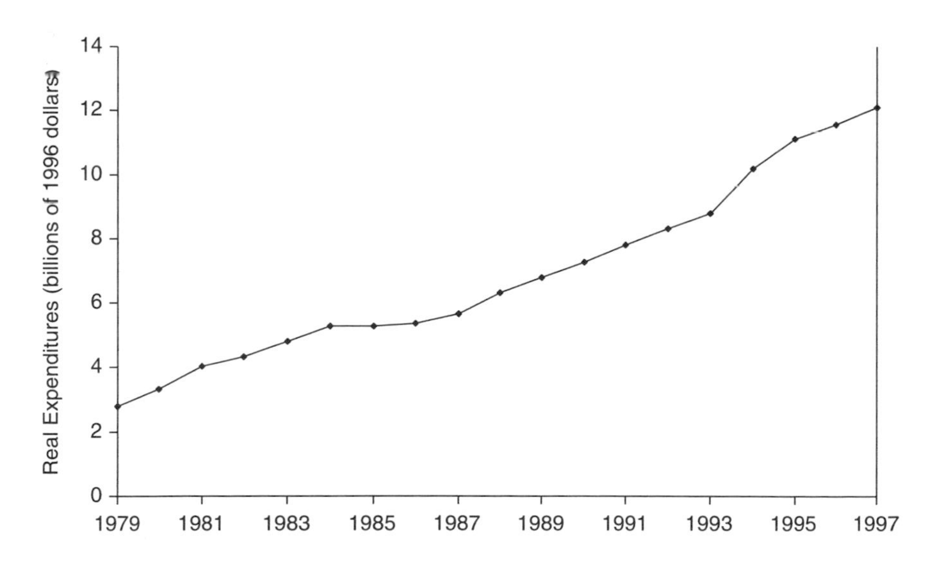

FIGURE 2-3 Real pesticide expenditures in the United States, 1979–1997. Source: Data from Aspelin and Grube (1999) and deflated by the implicit GDP price deflator (1996=100) from the Council of Economic Advisors, (2000).

lescence, creating markets for new pesticides by reducing the effectiveness of well-established ones and thus increasing expected returns on investment in new-product development. It gives pesticide developers a chance to increase their market share at their rivals' expense. It also decreases the window of opportunity to sell generic versions of established pesticides once their patent protection has lapsed, thereby increasing the share of time that developers receive patent protection.

Human Health Impacts

Occupational Effects and Risks: an Overview

Pesticides are designed to kill organisms that share many biochemical pathways and physiological processes with nontarget species in the agroecosystem, with domestic animals, and with humans. The biological commonalities make it difficult to develop pesticides that have ample margins of safety between the pest species and the nontarget organisms, including humans. Furthermore, the removal of "pest species", which might be enhanced by effects on nontarget species, will produce changes in the treated ecosystems. When one considers that the usual agricultural practice of cropping in monoculture involves major departures from "natural" ecosystems, the incremental impact of the use of pesticides on ecosystems becomes that much more difficult to estimate.

Candidate pesticides undergo extensive and rigorous laboratory and field testing. In most instances, the batteries of tests have been able to screen out chemicals that have undesirable characteristics. However, there are examples of failure of test protocols to warn against specific adverse effects that became apparent after pesticides came into common use. Two examples are the organophosphate insecticides Leptophos® and Mipafox®, which produce a delayed paralysis due to demyelination of motor axons (Johnson 1982). Workers exposed to the fumigant 1,2-dibromo-3-chloropropane unexpectedly experienced reduced or no sperm production (Thomas 1996). There have been many cases of poisoning and many more of suspected poisoning. However, most of the clinical cases have been due to gross overexposure in the course of misuse or suicidal use of the pesticide. (Levine 1991). In a survey of deaths from pesticide poisoning in England and Wales in 1945–1989, 73% of the recorded 1,012 deaths were due to suicides (Casey and Vale 1995).

It is difficult to generalize the toxicity of pesticides in a useful way. The target pest against which a pesticide is designed does not necessarily predict taxonomic risks of toxicity. For example, the fungicide methyl mercury has caused disproportionate mortality and morbidity in humans (Bakir et al. 1973, Hartung and Dinman 1972), and the herbicide paraquat has been associated with a disproportionate number of suicides (Staiff et al. 1975). Nor does the chemical structure necessarily predict risk. In the case of the insecticides, the organophosphate insecticides range from the essentially nontoxic Abate®, which has been used to control mosquito larvae in drinking-water supplies, and Ruelene® ,which has been used to control insect larvae that burrow into the skin of domestic animals, to the extremely toxic tetraethyl pyrophosphate, Systox®, Guthion®, and parathion, all of which are readily absorbed through the intact skin and by all other routes of exposures. The organophosphate insecticides differ substantially in toxicity and persistence. The basic mechanism of toxicity of all of the organophosphates is identical: inhibition of acetylcholinesterase (Mileson et al., 1998). However, that system is not the only one affected; organophosphates also inhibit other esterases to differing degrees, and some of these esterases are involved in the normal detoxification of organophosphate and carbamate insecticides. Consequently, combined exposures to several of these insecticides can result in synergistic, additive, and antagonistic responses (DuBois 1969).

Human exposures to pesticides occur through several routes (Ferrer and Cabral 1995). Exposures to pesticides in the general population tend to occur mainly through contact with residues in food or water but can also occur through accidental ingestion of seed prepared for sowing or through mistaken use of pesticides in food preparation because of their resemblance to food products. Occupational exposures to pesticides tend to occur mainly through dermal contact and inhalation. Occupational exposures constitute a distinct type and generally affect workers involved in the manufacture, transport, and application of these chemicals. Occupational epidemics tend to be more frequent in developing nations than in the United States, largely as a result of inappropriate technology transfer (for example, Senanayake and Peiris 1995). Nonetheless, occupational exposures have long been a source of concern in the United States.

The available statistics on poisoning by pesticides are difficult to compare because different agencies use different classifications and the classification schemes have changed over time. Reports on pesticide poisonings can be based on poison-control center reports, on hospital admission reports, and on reviewed clinical diagnoses. Even the quality of the information contained in death certificates is variable.

Manufacturing Risks

In general, the synthesis and formulation of pesticides are under better control than the later stages in the life cycle of a pesticide. Workers in manufacturing processes are under the direct or indirect supervision of occupational hygienists. Standards of permissible occupational exposures are promulgated by the Occupational Safety and Health Administration, and threshold limit values proposed by the American Council of Governmental Industrial Hygienists for individual pesticides are enforced.

However, there have been excessive exposures and adverse effects associated with manufacturing processes. The best known of these are related to the production and synthesis of 2,4,5-trichlorophenol, the precursor of a number of chlorinated herbicides. The synthesis of 2,4,5-trichlorophenol was associated with the unavoidable trace byproduct 2,3,7,8-tetrachlorodibenzo-p-dioxin (TCDD). The major effects noted were chloracne and occasional liver damage (Firestone 1977, Pocchiari, F. et al. 1979, Coulston and Pocchiari 1983, Kimbrough et al. 1984) and porphyria (Bleiberg et al. 1964). Furthermore, TCDD has been shown to be carcinogenic in laboratory bioassays of rats and mice, and there is epidemiological evidence of its human carcinogenicity (Gallo et al. 1991). The herbicides that were based on 2,4,5-trichlorophenol have been removed from the market, because of the difficulties encountered in removing TCDD and related dioxin impurities.

Most of the chlorinated-hydrocarbon insecticides—such as DDT, dieldrin, aldrin, toxaphene, and chlordane, as well as TCDD—have been found to be carcinogenic in rodent bioassays. That experience has generally not been duplicated in humans. However, human exposures tend to be much lower, even under occupational conditions, than those attained during bioassays (Leber and Benya 1994). Many of the chlorinated-hydrocarbon pesticides are very fat-soluble and exhibit a tendency to bioaccumulate. For these reasons, most of the chlorinated hydrocarbon pesticides are no longer on the US market.

Packaging, Distribution, and Application Risks

The Secretary's Commission on Pesticides (HEW 1969) called attention to a number of cases of poisoning due to improper storage or spillage during transportation involving the insecticides endrin, dieldrin, diazinon, mevinphos, and parathion. Another prominent case involved a shipment of dyed seed grain to Iraq; it had been treated with methyl mercury as a fungicide, was washed to remove the dye, and was used for food. The instructions on the seed bags were only in English; as a consequence there were 6,530 cases of poisoning, with 459 deaths. (Bakir et al. 1973).

In normal use, the potential for the highest exposures occurs during the handling of the concentrated pesticides in preparation for application to crops. The application process is the source of many exposure scenarios. The highest exposures are likely to occur during ground applications, especially for spot treatments when the plants being treated are taller than the applicator. Gardeners and nursery workers are likely to be exposured in enclosed environments. Exposures from broadcast or roto-misting can be relatively low or very high depending on the wind direction and strength in relation to the path taken by the applicator (Ecobichon 1996).

The exposure scenarios are only part of the assessment of the likely risk posed by pesticides. It is also important to know the extent to which specific pesticides are absorbed into the body and to know the quantitative and qualitative toxicity of the absorbed doses. The major routes of absorption of pesticides are inhalation, ingestion, and dermal absorption. For the applicator, all three routes can be important. The characteristics of the pesticide are also important. Most pesticides are readily absorbed when inhaled or ingested, and are also absorbed through intact skin. The rate of dermal absorption is influenced by molecular weight and by lipid solubility (Feldman and Maibach 1974, Moody et al. 1990). The problem is made more complex by the variance in permeability of the skin in different regions of the body. It is possible to absorb lethal doses of some organophosphate insecticides while walking through a freshly sprayed field, so intervals after spraying during which reentry into the field is not permitted have been established.

In many cases, incidents of acute poisoning are due to failures to observe safety regulations. In one conspicuous case, in California in 1989, workers harvesting cauliflower in a field sprayed 20 hours earlier with two organophosphate insecticides and one carbamate insecticide became ill after about 1 hour of exposure; at the time, state laws specified a 72-hour safety reentry interval (Ferrer and Cabral 1995). However, occupational exposures can occur even when safety regulations are enforced. In a review of pesticide poisonings of lawn-care and tree-service applicators reported to the New York State Pesticide Poisoning Registry in a 32-month period in 1990–1993, Gadon (1996) identified 28 unambiguous cases; 22 of the 27 people on whom information was available had been using protective equipment when the exposure occurred.

Residues in packaging materials can also be a source of substantial exposure and toxicity. One notable example involved a sack that had contained parathion. The sack was used to build a swing, and the percutaneous absorption was sufficient to kill two of five children using the swing (Hayes 1963).

Trends in Worker Safety Risk

The reporting of cases of poisoning due to pesticides is inconsistent by region, by circumstance, and over time. Nevertheless, trends begin to emerge at coarser scales of observation, especially when one concentrates on regions that have better reporting schemes, such as California and some Florida counties. From 1951 to 1967, 151 deaths due to agricultural chemicals were reported in California; 34 of them were occupational (Baginsky 1967). The number of reported poisoning cases is always much higher than the number of reported deaths. In 1966–1970, reported poisonings ranged from 1,347 to 1,493 cases per year; organophosphate insecticides were the most common causative agents (Hartung 1975).

Since then, a number of regulatory actions have decreased the number of deaths attributable to exposures to pesticides. Among them were the withdrawal of organomercurial fungicides and some of the more toxic organophosphate insecticides from commercial use. Perhaps most important, the use of many of the more potent pesticides has been limited to trained and licensed applicators.

California is one of a few states that enforce mandatory physician reporting of all occupational pesticide-poisoning incidents, including follow-up investigations to verify exposure and effects and to determine the likelihood that the pesticide was responsible (Blondell 1997). Statistics gathered by the California Environmental Protection Agency (1995, 1996, 1997) indicate that the agency received 1,995-2,401 reports of poisonings per year during the period of 1993–1995. The agency concluded that 656–1,332 of the reports could have involved exposures to pesticides in an occupational setting. Chlorine and hypochlorite were most frequently reported as causing illness. Throughout the 1993-1995 period, incidents of worker illness due to miscommunication of reentry provisions occurred. Blondell (1997) and Arne (1997) revealed potential undercounting problems in statistics collected on pesticide poisonings and accidental pesticide deaths.

The number of deaths related to pesticide exposures decreased dramatically compared with that in previous decades. For 1993 (California EPA 1995), one death was reported: that of a man who ingested strychnine-coated seeds, apparently with suicidal intent. Three deaths were reported for 1994 (California EPA 1996): two elderly men ingested malathion, and one person entered a building while it was being fumigated. For 1995 (California EPA 1997), one death was reported: a person broke into a motel room while it was being fumigated.

Special Focus: Health Risks of Farm Workers and EPA Worker Protection Standards

The impact of pesticides on the short- and long-term health of the agricultural workforce is poorly documented (Alavanja et al. 1996, Sanderson et al. 1995). Because agricultural workers are exposed to much higher levels of pesticides than the general public, studies of this group are more likely than studies of other groups to reveal relationships between health and exposure. Until recently the prevailing view has been that health studies of farm workers are very difficult to conduct because farm workers move often and are not generally interested in cooperating in studies. That view has been demonstrated to be inaccurate (e.g., Kamel et al. 1998). Some farm workers do move a lot, but others have home bases. Language barriers and concern over deportation are certainly issues,but it has been shown that these barriers can be overcome with appropriate use of camp aids by researchers who are sensitive to cultural issues.

Epidemiological studies conducted over the last 20 years point to some associations between particular pesticides and specific types of cancer (Zahm and Ward 1998) and subclinical neurological effects (Keifer and Mahurin 1997), but these studies have been limited in scope and technique. Most of the studies have focused on cancer incidence, have used case-control methods, and have been limited to small study populations. Recently initiated programs, such as the Agricultural Health Study (Alavanja et al. 1996)—which is being conducted cooperatively by the National Cancer Institute, EPA, the National Institute of Environmental Health and Safety, and the National Institute for Occupational Safety and Health (NIOSH)—are taking a more rigorous approach to the chronic health impacts of pesticides. The major results of these studies will not be available for many years and might not be able to point to specific pesticides that influence worker health. Nonetheless, the Agricultural Health Study is likely to produce important findings.

Some data exist on rates of acute, accidental pesticide poisonings, but problems with bias in reporting make it hard to estimate the true frequency of accidental poisonings (Arne 1997, Blondell, 1997). EPA and many farmworker advocacy groups view recent levels of such poisonings as unacceptable (EPA 1992, Davis and Schleifer 1998, Columbia Legal Services 1998).

In 1972, FIFRA was amended to broaden federal pesticide regulatory authority. EPA was given authority to ensure that registered pesticides were not used in a manner inconsistent with instructions on their pesticide labels. In 1974, EPA promulgated regulations under the title "Worker Protection Standards for Agricultural Pesticides" (40 CFR part 170). The regulations included

- Prohibition against spraying workers and other people
- Prohibition against reentry into a sprayed field before the pesticide had dried or dusts had settled, and a longer reentry period for 12 specific compounds

- Requirement of protective clothing for any person reentering a field before the designated reentry time
- Requirement for notifying workers regarding pesticide-related locations.

A 1983 EPA review of the 1974 FIFRA regulations designed to protect agricultural workers concluded that 40 CFR part 170 was inadequate. A major concern was the lack of enforceability. In 1992, after a long process of rule-making and public participation, EPA promulgated revisions to the 1974 rules (EPA 1992). The risk-benefit analysis in the 1992 *Federal Register* report stated that "EPA estimates that at least tens of thousands of acute illnesses and injuries and a less certain number of delayed onset illnesses occur annually to agricultural employees as the result of occupational exposures to pesticides used in the production of agricultural plants. These injuries continue to occur despite the protections offered by the existing part 170 and by product-specific regulation of pesticides." Additionally, The National Agricultural Statistics Service (NASS) estimated that 32,800 agricultural-related injuries occurred to children or adolescents under the age of 20 who lived on, worked on, or visited a farm operation in 1998. Many of these included exposures to pesticides (NASS 1999).

The 1992 regulations stated further that "the Agency believes that this (new 1992) rule will reduce substantially the current illness and injury incidents at modest cost to agricultural employers, pesticide handler employers, and registrants." The estimated cost after the first year was predicted to be $49.4 million per year. "Assuming that the majority of the current acute illnesses and injury incidents...are prevented through compliance with this new rule, there will be significant benefits to agricultural workers and pesticide handlers." In developing the new regulations, EPA indicated that the "minor use crops are the ones this Worker Protection Standard will impact the most. Much of agricultural labor is used in minor use crops, and it is in the production of these crops where the greatest chance of pesticide exposure to agricultural workers occurs."

The rules in the 1992 Worker Protection Standard (WPS) apply to all farm, forest, nursery, and greenhouse operations. The WPS is not applicable to rangelands, pastures, livestock operations, postharvest operations, structural pest control, nonagricultural plants or noncommercial plants.

The 1992 WPS and amendments to it are complex, but the general guidelines are as follows:

- General information on safe use of pesticides and facts about each pesticide application on the farm (defined broadly) must be posted in a central, accessible location.

- Pesticide-safety training must be provided to each pesticide handler and agricultural worker at least once every 5 years.
- Decontamination sites must be established within .025 mile of all workers and handlers.
- Commercial pesticide operators must inform the agricultural establishment of all facts regarding pesticide applications on the agricultural establishment property.
- Emergency assistance must be given to any employee who is poisoned or injured by a pesticide, and facts about the exposure must be given to medical personnel.
- Only appropriately trained and equipped handlers may enter treated areas during the restricted-entry interval (REI). The REIs for class I, II, III and IV pesticides are 48, 24, 12, and 12 hours, respectively.
- Workers must receive oral or written warnings of pesticide application, depending on the pesticide label.
- Warning signs must be posted before pesticide application and must be removed within 3 days after the end of the REI.
- Spray equipment and personal protective equipment (PPE) must be cared for and be disposed of in accordance with specific rules (for example, all PPE must be stored and washed separately from other clothing).

Further discussion of the WPS guidelines, and of issues of compliance and the potential for improvement of the guidelines can be found in the last major section of Chapter 3.

Food Residues

Although the important contribution of pesticides to world food production cannot be ignored (Klassen 1995), exposure to residues from their use is a continuing concern. Pesticide residues can occur in food as a result of treatment of a food crop or food animal with pesticides, of inadvertent contact with the chemical through exposure to air or water contaminated with the chemical, or in the case of food animals, of consumption of feed contaminated with the chemical. The amount of residue encountered in these situations is a complex function of such factors as the treatment rate or contact level, the physical and chemical properties of the pesticide, the time between exposure to the pesticide and harvest of the crop or food animal, and the processing or other treatment of the food commodity prior to its consumption as human food. A complicating factor is the decomposition of the parent pesticide to one or more metabolites or breakdown products; decomposition products can be toxic and so need to be considered when dietary exposures are being calculated.

The occurrence and concentrations of pesticide residues in foods are monitored by several organizations (OTA 1988). USDA annually performs one of the most important surveys of fresh and processed fruit, grain, vegetables, and milk in the Pesticide Data Program as part of its efforts to meet the mandates of the FQPA. This effort includes collecting data on pesticide residues in foods most likely to be consumed by infants and children. FDA monitors residues in crops and other foods, harvested in and imported into the United States except meat, milk, and eggs, which fall under the purview of USDA. Several states, including California, Florida, and Texas, conduct random monitoring of residues in foods. And several food industries, most notably those dealing with infant food or such high-value products as wine and some packaged and canned foods, analyze food ingredients for pesticide residues.

The percentage of food that tests positive for pesticide residues is usually 20% or less for all pesticides (Archibald and Winter 1990). The frequency varies widely with the pesticide and the food commodity. For example, ethylene bisdithiocarbamate (EBDC) fungicides were found in 49 of 124 samples of succulent garden-variety peas but in only four of 100 samples of bananas (NRC 1993). Chlorpyrifos was found in 116 of 968 samples of fresh apples but in only nine of 751 samples of succulent peas (NRC 1993). The differences are in part a result of the specifics of registration (chlorpyrifos is registered for use on apples but not on succulent peas) and of the percentage of acreage treated even when registration and a tolerance exists for a chemical-commodity combination.

When residues are detected, they rarely exceed established tolerance limits. Of all FDA surveillance samples, both domestic and imported, 3% or fewer were found to be in violation (Archibald and Winter 1990). Violations are very rare because tolerances are set to take into account the maximal residue expected at harvest when the pesticide is used according to its label. Violations also occur when a residue is found on a commodity for which there is no registration for the pesticide found—often the result of illegal use or inadvertent residue contamination.

The concern over pesticide residues in foods has many dimensions. Some pesticides, such as benomyl and EBDCs, or their metabolites (ethylene thiourea-ETU is a metabolite of EBDC) are animal carcinogens or suspected carcinogens (NRC 1987, Winter 1992). Their presence in foods can increase the risk of death due to cancer. Risk is proportionate to the frequency with which a residue is encountered, the concentration of the residue, and the frequency of consumption of foods containing the residue. But because carcinogens do not appear to act through a threshold mechanism (NRC 1987), any residue consumed can produce a statistical increase in lifetime dietary cancer risk.

Population subgroups differ in exposures to residues. Infants and children, for example, can receive higher exposures to pesticides that occur regularly as residues on such foods as apples, bananas, and pears because they consume more of these foods than adults and because they are smaller than adults. Their exposure to a pesticide in these foods is higher, on a bodyweight basis, than adults' (NRC 1993). The higher exposure, combined with the greater susceptibility of infants and children to some pesticides, can increase risks for these population subgroups.

People can be exposed through water or air or by dermal contact, in addition to food intake. Thus, their total, or aggregate, exposure can be considerably higher than their exposure via food intake. Exposure can be to several chemical residues simultaneously because some commodities are treated with more than one pesticide and because people consume many foods, any or all of which can contain residues. The effect of these cumulative residues and whether they act independently of each other or in an additive or synergistic manner, are subjects of active debate.

Other acute or chronic effects in people can be associated with residues in food, particularly in the context of illegal use of pesticides. Although aldicarb is still used legally on some food crops, thousands of people in the western United States became ill in 1985, some seriously, from consuming watermelons contaminated by the illegal use of this insecticide. Although random, composite sampling might show the remaining residues to be well within tolerance, there is some concern that individual samples or other commodity units could contain above-tolerance, possibly harmful residues (NRC 1993). Suspicions regarding potential teratogenic, neurotoxic, or hormone-disrupting effects also exist for some chemicals or residue mixtures; these are additional subjects of current research and public interest.

The concentration of pesticide residues in foods and the frequency with which they occur have decreased substantially in recent years. Data on pesticide residues in foods are generally available from field trials conducted by the registrant, monitoring programs of FDA and state agencies, and marketbasket surveys of FDA and USDA. Marketbasket surveys—such as FDA's Total Diet Study, in which foods purchased from supermarkets and prepared for consumption are analyzed—provide perhaps the best record of dietary intake, although the data are few. Marketbasket surveys support the idea that, generally, very low exposure to pesticide residues occurs through foods in the United States.

In part, the low exposure is due to the effects of commercial processing and food preparation (NRC 1993, Elkins 1989, Gelardi and Mountford 1993). The decline in food residues has occurred in large part, however, because of changes in pesticide chemistry and improved monitoring of the food supply. Pesticide chemists now favor the use of chemicals of lower stability, which break down substantially before harvest and before human consumption of foods. The stable organochlorine insecticides in wide use in the 1940-1980 era (such as DDT, toxaphene, and chlordane) have been replaced with less-persistent, less-bioaccumulative chemicals. And, many modern chemicals are effective at much lower dosage, so the absolute amount of residue on treated food is one-tenth or less of the amounts of the older chemicals that they replace. That is not in itself a guarantee of less risk to consumers, but it turns out that modern chemicals generally have higher margins-of-safety than the older chemicals. Furthermore, current patterns of pesticide registrations (Figure 2-4) suggest that a greater proportion of the modern pesticides are considered reduced risk chemicals (EPA, 1999).

The FQPA (1996) poses even more challenges to industry to decrease risks from exposure to pesticide residues (FQPA 1996, McKenna and Cuneo 1996). Both aggregate-exposure chemicals and cumulative-exposure chemicals must be considered in risk assessment. Exposures of sensitive subpopulations, such as infants and children, are to be given even more attention than that afforded as a result of the National Research Council's 1993 report (NRC, 1993). Risks due to disruption of endocrine systems must be addressed with new tests in the chemical evaluation phase. A single health-based standard for pesticides in foods was established, replacing the paradox created by application of the "Delaney clause" (which was eliminated by the FQPA) to processed foods (NRC 1987).

Soil, Air, and Water Exposures

An understanding of the transport and fate of pesticides in the environment is required to evaluate the potential impact of pesticides on human health and the environment; for example, it is needed to conduct risk assessments, such as evaluating the probability that a pesticide application will contaminate groundwater. Such knowledge is also required to develop and evaluate methods for remediating environmental contamination. Just as important, knowledge of contaminant transport and fate is necessary to design pesticides and application strategies that minimize adverse impacts.

The four general processes that control the movement of chemicals in the environment are advection, dispersion, interphase mass transfer, and transformation reactions. Advection, also referred to as convection, is the transport of matter by the movement of a fluid responding to a gradient in fluid potential. For example, a pesticide dissolved in water will be

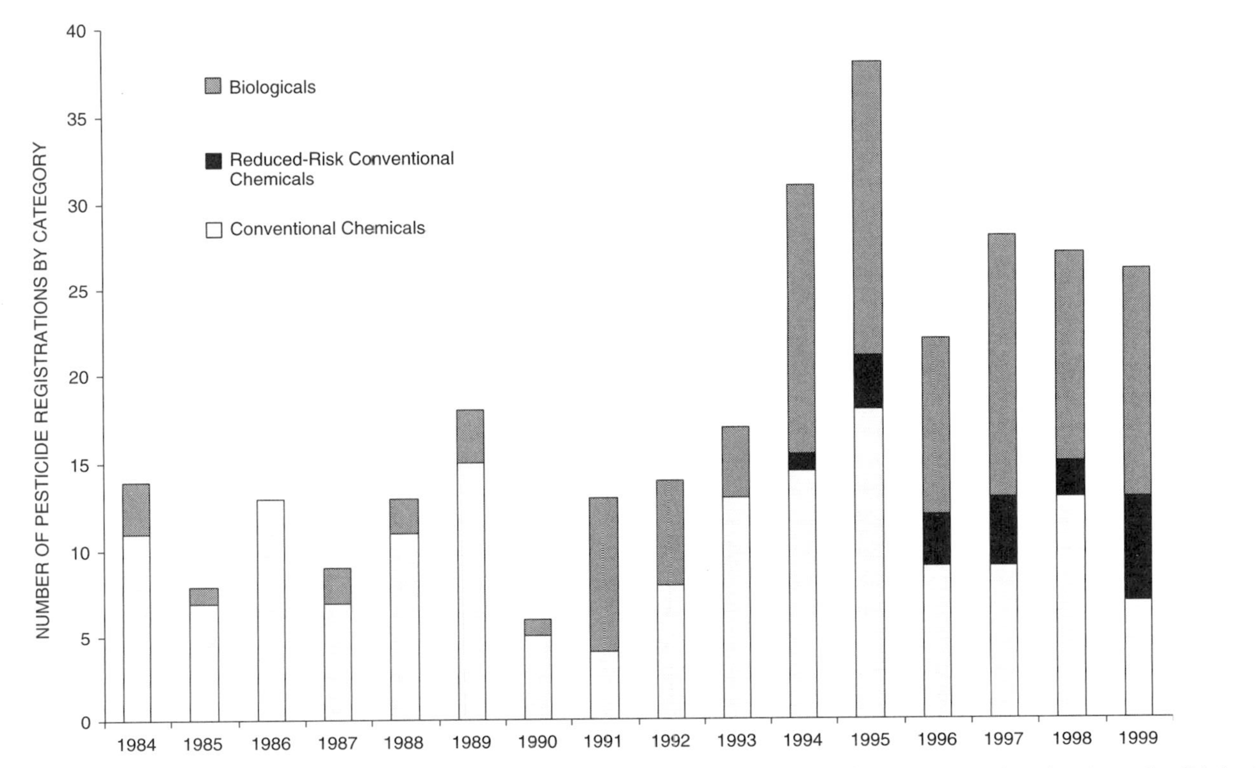

FIGURE 2-4 Registration of safer chemicals. Proportion of pesticide active ingredients that are considered to be safer (biological chemicals and reduced-risk conventional chemicals) than conventional chemical pesticides has steadily increased over the last several years.
Source: EPA, 1999.

carried along by the water as it flows through (infiltration) or above (runoff) the soil. Dispersion is the spreading of matter about a mean position, such as the center of mass; spreading is caused by molecular diffusion and nonuniform flow fields. Interphase transfers—such as sorption, liquid-liquid partitioning, and volatilization—involve the transfer of matter in response to gradients in chemical potential or, more simply, concentration gradients. Transformation reactions include any process by which the physicochemical nature of a chemical is altered; examples are biotransformation (metabolism by organisms) and hydrolysis (interaction with water molecules). Additional information regarding the factors and processes that influence the transport and fate of contaminants in the environment can be found in Pepper et al. (1997).

The fate of a specific pesticide in the environment is a function of the combined influences of those four processes. The combined impact of the four processes determine the "pollution potential" and "persistence" of a pesticide in the environment. The pollution potential characterizes, in essence, the "ability" of a pesticide to contaminate the medium of interest (soil, water, or air). Pesticides with larger pollution potentials are generally transported readily (for example, low sorption) and not transformed to any great extent (they are persistent). High rates of transport mean that a pesticide readily moves away from the site of application. A low transformation potential means that a chemical will persist, and thus maintain its hazard potential, for a longer time than one with a high potential. The risk posed by a specific pesticide to humans or other receptors is, of course, a function of its toxicity, as well as its pollution potential. It is therefore important to understand both types of properties. For example, pesticides that are very mobile, persistent, and highly toxic will generally be associated with the greatest risk.

Once a pesticide is applied to or spilled onto soil, it can remain in place, or transfer to the air, surface runoff, or soil-pore water. Transfer of pesticides to surface runoff during precipitation or irrigation is a major concern associated with non-point-source pollution. Once entrained into surface runoff, a pesticide can be transported to surface-water bodies. Consumption of contaminated surface water is a major potential route of exposure to pesticides. A recent discussion of pesticides and non-point-source pollution is presented in Loague et al., (1998).

The other major potential soil-water route of pesticide exposure of humans is consumption of contaminated groundwater. Once applied to soil, a pesticide can partition to the soil-pore water. It then has the potential to move down to a saturated zone (aquifer), thereby contaminating the groundwater. Whether that occurs, the time it takes, and the resulting degree of contamination depend on numerous factors. Major factors are the magnitude and rate of infiltration and recharge, soil type, depth to the aquifer, quantity of pesticide applied, and its physiochemical properties (such as solubility, degree of sorption, and transformation potential). Volatile pesticides can move by advection and diffusion in the soil-gas phase in addition to the pore-water phase, and these provide additional means to travel to an aquifer.

The transport and fate of pesticides in porous media (such as soil) and the potential for pesticides to contaminate groundwater have been the subjects of an enormous research effort. The initial paradigm for transport of chemicals in porous media was based on assumptions that the porous medium is homogeneous and that the rates of inter-phase mass transfer and reaction are linear and essentially instantaneous. However, it is well known that the subsurface is in fact heterogeneous, and that many phase transfers and transformation reactions are not linear or instantaneous. In addition, the transport of many contaminants is influenced by multiple factors and processes (Brusseau 1994). The transport and fate of pesticides in the subsurface are complex and generally difficult to predict accurately.

An example of the complex behavior of chemicals in soil is the long-term persistence observed in many field studies. The results of several studies have shown that chemicals, including pesticides, that have been in contact with soil for long times are much more resistant to desorption, extraction, and degradation. For example, contaminated samples taken from field sites exhibit solid:aqueous distribution ratios that are much larger than those actually measured or estimated on the basis of spiking the porous media with the same contaminant (adding contaminant to an uncontaminated sample) (Steinberg et al. 1987, Pignatello et al. 1990, Smith et al. 1990, Scribner et al. 1992). In addition, the desorption rate coefficients determined for previously contaminated media collected from the field are much smaller than those obtained from spiked samples (Steinberg et al. 1987, Connaughton et al. 1993). The confounding aspect of those results is that they are often observed for chemicals that would not necessarily be expected to exhibit substantial persistence, because they are, for example, volatile or readily degradable. The mechanisms for such behavior are unclear and are the subject of current investigation. Additional information on the transport and fate of pesticides in soil and groundwater can be found in several books (Sawhney and Brown 1989, Cheng 1990, Linn et al. 1993).

As discussed in a recent NRC (1994), it is difficult to remediate contaminated soil and groundwater. There are three general approaches: control, removal, and in situ treatment. The purpose of control-based approaches, which include the use of well fields (hydrodynamic control) and physical barriers (slurry walls), is to prevent further migration of the contamination. The purpose of removal-based approaches—which include excavation, the pump-and-treat method, and soil-vapor extraction—is to remove the contamination from the subsurface. The purpose of in situ treatment, which includes in situ bioremediation and in situ chemical transformation, is to reduce or eliminate the mass of the hazardous chemical in place. Any of those methods can be successful for a specific site under specific circumstances. However, because of the complexity of the subsurface and the physiochemical nature of many contaminants, the effectiveness of any remediation method will be constrained in many cases. As a result, subsurface remediation is expensive, and there is no guarantee of complete success. Although advances in remediation and characterization technologies continue to be made, it remains true that pollution prevention is much cheaper than pollution remediation in the long term.

Pesticides enter the air by a variety of processes. Drift during application can result in the entry of small aerosol particles and vapors into the air and their movement downwind from the spraying operation. After application is completed, additional residue can enter the air by volatilization and wind erosion. The resulting residue, both particles and vapors, can travel downwind, where deposition can occur by the wet processes of rainfall, snowfall, and fog coalescence washout or the dry processes of vapor exchange with surfaces and fallout of particles. Degradation can accompany the downwind processes, and the products of decomposition can join the parent-chemical residue in further transport and deposition.

Airborne pesticides can undergo long-range transport to remote environments (Kurtz 1990). The presence of residues of organochlorine chemicals in the earth's polar regions, where they accumulate in the body fat of Inuits and in seals and polar bears, is evidence of the operation of a "cold condensation" mechanism that moves residues from temperate and tropical areas of use to colder regions (Wania and Mackay 1993). The finding of current-use chemicals in the alpine and subalpine regions of the Sierra Nevadas (owing to airborne movement from California's farming valleys) (Zabik and Seiber 1993) and in the polar regions is a cause of continuing concern about long-range movement of pesticides.

There are several reasons to be concerned about airborne residues (Seiber and Woodrow 1995). Airborne residues can represent a direct hazard to humans, wildlife, and vegetation. The human hazard might be most prominent for farm workers who apply chemicals or work in and around treated fields. But people who live, work, or play downwind can also be exposed. The recent hypothesis of possible endocrine system disruption by exposure to "background" chemicals has raised concern about the potential of adverse population-level effects of ecosystem-scale contamination, such as can be caused by airborne residues (Arnold et al.

1996). Atmospheric residues might cause large-scale, long-term effects that are only slowly reversible. The entry of methyl bromide, a fumigant (see Box 2-1) into the stratosphere and its contribution to the depletion of the earth's protective ozone layer have resulted in the planned phaseout of this chemical by the year 2001.

Generally, the dermal route of contact is much greater than the respiratory route in occupational exposures and those immediately downwind. But important exceptions occur for very volatile pesticides, particularly fumigants, such as methyl bromide, ethylene oxide, and telone (Ecobichon 1991). For example, accidental deaths are reported each year that are due to occupational and nonoccupational exposure to fumigants used in confined spaces (Mehler et al. 1992). Downwind exposures of nontarget plants and animals can also occur, with herbicide damage the most obvious example. Downwind exposures can involve contact with airborne residues or with deposited residues in soil, vegetation, or water. Multimedia exposures might be more important for plants and wildlife than for people, although certainly people can be exposed unintentionally to deposited residues.

Atmospheric residues constitute a missing element in the chemical mass balance of a pesticide. Residues that leave the target area in the air pose an economic loss to the grower because they consist of chemical that is no longer available for pest control. From a broader perspective, losses to the atmosphere consist of mass that can no longer be accounted for or controlled. Given the large losses to the air that occur in some cases of pesticide use, it is difficult to convince an already wary public that pesticides are safe when substantial portions leave the target zone and end up where they are not needed or wanted. Few instances are known or reported in which downwind airborne residues have resulted in measurable harm to people under normal agricultural pesticide-use conditions. The use of application buffer zones and other restrictions on the use of such chemicals as methyl bromide, which has a tendency to volatilize, or paraquat, whose toxic action is on the lung, has minimized exposures (Baker et al. 1996). But concerns still exist among downwind residents, and these concerns are driving attitudes toward pesticide usage that cannot be dismissed. It might become more important in the future to pay special attention to ways to reduce off-target airborne loss of chemicals by improving application efficiency, imposing additional environmental-use conditions on applicaters, and encouraging the use of pesticides that experience minimal loss to the atmosphere (because of low vapor pressure, effectiveness at low dosage, and rapid environmental degradation) (Seiber and Woodrow 1998).

As discussed, the release and transport of pesticides in the environment are complex and can have many adverse effects. Thus, a question of

BOX 2-1
Fumigants

Fumigants are a loosely defined group of pesticide active ingredients and formulations that act through partial or sole distribution of the chemical and delivery to the target in the vapor form (Ware, 1983). Fumigants include chemicals applied to the soil before harvest to control nematodes, insects, and undesirable microorganisms (soil fumigants). They also include chemicals used after harvest to control pests in harvested commodities before storage or transport to markets and chemicals used to rid infested buildings of termites, cockroaches, and other pests. Fumigation is a specialized use that has recently attracted unusual interest because of the visibility of some fumigation operations and the past contribution of fumigants to human health and environmental problems.

Several fumigants have been withdrawn or banned, and several current fumigants are undergoing reevaluation or phaseout. Dibromochloropropane (DBCP), a soil nematicide, was widely used because it could be used after planting on numerous fruit and vegetable crops susceptible to nematode root infestation. In 1977, it was withdrawn after it was found to cause sterility in male workers exposed during manufacture. Coming from areas of manufacture, formulation, and use, it was also a contaminant in groundwater. Its registrations, except for use on pineapples in Hawaii that was deemed essential for production, were canceled by EPA in 1979. Ethylene dibromide (EDB) enjoyed widespread use as a preharvest soil fumigant to control nematodes, eggs, and soil insects. EDB was a widely distributed groundwater contaminant, partly because of its water solubility, method of use (injection into soil followed by irrigation to improve penetration), high rate of use, and stability. In one summary, wells in six states were reported to be contaminated with EDB; this resulted in contamination of 520 wells (of 5,133 tested) above the health advisory level for drinking water. Health concern resulted from positive carcinogenic testing in experimental animals. The use of EDB was cancelled by EPA in the early 1980s.

Methyl bromide is a highly effective preplant fumigant with wide use on high-value crops—such as citrus, other fruit, and nut orchard crops; grapes; and strawberries—and in nurseries. After the cancellation of EDB and DBCP, it became the fumigant of choice because of its effectiveness against a wide spectrum of pests, including arthropods, nematodes, fungi, bacteria, and weeds (Noling, 1997). Methyl bromide is an effective postharvest fumigant for stored nuts and other commodities, and its use is required in some cases to circumvent quarantine of produce that might, through shipping, transport pests into uninfested areas. It is also used to fumigate structures; this use has resulted in a few fatal accidents when reentry regulations were not followed after treatment with the chemical. Methyl bromide is a gas at ambient conditions and leaves no organic residue in the treated soil or commodity, because it breaks down to bromide and eventually carbon dioxide. But methyl bromide is stable in the atmosphere, creating a potential for exposure among workers and downwind residents and for diffusion to the stratosphere, where its interaction with high-energy ultraviolet radiation sets off a series of reactions that lead to depletion of stratospheric ozone (Methyl Bromide Global Coalition, 1992).

Because of its ozone-depleting potential, methyl bromide is scheduled for phaseout in the United States by the year 2001 as a consequence of rules in the 1990 Clean Air Act Amendments and Montreal Protocol agreements (Honagonahalli and Seiber, 1997). In 1993, nearly 15 million pounds of methyl bromide were used in California (California Environmental Protection Agency, 1994), partly because of the unusually high application rate—up to 500 lb/acre in some preplant uses.

There is some anxiety regarding the availability or suitability of alternatives to methyl bromide if the scheduled phaseout occurs. Current registered alternatives are 1,3-D (1,3-dichloropropane, Telone II) and metam sodium, which generates the active fumigant methylisothiocyanate when the parent chemical breaks down in soil or water. Telone was placed in the reregistration and special-processes review by EPA in 1986; this requires EPA to conduct a risk–benefit analysis leading to a regulatory decision (Roby and Melichar, 1996). Telone was also suspended by the California Department of Pesticide Regulation (DPR) in 1990 after air monitoring showed residues above the state's acceptable air concentration limit around fumigated fields in California's San Joaquin Valley. The suspension was lifted in 1994 on the basis of changes in application practice and additional monitoring studies conducted by the registrant and further review by DPR.

MITC received adverse publicity after a 1991 spill of metam sodium in the Sacramento River that caused widespread intoxication of fish and other aquatic organisms and loss of vegetation. But metam sodium registration continues, and the product is finding expanding markets as the regulatory status of methyl bromide and Telone remains in some question. Another alternative is chloropicrin, which is often used as an odorant or residual pesticide in combination with methyl bromide and other mainstream fumigants. Other fumigants include phosphine, which is released from aluminum and other phosphide salts for control of rodents and other vertebrate pests, and sulfuryl fluoride (Vikane). Nonfumigant pesticides, such as carbofuran and methomyl, can be substituted for fumigants in some soil pest-control situations. Nonsynthetic alternatives to fumigants are also receiving increasing attention. These include solar heating and natural nematicide chemicals present in some plants in the case of soil fumigation and controlled atmospheres and heating in the case of postharvest and building-space fumigation.

The fumigant situation is a highly visible, somewhat controversial dilemma that faces conventional pesticides on many fronts. Long-standing, well-accepted chemicals can come under increasing scrutiny as a result of some adverse characteristic (chronic toxicity in experimental animals, groundwater contamination, stratospheric ozone depletion) or changes in federal or state regulations. Registrants conduct additional studies to preserve registration of their products. Agriculturists worry about whether loss of productivity will result if particular pesticides are no longer available. Some in the public sector call for swift action in removing or restricting offending chemicals. And researchers go back to the drawing board to devise new solutions to pest problems.

critical concern is, to what extent have pesticides contaminated the environment? One way to address the question is to examine the occurrence of pesticides in surface water, groundwater, and air. An analysis of such occurrences in the United States has recently been reported under the auspices of the National Water-Quality Assessment Program of the US Geological Survey (USGS) (Majewski and Capel 1995, Barbash and Resek 1996, Larson et al. 1997). The reports compile and synthesize the many occurrence studies that have been conducted in the United States. In addition, the results of a recent data-collection program for groundwater conducted by USGS have been reported (Kolpin et al. 1998). The results from the latter study are used below to illustrate the occurrence of pesticides in the environment.

To examine pesticide occurrence in groundwater, samples were collected from recently recharged groundwater (generally less than 10 years old) in 41 land-use areas scattered throughout the United States. As reported in Kolpin et al. (1998), pesticides were detected at 54.4% of the 1034 sampling sites. Many pesticides were found; 39 of the 46 that were looked for were detected. Pesticide concentrations were generally low; more than 95% of the detected concentrations were less than 1 μg/L. Maximal contaminant concentrations developed by EPA for drinking water, which have been established for 25 of the 46 pesticides, were exceeded for only one pesticide at a single location. Pesticides were detected in both agricultural (56.4%) and urban (46.6%) locations.

Those results indicate that pesticides are widespread in shallow groundwater. As noted in Larson et al. (1997), they are also widespread in surface waters. The potential impact on human health is uncertain, given the relatively low concentrations and the lack of understanding of the impact of low concentrations on human health. In addition, more than one pesticide is often found at a site. For example, two or more pesticides were detected in shallow groundwater at 73% of the sites where pesticides were detected (Kolpin et al. 1998). The impact of chemical mixtures on human health is unclear. The widespread detection of pesticides in urban areas illustrates the fact that pesticide use and its associated problems are not confined to agricultural applications. Clearly, strategies for minimizing the impact of pesticides on human health and the environment will be successful only if all uses of pesticides are considered.

Ecological Problems: Impacts on Nontarget Organisms

The basic intent in the design of pesticides is to produce substances that are highly toxic to pest species, but much less toxic to the nonpest species so that there can be a useful margin of safety. The toxicity to each species is related to the characteristics of the substance and to the dose absorbed by the organism. There appear to be no substances that are nontoxic to all species regardless of dose.

Because the status of any particular organism as a pest depends not on its taxonomy or physiology, but rather on whether its ecological role brings it into conflict with humans, it is to be expected that pest species often share many biochemical pathways, physiological functions, anatomical features and life history attributes with nontarget species. Therefore, it should be obvious that it is difficult to design pesticides that exhibit ample margins of safety with respect to all nontarget species. Given the large number of potential pest species and the plethora of commercial and nontarget species, it is also very rare to have pesticides available containing the desired specific activity against target pests.

However, degrees of specificity are readily apparent. Specificity varies with likelihood of encounter or exposure. A general difference between humans and insects, for example, is that insects are much smaller. The basic body plans of animals can be summarized as a collection of spheres and cylinders whose surface areas increase according to the squares of the applicable radii, and whose volumes increase according to the cubes of the same radii. That translates into disproportionately greater surface area in relation to body volume or mass as organisms become smaller. The ratio of surface area to body mass in insects relative to their body mass can easily be 100 times the ration in humans and common domestic animals. That has the direct effect of proportionally increased absorption rates in case of the insects. The combination of rapid absorption rate and small volume into which the pesticides are being absorbed results in a rapid attainment of the maximal concentration (or dose). Ease of absorption associated with proportionately large surface areas is also exploited in the case of pesticides used with other arthropods (acaricides, nematicides, and so on). The comparatively large surface area of plant leaves has also been used with some herbicides to provide selectivity against broad-leaved plants.

Probability of exposure notwithstanding, perhaps the most important consideration in determining the probability of nontarget effects is the degree of biochemical or physiological similarity to the target species. Nontarget effects of broad-spectrum pesticides are thus to be expected; these compounds are aimed at target sites that are widely distributed among organisms. The degree of risk often correlates highly with the degree of resemblance to the target pest. There is ample documentation of the nontarget effects of the early chlorinated hydrocarbon insecticides on nontarget insects; the phenomena of secondary pest problems and pest resurgence are attributable in large part to the greater toxicity of these insecticides to arthropod natural enemies than to arthropod target species. However, pesticides with even narrower modes of action can have nontarget effects that reflect degree of similarity. Insecticides that act as chitin-synthesis inhibitors have historically caused problems among aquatic crustaceans, which as arthropods synthesize chitin for exoskeleton production, when residues contaminate watersheds.

Shifts to more narrow-spectrum compounds administered at lower rates were motivated partly by the desire to reduce nontarget effects; reducing environmental persistence also contributes to reducing the likelihood of encounter by nontargets. However, according to some sources (Benbrook 1996), these shifts might not have reduced nontarget effects as much as intended. Croft (1990) summarized data on the impact of 400 pesticide active ingredients on over 600 species of beneficial arthropods. His basic finding, as might have been predicted, was that insecticides have greater impacts than either herbicides or fungicides. However, calculation of selectivity ratio (the ratio of LD 50 for target species to LD 50 for nontarget species) suggested that toxicity to nontarget arthropods, partly because of the toxicity of synthetic pyrethroids, has increased (Benbrook 1996), although statistical analysis of such data presents challenges.

A major concern historically about the use of insecticides has been nontarget effects on vertebrates. Organophosphates and pyrethroids continue to present toxicity problems in fish and have been associated with major fish kills on several occasion after aerial spraying. Data compiled by the US Fish and Wildlife Service (Mayer and Ellersieck 1986) on over 400 chemicals suggest that sensitivity varies with species, age, and water temperature. Concerns also exist about nontarget effects on terrestrial vertebrates; indeed, food-chain effects culminating in reproductive failures in birds contributed substantially to raising public awareness and activism with respect to pesticide regulations (for example, Carson 1962). Many of the chlorinated hydrocarbon pesticides exhibited very high lipid solubility. Consequently, they were readily absorbed, bioconcentrated, and biomagnified, especially in food webs that terminated with top predators (Peterle 1991). That exposure altered calcium disposition, especially in some species of predatory birds, which led to reproductive failure due to the thinning and breaking of the eggshells (Cramp 1963) and to serious declines in populations, especially of peregrine falcons, ospreys, and brown pelicans. Since the banning of DDT, most of the predatory-bird populations have recovered substantially (Peterle 1991, Robbins et al. 1986, Greig-Smith 1994).

Although many of the older compounds with such properties have been banned, organophosphates and carbamates still in use have been implicated in nontarget effects on songbirds, waterfowl, and game birds. Indirect effects on terrestrial birds and mammals have been documented; insecticide use can reduce cover and prey abundance and thus lead to greater mortality (Blus and Henry 1997). Also recorded are secondary and tertiary poisonings resulting from ingestion of contaminated bird prey by raptors and other birds of prey.

Developing pesticides with specific modes of action has been given a high priority to reduce nontarget effects. Very high degrees of specificity in the control of some animal pests can be gained through such techniques as the use of specific pathogens, sterile-male techniques, and the use of species-specific pheromone blends, although the utility of these approaches depends on the availability of specific pathogens and the appropriateness of the mating habits of the pest species. However, unexpected nontarget effects have arisen even when metabolic resemblances seemed unlikely. A number of herbicides derive their potency against plants from their ability to inhibit photosynthesis. That generally affords a greater margin of safety toward animals. The phenoxy acetic acid herbicides 2,4-D (2,4-dichlorophenoxy acetic acid), 2,4,5-T (2,4,5-trichlorophenoxy acetic acid), and Silvex (2,4,5-trichlorophenoxy propionic acid) are specific to broad-leaved plants and woody plants by mimicking plant growth hormones. The trichloro compounds were removed from the market because they could not be produced without the highly toxic dioxin or TCDD as an impurity (Leung and Paustenbach 1989). Indeed, many of the pesticides with putative endocrine-disrupting effects in humans are herbicides or fungicides (Clement and Colborn 1992), and some herbicides might have unexpected toxic effects by virtue of their ability to inhibit detoxification enzyme systems (Benbrook 1996).

Specificity is not always the goal of pest management. It is often desirable to remove all plants from a particular plot. Such intentionally nonspecific applications make use of fundamental biochemical processes, but are still subject to the surface area-volume limitations. The herbicides 2,4-dinitrophenol and 2,4-dinitro-*o*-cresol inhibit the organic phosphorylation of adenosine diphosphate to adenosine triphosphate, which is centrally important in all higher plants and animals. Consequently, those two herbicides have been widely used to clear rights of way, and they have had to be used with caution by the applicater, who is also susceptible to the inhibition of organic phosphorylation.

The application of pesticides results in indirect effects on ecosystems by reducing local biodiversity and by changing the flow of energy and nutrients as the biomasses attributable to individual species are altered. Numerous studies have shown that pesticides decrease crop loss, but the potential indirect environmental impacts of pesticides have not been widely studied. An ecological perspective of pesticide effects include direct and indirect effects on all associated organisms, not just the pests in question. Thus, ecosystem impacts of pesticides can include effects on human health, domestic and wild animal response, effect on natural enemies of pests, crop pollinators and honey bee response (Buchmann and Nabhan 1996), and soil microorganism community response. Domestic and nontarget wild animals can be poisoned by direct contact with pesticides (particularly rodenticides) and by feeding on contaminated pests (Barton and Oehme 1981, Beasley and Trammel 1989). Deleterious effects of pesticides on wild birds include death from direct exposure or from eating contaminated prey (Flickinger et al. 1991, Stone and Gradoni 1985, White et al. 1982).

Indirect effects come about as a result of alterations by pesticides in ecosystem trophic dynamics. Birds, for example, experience reduced survival, growth and reproductive rates, and habitat through elimination of food sources and refuges (McEwen and Stephenson 1979, Potts 1986). It has been suggested that native wild bird populations or communities can be good indicators of ecosystem integrity because they tend to be sensitive to vegetation structural changes and disruption of food webs over several trophic levels (Blus and Henery 1997, Fry et al. 1986, Landres et al. 1988). Many birds feed on diverse insects, particularly at early developmental stages, and on a wide array of seeds as they mature (Ehrlich et al. 1988). Further research on bird population responses to pesticides and as ecosystem- integrity indicators may be warranted.

Weed species shifts can result from intensive and consistent use of herbicides. In this case, herbicides cause local extinction of some weed species and thus leave an ecological niche for species that have greater tolerance of, can avoid, or are resistant to the herbicide. Fryer (1982) showed that the frequency of many broadleaf weed species in cereal production fields of Great Britain declined and the grass weeds (such as *Avena fatua* and *Alopecurus myosuroides*) increased as a result of 40 years of routine use of the selective broadleaf herbicide 2,4-D. Weeds that are taxonomically closely related to the crop are often selected because of common physiology and therefore similar response to a herbicide. Weeds in the Solanaceae family are common in potato and tomato crops (same family) in which trifluralin is commonly used as a selective herbicide. Most solanaceous weeds are tolerant of trifluralin.

Pesticide effects on nontarget beneficial organisms have not been extensively studied for most agricultural systems. However, in agroecosystems where some studies have been conducted, maintenance of crop monocultures with pesticides has created simplified plant and insect communities dominated by a few pests (Ryszkowski et al. 1993). Several studies have shown that as structural diversity is increased with intercropping or weeds, the diversity of pest and beneficial insects increases, but damage by insect pests is generally reduced (Altieri et al. 1978, Dempster 1969, Dempster and Coaker 1974, Perrin 1977, Foster and Ruesink 1984, Risch et al. 1983, Tahvanainen and Root 1972, Van Emden 1990). There is some evidence biological control of insect pests can be maximized by increasing the availability of alternate hosts such as weeds (Debach and Rosen 1991, Van Emden 1965, vanEmden 1990, Powell 1986, Rabb et al. 1976). Fungicides could contribute to weed or insect outbreaks by reducing natural pathogenic fungi on these pests (Johnson et al. 1976).

Honey bees and wild bees are important as pollinators in many crops (Robinson et al. 1989), as well as for production of honey. Yearly costs of direct honey bee losses and of yield reductions resulting from reduced pollination in response to off-target response to insecticides are high (Pimentel et al. 1992). In addition, weed control by herbicides can have negative effects on pollinators by selectively removing important nectar sources.

Microorganisms and invertebrates play an important role in biogeochemical cycling of carbon, energy, water and nutrients essential for crop growth. Maintenance of these cycles is critical for ecosystem integrity (Brock and Madigan 1988). Pesticides can disrupt these processes with even sublethal effects on the microorganisms. Little is known about the dynamics of soil microorganism communities and their response to pesticides. The potential for disruption—though interference with mycorrhizal associations, nutrient uptake, or susceptibility to disease—is great (see references cited in Benbrook 1996).

PUBLIC PERCEPTION OF PESTICIDES

It is widely believed by the US public that pesticides pose substantial dangers to the population at large through residues on foods and groundwater contamination, to farmworkers through occupational exposure, and to wildlife and the environment. Numerous surveys conducted in 1984–1990 scattered across the United States showed that most Americans had serious concerns about pesticide residues on foods (Sachs et al. 1987; Jolly et al. 1989, Food Marketing Institute 1989, Porter Novelli 1990, Dunlap and Beus 1992, Weaver et al. 1992). The concern might have diminished since then. A national poll conducted in 1994 found that about 35% of Americans believed that pesticides were very dangerous for themselves and 38% believed that pesticides were very dangerous for the environment (National Opinion Research Center 1994)—about half the percentages reporting such concerns only a few years earlier.

Whether scientifically valid or not, public perception of pesticide risks has an important effect on the development of pesticide regulations and on food purchases (Jolly 1991, Bruhn et al. 1992). According to a report sponsored by the Council for Agricultural Science and Technology (van Ravenswaay 1995), research on public perception of pesticide risks "is in its infancy, and more research is needed to develop valid and reliable theories, methods, and conclusions about public perception of agrichemicals and other agricultural technologies." Although there are important limitations to available research results, existing studies offer useful insights into both the perceptions themselves and ways to improve research on public perception.

A number of studies have emphasized the great diversity in perception of pesticide risk among segments of the US population (for example, Jolly 1991, Baker and Crosbie 1993, van Ravenswaay and Hoehn 1991a). Van Ravenswaay and Hoehn (1991a) found that about 25% of over 900 participants in a nationwide survey felt that there was a 1% or greater chance that someone in their household would "have health problems someday because of the current level of pesticide residues on their food"; 4.4% of the participants felt that such residue-related health problems were certain to occur. In contrast, almost 25% of the participants felt that the chance of such health problems was less than 0.0001%, and 4.1% thought that there was "no chance" of such problems. People seem to vary widely in their assessment of how harmful pesticide residues are for their health. However, the diverse responses could also reflect differences in how people in the survey defined "health problems".

A survey of households in Seattle, Washington, and Kobe, Japan (Jussaume and Judson 1992), assessed the affect of household characteristics on concern about food safety. In general, the Kobe households were more concerned about food safety than the Seattle households. Households with children under 18 years old and households with incomes below $50,000 per year (10,000,000 Yen) were less likely to trust the safety of government-inspected foods. Other studies have found a negative relationship or no relationship between education and concerns about pesticides (Misra et al. 1991, Dunlap and Beus 1992). Jussaume and Judson (1992) found that households in which more vegetables were consumed were more concerned about food safety and had less trust in government, business and farmers than other segments of the sample.

It is difficult to compare results of those studies because they sampled different geographic areas and presented different questions (or posed similar questions in a different way). Sachs et al. (1987) tried to address that problem in assessing whether public concern about pesticide use had increased between the 1960s and the 1980s. They essentially replicated a study of Bealer and Willits (1968) conducted 19 years earlier. The results indicated that in 1965 94% of the public felt that the "government adequately regulates chemical use in or on food." In 1984, only 48.9% of respondents agreed with that statement. Other measures of risk posed by pesticide residues also increased dramatically.

Instead of simply documenting the level of concern about pesticide risks by direct inquiry, a few surveys were designed to incorporate a method called "contingent valuation" to assess how much a consumer is willing to pay for implementation of measures that could decrease the risks posed by pesticide residues on food. A commonly used measure of potential consumer response is the impact of concerns on demand—in this case, the extent to which the presence of residues affects the unit price that consumers are willing to pay for any given quantity purchased. If the value of avoiding residues is large, consumers should be willing to pay a substantial premium for produce that is free of residues. A number of studies conducted in 1988–1990 attempted to estimate the consumers' willingness to pay for produce certified to be free of pesticide (Gallup 1989, Ott 1990, Misra et al. 1991, Weaver et al. 1992). In those studies, respondents who reported being willing to pay much more for such produce were asked to choose which percentage markup represented the most extra that they would be willing to pay. This approach does not replicate an actual choice situation; that is, respondents knew their answer would have no direct financial consequences, such as actually paying more. Survey formats of this kind tend to generate excessively high reports of willingness to pay. Even so, few consumers reported being willing to pay more than 5% extra more for certified residue-free produce. Even fewer reported being willing to buy certified residue-free produce that had lower cosmetic quality or more surface defects. Misra et al. (1991) found that 56% of their respondents felt that fresh produce should be tested and certified as free of pesticide residue. When respondents were asked whether they were willing to pay extra for residue-free foods, 46% said "yes" and 26% said "no". Of those who said "yes", only 14% were willing to pay more than 10% extra, and only 1% were willing to pay more than 20% extra. In another study, which surveyed people visiting food stores that sold both organic and conventional produce, Jolly (1991) found that people who had purchased organic produce at least once in 3 months were willing to pay, on the average, an extra 37%, 40%, 61%, and 68%, respectively, for organic apples, broccoli, peaches, and carrots. The typical premiums on organic produce are higher than those percentages and in part explain the small market for such produce. Only 3.1% of consumers in the Jolly study who had purchased organic produce in the preceding 3 months were willing to buy organic apples when the premium was 147%.

Van Ravenswaay and Hoehn (1991b) used market data from the New York City area in assessing how much consumers would have been willing to pay to expedite removal of Alar from agricultural use. They estimated that New Yorkers would have been willing to pay an extra 30% for apples to avoid perceived risks associated with Alar in 1989. In nationwide household surveys by the same authors (van Ravenswaay and Hoehn 1991b, c), the general public was willing to pay 23.6 cents per pound extra for apples with no detectable pesticide residues and 37.5 cents per pound extra for apples with absolutely no pesticide residue; (the normal cost of apples was 79 cents per pound. Of special interest was the fact that respondents who considered pesticide residues to pose "high risk" were willing to pay only 1 cent more per pound than the average respondent. Results of a separate study by Yarborough and Yarborough (1985, cited in Goldman and Clancy 1991) also found that the relationship between a person's "pesticide concern" and altered purchasing patterns was weak.

Decreased pesticide use might be accompanied by increased cosmetic damage to produce. Brunn (1990) examined consumer response to such cosmetic damage by showing photographs of oranges to shoppers in the Los Angeles and San Francisco areas. Shoppers were shown photographs of oranges with 0%, 10%, and 20% damage by thrips. The shoppers indicated that they would be willing to buy the oranges with the 10% and 20% scarring if the prices were 88% and 78% of the price of unblemished oranges. When shoppers were told that the blemished oranges were grown with 50% less pesticide, most of them said that they preferred the blemished over the nonblemished oranges; 63% preferred 10% scarred oranges and 58% preferred 20% scarred over the nonblemished oranges.

Baker and Crosbie (1993) criticized the contingent-valuation method used in many of those studies because it assesses only one variable and includes all consumers in each estimate. They used a technique called conjoint analysis on a small sample of shoppers at two San Jose, California supermarkets in 1992 to establish consumer groups that differ in behavior. In their analysis, they were able to examine the tradeoff between decreased pesticide use and increased pest damage. Their results indicated, for example, that, if a program banning carcinogens raised the price of produce by 20 cents per pound, the publicly acceptable increase in pest damage due to the ban could not exceed 14%. Cluster analysis indicated that the shoppers could be divided into three groups. About 30% cared about price and quality, but not pesticide residue. A majority, 55%, cared about price and quality and whether the produce met government standards for residue. The remainder, about 15%, wanted stricter government regulation of pesticide use on the farm.

Most of the data on health effects of pesticides are focused on carcinogenicity. Surveys of public opinion indicate that consumers are concerned about pesticide residue because of fear of cancer, but the public attitude toward pesticide residue is shaped by a much broader array of factors. Other health factors such as allergies and nervous system disorders, are of concern (van Ravenswaay 1995); but a number of studies indicated that consumers who were concerned with pesticide residue associated it with environmental problems (Hammitt 1986, Higley and Wintersteen 1992).

It is important that scientists, policy makers, and companies gain a realistic understanding of public attitudes toward pesticides and of the public actions that are likely to result from these attitudes. Van Ravenswaay (1995) and others (such as Baker and Crosbie 1993, Jussaume and Judson 1992) have been critical of the state of research in public perception of pesticides. Van Ravenswaay (1995) concludes that there is an immediate need for more basic research that includes development of valid and reliable theories and methods for quantitatively assessing public concerns. There is also a need for basic research that will improve communication channels between scientists, policy- makers, and the public.

One possible reason for the apparent low willingness to pay for residue-free produce is that high levels of concern about residue did not necessarily indicate belief in immediate individual danger. For example, one study found that while 72% of respondents believed pesticides in food to be a major health risk but only 47% believed that pesticides made the US food supply unsafe (Dunlap and Beus 1992).

Some have argued that the growth of organic-food sales is an indication of the public's willingness to pay to avoid both residue on foods and environmental problems associated with the use of pesticides in food production. According to data published by the *Natural Foods Merchandiser* (1997), organic-food sales have seen an annual growth rate of approximately 21% between 1980 and 1996. Organic-food sales have risen far faster than total food sales, which were only a little over twice as large in nominal terms in 1996 as in 1980. Prices of organic produce average 25-35% higher than prices of comparable conventional produce (Hammitt 1986, Morgan and Barbour 1991); purchasers of organic foods seem willing to pay a substantial premium for them. However, concerns about pesticides constitute only part of the motivation for buying organic foods. Most purchasers of organic foods believe that they are more nutritious and flavorful than conventionally grown foods (Hammit 1986, Jolly et al. 1989), and certification as residue-free is not the sole criterion of demand for organic produce.

The *Hartman Report* (Hartman Group 1996), phases I and II, summarizes an extensive survey of 1,800 consumers to assess attitudes toward food and the environment. Some key points follow.

- Concern for the environment is vague and diffuse. About 71 % of survey respondents indicate interest in purchasing earth-sustainable grocery products; this percentage drops to 46% if there is a premium on the price of products.
- Concerns about water pollution are greater than concerns about chemical residues in food; 40–50% of those surveyed were knowledgeable about groundwater issues.
- About 68 % of those surveyed indicate a preference for the substi-

tution of naturally occurring substances or organic methods for synthetic pesticides.

- Consumers are more likely to buy food products with many environmental benefits than with just a single environmental benefit.
- Environmental claims certified by a third party, such as a government agency, are stronger than claims of a grocer or manufacturer.
- Consumers would not prefer to purchase products labeled "IPM" (integrated pest management) *unless the label indicated that the product was produced with less pesticide*; about 50% of consumers surveyed would be motivated to buy food produced with IPM methods (reduced chemical pesticides); this fraction increases to more than 60% if IPM is used with pesticide reduction in combination with such other practices as soil and water conservation.

At an increasing rate, consumers in Europe, Asia, and United States, are objecting to genetically modified foods. Consumers are demanding the right to know if their food purchases originate in genetically modified organisms (GMOs). In Europe, where the attacks on GMOs are the most extreme (Gaskell et al. 1999), experimental field plots containing genetically modified crops have been destroyed by anti-GMO activists. Major food retailers (such as Tesco, Marks and Spencer, and Salisbury) are responding and have chosen not to sell GMO foods. Major grain processors in the United States (such as ADM and Cargill), fearing refusal of their GMO grain at European ports, are shipping only non-GMO grain. However, a recent study (CGIAR, 2000) does suggest that, despite current levels of concerns, biotechnology is expected to play a major role in enhancing food security in developing nations. Furthermore, recommendations to strengthen regulations surrounding GMO products (NRC, 2000) will probably assuage public concerns about the safety of these products. Policy-makers and nongovernment groups expect the GMO issue to continue to be a key issue in global trade negotiations (Helmuth, 2000).

To the extent that concern about pesticide residue does influence organic food-purchases, it is limited to a small segment of the population. Despite its rapid growth, organic-food demand makes up a miniscule share of total food demand. In 1996, for example, organic food sales were $3.5 billion, only 0.46% of total food expenditures of $756.1 billion (Natural Foods Merchandiser 1997, Council of Economic Advisers 1998). That share is consistent with the finding that relatively few consumers are willing to pay a large premium for residue-free produce.

Although there has been some moderation in attitudes, public perceptions of pesticides today are overwhelmingly negative relative to attitudes toward many other classes of chemicals. According to Petty (1995), attitudes toward pesticides are "more negative than attitudes toward other potentially 'unsafe' substances" for several reasons. Unlike other potentially dangerous components of the diet, such as saturated fats and excess salt, pesticide residues offered no tangible positive consequences for consumers. That pesticide use might play a role in producing blemish-free fruits and vegetables is not readily apparent to a population that is increasingly removed from agriculture. Whereas in 1900, 80% of the US public was involved in food production, the percentage has shrunk to less than 2% in 1990. Moreover, the amelioration of pesticides as a risk factor is perceived as a responsibility beyond that of the individual; responsibility for correction lies with producers rather than consumers (in contrast with other forms of risky substances in the diet).

In urban settings, pesticides remain a source of concern. A telephone survey conducted in Kentucky in 1994 (Potter and Bessin, 1998) revealed that, although a substantial majority of respondents (over 90%) were concerned about the presence of insects in their homes the same respondents were collectively (77%) either very or somewhat concerned about pesticide use in their homes. Those respondents also were very to somewhat concerned about pesticide use in workplaces (about 73%) and in their children's schools (about 87%). Those concerns were expressed by respondents independently of their socioeconomic level. As to the nature of the concerns, close to two-thirds of respondents who had opinions expressed the belief that pesticides can cause cancer. A similar majority expressed a preference for biologically based alternatives to chemicals if such were available, on the basis of a belief that these products are "safer"; and 82.8% indicated willingness to pay more for pest-control operators who use less pesticide. Surprisingly, 98.6% of respondents could not define the meaning of "IPM", the pest-management philosophy that has prevailed in the field for close to 4 decades. Over 96% could not define the philosophy even after the abbreviation was identified. Such overwhelming majorities indicate that efforts on the part of the EPA, USDA, and FDA to educate consumers and promote acceptance of IPM approaches have not been particularly effective (Potter and Bessin, 1998).

Another factor influencing public attitudes toward pesticides is perceptions of the reliability of information providers. In general, information derived from sources that stand to benefit from dissemination of that information is perceived as less trustworthy than information derived from sources that enjoy no apparent benefit from dissemination. Thus, information from proponents of pesticide use—chemical companies and food growers and distributors—is perceived as less trustworthy. Another potential factor in public perception is the accessibility of the information provided from different sources and the accessibility of that information to a general public with little science education (Augustine 1998).

REFERENCES

Agrios, G.N. 1988. Plant Pathology. San Diego: Academic Press.

Alavanja, M.C.R., D.P. Sandler, S.B. McMaster, S.H. Zahm, C.J. McDonnell, C.F. Lynch, M. Pennybacker, N. Rothman, M. Dosemeci, A.E. Bond, and A. Blair. 1996. The agricultural health study. Env. Health Pers. 104:362–69.

Altieri, M.A., C.A. Francis, A. Schoonhoven, and J. Doll. 1978. Insect prevalence in bean (*Phaseolus vulgaris*) and maize (*Zea mays*) polycultural systems. Field Crops Research 1:33–49.

Archibald, S.O., and C.K. Winter. 1990. Pesticides in our food: Assessing the risks. Pp. 1-50 in Chemicals in the Human Food Chain, C. K. Winter, J. N. Seiber and C .F. Nuckton, eds. New York: Van Nostrand.

Arnold S. F., D. M. Klotz, B. M. Collins, P. M. Vonier, L. J. Guillette, Jr. and J. A. McLachan. 1996. Synergistic activities of estrogen receptor with combinations of environmental chemicals. Science 272:1489–1491.

Aspelin, A. L., and A. H. Grube. 1999. Pesticide Industry Sales and Usage. 733-R-99-001. Washington, DC: US Environmental Protection Agency, Office of Pesticide Programs.

Augustine, N. 1998. What we don't know does hurt us: How scientific illiteracy hobbles society. Science 279:1640-1641.

Babcock, B. A., E. Lichtenberg, and D. Zilberman. 1992. Impact of damage on the quality and quantity of output. Am. J. of Ag. Ec. 74:163–172.

Baginsky, E. 1967. P. 35 in Occupational Disease in California Attributed to Pesticides and Other Agricultural Chemicals. Berkeley: State Department of Public Health.

Baker, G. A., and P. J. Crosbie. 1993. Measuring food safety preferences: identifying consumer segments. J. of Ag. and Res. Ec. 18:277–287.

Baker, L. S., D. L. Fitzell, J. N. Seiber, T. R. Parker, T. Shibamoto, M.W. Poore, K. E. Longley, R. P. Tomlin, R. Propper and W. Duncan. 1996. Ambient air concentrations of pesticides in California. Env. Sci. Tech. 30:1365–1368.

Bakir, R., S. F. Damluji, L. Amin-Zaki, M. Murtadha, A. Khalidi, N. Y. Al-Rawi, S. Tikriti, H. I. Dhahir, T. W. Clarkson, J. C. Smith, and R. A. Doherty. 1973. Methylmercury poisoning in Iraq. Science 181:230–241.

Ball, V. E. 1985. Output, input and productivity measurement in US agriculture. Am. J. of Ag. Ec. 67:476–486.

Ball, V. E. 1988. Modeling Supply Response in a Multiproduct Framework. Am. J. of Ag. Ec. 70:813–825.

Ball, V. E., J.C. Bureau, R. Nehring, and A. Somwaru. 1997. Agricultural productivity revisited. Am. J.of Ag. Ec. 79:1045–1063.

Barbash, J. E., and E. A. Resek. 1996. Pesticides in Ground Water: Distribution, Trends, and Governing Factors. Chelsea, Mich: Ann Arbor Press.

Barton, J., and F. Oheme. 1981. Incidence and characteristics of animal poisonings seen at Kansas State University from 1975 to 1980. Vet. Human Toxicol. 23:101–102.

Bealer, R. C., and F. K. Willits. 1968. Worriers and non-worriers among consumers about farmers' use of pesticides. J. Cons. Affairs 2:188–204.

Beasley, R., and H. Trammel. 1989. Insecticide. Pps. 97–107 in Current Veterinary Therapy: Small Animal Practice, R. W. Kirk, ed. Philadelphia: W. B. Saunders.

Benbrook, C. M. 1996. Pest Management at the Crossroads. Yonkers, N.Y.: Consumers Union.

Bleiburg, J., M. Wallen, R. Brodkin, and I. Applebaum. 1964. Industrially acquired porphyria. Arch. Dermatol. 84:793-797.

Blondell, J. 1997. Epidemiology of pesticide poisonings in the United States, with special reference to occupational cases. Occ. Med.: State of the Art Reviews 12(2):209–220.

Blus, C. .J., and C. J. Henery. 1997. Field studies on birds: unique and unexpected relations. Ecol. App. 7:1125–1132.

Brattsten, L. B. and S. Ahmad, eds. 1986. Molecular Aspects of Insect-Plant Interactions. New York: Plenum Press.

Bridges, D. C., and R. L. Anderson. 1992. Crop Losses Due to Weeds in the United States. Champaign, Ill.: Weed Science Society of America.

Brock T. D., and M. T. Madigan. 1988. Biology of Microorganisms, 5th edition. Englewood, N.J: Prentice-Hall.

Bruhn, C., M. K. Diaz-Knauf, N. Feldman, J. Harwood, G. Ho, E. Ivans, L. Kubin, C. Lamp, M. Marshall, S. Osaki, G. Stanford, Y. Steinbring, I. Valdez, E. Williamson, and E. Wunderlich. 1992. Consumer food safety concerns and interest in pesticide-related information. J. of Food Safety. 12:253–262.

Brunn, D. 1990. Consumer acceptance of cosmetically imperfect produce. J. of Consumer Affairs. 24(2):268–279.

Brusseau, M. L. 1994. Transport of reactive contaminants in porous media. Rev. of Geophys. 32(3):285–314.

Buchmann, S. L., and G. P. Nahan. 1996. The Forgotten Pollinators. Washington, DC: Island Press.

California EPA 1994. Pesticide Usage Report, 1993. Sacramento: California Department of Pesticide Regulation,. [Online]. Available: HtmlResAnchor http://www.cdpr.ca.gov/docs/dprdocs/docsmenu.htm.

California EPA 1995. California Pesticide Illness Surveillance Program Summary Report – 1993. Sacramento: California Environmental Protection Agency, Dept. of Pesticide Regulation.

California EPA 1996. California Pesticide Illness Surveillance Program Summary Report – 1994. Sacramento: California Environmental Protection Agency, Dept. of Pesticide Regulation.

California EPA 1997. Overview of the California Pesticide Illness Surveillance Program – 1995. Sacramento: California Environmental Protection Agency, Dept. of Pesticide Regulation.

Capalbo, S. M., and J. M. Antle, eds. 1988. Agricultural Productivity: Measurement and Explanation. Washington, DC: Resources for the Future.

Carlson, G. A. 1977. Long run productivity of insecticides. Am. J. of Ag. Ec. 59:43–548.

Carlson, G. A., and C. E. Main. 1976. Economics of disease loss management. Ann. Rev. of Phyt. 17:149–161.

Carlson, G. A., and L. Suguiyama. 1985. Economic evaluation of area-wide cotton insect management: boll weevils in the Southeastern United States. Bulletin 473. Raleigh: North Carolina State University, North Carolina Agricultural Research Service.

Carlson, G. A., G. Sappie, and M. Hammig. 1989. Economic Returns to Boll Weevil Eradication. Agricultural Economic Report No. 621, Washington, DC: US Department of Agriculture, Economic Research Service.

Carrasco-Tauber, C., and L. J. Moffitt. 1992. Damage control econometrics: functional specification and pesticide productivity. Am. J. of Ag. Ec. 74:158–162.

Casey, P., and J. A. Vale. 1995. Deaths from pesticide poisoning in England and Wales 1945-1989. Human and Exptl. Tox. 13:95–101.

CGIAR (Consultative Group on International Agricultural Research) and US National Academy of Sciences. 2000. Agricultural Biotechnology and the Poor, Conference Proceedings, G. J. Persley and M. M. Lantin, eds. Washington, DC: CGIAR.

Chambers, R. G., and E. Lichtenberg. 1994. Simple econometrics of pesticide productivity. Am. J. of Ag. Ec. 76:407–418.

Chambers, R. G., and E. Lichtenberg. In press. Pesticide productivity and pest damage in the United States: an aggregate analysis. In The Economics of Pesticides and Food Safety: Essays in Memory of Carolyn Harper. C. E. Willis and D. Zilberman, eds. Boston: Kluwer Academic Publishing.

Chambers, R. G., and R. D. Pope. 1994. A virtually ideal production system: specifying and estimating the vips model. Am. J. of Ag. Ec. 76:105–113.

Chavas, J. P., and T. Cox. 1988. A nonparametric analysis of agricultural technology. Am. J. of Ag. Ec. 70:303–310.

Cheng, H. H., ed. 1990. Pesticides in the Soil Environment: Processes, Impacts, and Modeling. Soil Science Society of America Book Series, No. 2. Madison, Wis: Soil Science Society of America.

Clement, C.R. and T.. Colborn. 1992. Herbicides and fungicides: a perspective on potential human exposure. Pp. 368–384 in Chemically-induced Alterations in Sexual and Functional Development: The Wildlife/Human Connection. T. Colborn and C. Clement, eds. N.J.: Princeton Scientific Publishing Company.

Columbia Legal Services. 1998. Enforcement of Farm Worker Pesticide Protection in Washington State. Report, Nov. 1998. Seattle, Wash.: Columbia Legal Services.

Connaughton, D. F., J. R. Stedinger, L. W. Lion, and M. L. Shuler. 1993. Description of time-varying desorption kinetics: Release of naphthalene from contaminated soils. Enviro. Sci. Tech., 27(11):2397–2403.

Coulston, F., and F. Pocchiari. 1983. Accidental Exposure to Dioxins – Human Health Aspects. New York: Academic Press.

Council of Economic Advisers. 1998. Economic Report of the President. Washington, DC: Government Printing Office.

Council of Economic Advisers. 2000. Economic Report of the President. Washington, DC: Government Printing Office.

Cramer, H. H. 1967. Plant Protection and World Crop Production. Leverkusen, Germany : Bayer Pflanzenschutz.

Cramp, S. 1963. Toxic chemicals and birds of prey. Brit. Birds 56(4):124.

Croft, B. 1990. Pesticide effects on arthropod natural enemies: a database summary. Chapter 2 in Arthropod Biological Control Agents and Pesticides. New York: J. Wiley.

CTIC (Conservation Technology Information Center). 1998a. 1998 Crop Residue Management Executive Summary (1998 Conservation Tillage Report). [Online]. Available: HtmlResAnchor http://www.ctic.purdue.edu/Core4/CT/CT.html.

CTIC (Conservation Technology Information Center). 1998b. 1990–1998 Conservation Tillage Trends. [Online]. Available: http:/www.ctic.purdue.edu/Core4/CT/CTSurvey/NationalData.html

Curtis, C. G., L. M. Cook and R. J. Wood. 1978. Selection for and against insecticide resistance and possible methods of inhibiting the evolution of resistance in mosquitoes. Ecol. Ent. 3:273–287.

Davis, S., and R. Schliefer. 1998. Indifference to Safety: Florida's Investigation into Pesticide Poisoning of Farmworkers. Washington, DC: Farmworker Justice Fund, Inc.

Debach, P., and D. Rosen. 1991. Biological Control by Natural Enemies. New York: Cambridge University Press.

Dempster, J .P. 1969. Some effects of weed control on the numbers of the small cabbage white (*Pieris rapae*) on Brussels sprouts. J. Appl. Ecol. 6:339–345.

Dempster, J. P., and T.H. Coaker. 1974. Diversification of crop ecosystems as a means of controlling pests. Pp. 59–72 in Biology in Pest and Disease Control. D. Price-Jones and M. E. Solomon, eds. Oxford: Blackwell Scientific.

DuBois, K. P. 1969. Combined effects of pesticides. Can. Med. Assoc. J. 100:173–179.

Duffy, D. L., C. A. Mitchell, and N. G. Martin, 1998. Genetic and environmental risk factors for asthma. Am. J. Respir. Crit. Care Med. 157:840–845.

Dunlap, R. E., and C. E. Beus. 1992. Understanding public concerns about pesticides: an empirical examination. J. Cons. Aff. 26(2):418–438.

Dyte, C. E. 1972. Resistance to synthetic juvenile hormone in a strain of flour beetle, *Tribolium castaneum*. Nature. 238:48.

Ecobichon, D. J. 1991. Toxic Effects of Pesticides. Pp. 610–611 in: Casarett and Doull's Toxicology, M. O. Amdur, J. Doull and C.D. Klassen, eds. New York: Pergamon Press.

Ecobichon, D. J. 1996. Toxic Effects of Pesticides. Pp. 643-689 in: Casarett and Doull's Toxicology, 5th Edition, C. D. Klaassen, ed., New York: McGraw Hill.

Ehrlich, P. R., D. S. Dobkin and D. Wheye. 1988. The Birder's Handbook. New York: Simon and Schuster.

Elkins, E. R. 1989. Effect of commercial processing on pesticide residues in selected fruits and vegetables. J. Assoc. Off. Anal. Chem. 72(3):533–535.

EPA (US Environmental Protection Agency). 1992. Worker Protection Standard, Hazard Information, Hand Labor Tasks on Cut Flowers and Ferns Exception; Final Rule, And Proposed Rules. Federal Register 57(163):38102–38176.

EPA (US Environmental Protection Agency). 1998. Status of Pesticides in Registration, Reregistration, and Special Review (Rainbow Report). Washington, DC: EPA, Office of Pesticide Programs.

EPA (US Environmental Protection Agency). 1999. Office of Pesticide Programs 1999 Programs Biennial Report for FY 1998 and 1999. Washington, DC: US Government Printing Office.

EPA (US Environmental Protection Agency). 2000. Clinton-Gore administration acts to eliminate major uses of pesticide dursban to protect children and public health. Headquarters Press Release June 8.

ERS (Economic Research Service). 1997. Agricultural Resources and Environmental Indicators 1996-97. Agricultural Handbook No. 712. Washington, DC: US Department of Agriculture

Farnsworth, R. L., and L. J. Moffitt. 1981. Cotton production under risk: an analysis of input effects on yield variability and factor demand. W. J. Ag. Ec. 6:155–163.

Feldmann, R. J., and H. I. Maibach. 1974. Percutaneous penetration of some pesticides and herbicides in man. Toxicol. & Appl. Pharmacol. 28:126–132.

Ferrer, A., and R. Cabral. 1995. Recent epidemics of poisoning by pesticides. Tox.. Let. 82(3):55–63.

Firestone, D. 1977. The 2,3,7,8-tetrachlorodibenzo-p-dioxin problem: A review. In., Chlorinated phenoxy acids and their dioxins: Mode of action, health risks, and environmental effects, C. Ramel, ed. Ecol. Bull. (Stockholm) 17.

Flickinger, E. L., G. Juenger, T. J. Roffe, M. R. Smith, and R. J. Irwin. 1991. Poisoning of Canada geese in Texas by parathion sprayed for control of Russian wheat aphid. H. Wildl. Dis. 27:265–268.

Food Marketing Institute. 1989. 1989 Trends: Consumer Attitudes and the Supermarket. Washington, DC: Food Marketing Institute.

Foster, M. S., and W. G. Ruesink. 1984. Influence of flowering weeds associated with reduced tillage in corn on a black cutworm (*Lepidoptera: Noctuidae*) parasitoid, *Meteorus rubens* (Nees von Esenbeck). Env. Ent. 13:664–668.

FQPA (Food Quality Protection Act). 1996. US Public Law 104-170. 104th Congress, August 1996.

Fry, M. E., R. J. Risser, H. A. Stubbs and J. P. Leighton. 1986. Species selection of habitat evaluation procedures. In Wildlife 2000: Modeling Habitat Relationships of Terrestrial Vertebrates, J. Verner, M.L. Morrison and C.J. Ralph, eds. Madison: University of Wisconsin Press.

Fryer, J. D. 1982. The British Crop Production Council and Weed Research Organization: an interwoven history Great Britain. Pp. 78–83 in Report – Weed Research Organization. Brighton, England : British Crop Protection Council,

Gadon, M. 1996. Pesticide poisonings in the lawn care and tree service industries. J. Occ. and Env. Med. 38:794–799.

Gallo, M. A., R. J. Scheuplein, and K. A. van der Heijden, eds. 1991. Biological basis for risk assessment of dioxins and related compounds. Banbury Report 35. Cold Spring Harbor, N.Y.: Cold Spring Harbor Laboratory Press.

Gallup. 1989. Gallup Poll Report 286. Princeton, N.J: Gallup Poll Publishing.

Gaskell, G., M. W. Bauer, J. Durant, and N. C. Allum. 1999. Worlds apart? The reception of genetically modified foods in Europe and the US. Science 285:384–387.

Gelardi, R. C.,and M. K. Mountford. 1993. Infant Formulas: evidence of the absence of pesticide residues. Reg. Tox.. Pharm. 17(2,pt.1):181–192.

Georghiou, G. P. 1972. The evolution of resistance to pesticides. Ann. Rev. Ecol. Syst. 3:133–168.

Georghiou, G. P., and C. E. Taylor, 1976. Pesticide resistance as an evolutionary phenomenon. Proc. XV Int. Cong. Ent. 1976:759–785.

Georghiou, G.P., and R. Mellon. 1983. Pesticide resistance in time and space. Pp. 1–46 in. Pest Resistance to Pesticides, G. Georghiou and T. Saito, eds. New York: Plenum Press.

Gergen, P. J., K. M. Mortimer, P. A. Eggleston, D. Rosenstreich, H. Mitchell, D. Ownby, M. Kattan, D. Baker, E. C. Wright, R. Slavin, and F. Malveaux, 1999. Results of the National Inner-City Asthma Study (NCICAS) environmental intervention to reduce allergen exposure in inner-city homes. J. Allergy Clin. Immunol. 103:501–506.

Goldman, B. J., and K .L. Clancy. 1991. A survey of organic produce purchases and related attitudes of food cooperative shoppers. Am. J. Alt. Ag. 6(2):89–92.

Gotsch, N., and U. Regev. 1996. Fungicide use under risk in swiss wheat production. Ag. Ec. 14:1–9.

Greig-Smith, P.W. 1994. Understanding the impact of pesticides on wild birds by monitoring incidents of poisoning. Pp. 301-319 in Wildlife Toxicology and Population Modeling, R.J. Kendall and T.E. Lacher, eds. Boca Raton, Fla.: CRC Press.

Griliches, Z. 1963. The sources of measured productivity growth: US agriculture, 1940-1960. J. Pol. Ec. 71:331–364.

Hammitt, J. K. 1986. Estimating Consumer Willingness to Pay to Reduce Food-borne Risk. (R-3447-EPA). Santa Monica, Calif.: Rand Corporation.

Harper, J. L. 1957. Ecological aspects of weed control. Outlook Agric. 1:197–205.

Hartman Group. 1996. The Hartman Report: Food and the Environment: A Consumer's Perspective. Bellevue, Wash.: Hartman Group.

Hartung, R. 1975. Occupational hazards of selected pesticides. Pp. 833–842 in Occupational Medicine, C. Zenz, ed. Chicago, Ill.:Yearbook Medical Publishers.

Hartung, R. and B. D. Dinman, eds. 1972. Environmental Mercury Contamination. Ann Arbor, Mich.: Ann Arbor Science Publications.

Hayes, W. J., Jr. 1963. Clinical Handbook on Economic Poisons: Emergency Information for Treating Poisoning. Public Health Service Publication No. 476. Washington, DC: US Government Printing Office.

Headley, J. C. 1968. Estimating the productivity of agricultural pesticides. Am. J. of Ag. Ec. 50:13–23.

Heap, I. M. 1997. The occurrence of herbicide-resistant weeds worldwide. Pesticide Science. 51:235–234.

Helm, R. M., A. W. Burks, L. W. Williams, D. E. Milne and R. J. Brenner. 1993. Identification of cockroach aeroallergens from living cultures of German or American cockroaches. Int. Arch. Allergy & Appl. Immunol. 101:359–363.

Helmuth, L. 2000. Both sides claim victory in trade pact. Science 287:782–783.

HEW (US Department of Health, Education and Welfare). 1969. Report of the Secretary's Commission on Pesticides and Their Relationship to Environmental Health. Washington, DC: US Government Printing Office.

Higley, L. G., and W. K. Wintersteen. 1992. A novel approach to environmental risk assessment of pesticides as a basis for incorporating environmental costs into economic injury levels. Am. Entomol. 38:34–39.

Honaganahalli, .P. S., and J. N. Seiber. 1997. Health and environmental concerns over the use of fumigants in agriculture: the case of methyl bromide. Pp. 1–13 in: Fumigants: Environmental Fate, Exposure, and Analysis, J. N. Seiber, J.A. Knuteson, J.E. Woodrow, N.L. Wolfe, J. V. Yates, and S. R. Yates, eds. ACS Symposium Series 652. Washington, DC: American Chemical Society.

Horowitz, J. K., and E. Lichtenberg. 1993. Insurance, moral hazard, and chemical use in agriculture. Am. J. of Ag. Ec. 75(4):926–935.

Horowitz, J.,K., and E. Lichtenberg. 1994. Risk-increasing and risk-reducing effects of pesticides. J. Ag. Ec. 45(1):82–89.

Hoy, C. W., G. P. Head, and F. R. Hall. 1998. Spatial heterogeneity and insect adaptation to toxins. Ann. Rev. of Ent. 43:185–205.

Hueth, D., and U. Regev. 1974. Optimal pest management with increasing pest resistance. Am. J. of Ag. Ec. 56:456–465.

Jacobsen, B. J. 1997. Role of plant pathology in integrated pest management. Ann. Rev. Phyto. 35:373–391.

Jaenicke, E. C. 1997. The Myths and Realities of Pesticide Reduction: A Readers's Guide to Understanding the Full Economic Impacts. Greenbelt, Md.: Henry A. Wallace Institute of Alternative Agriculture.

Johnson, D. A., T. F. Cummings, P. B. Hamm, R. C. Rowe, J. S. Miller, R. E. Thornton, and E. J. Sorensen. 1997. Potato late blight in the Columbia Basin: An economic analysis of the 1995 epidemic. Plant Disease. 81:103–106.

Johnson, D. W., L. P. Kish, and G. E. Allen. 1976. Field evaluation of selected pesticides on the natural development of the entomopathogen, *Nomuraea rileyi*, in the velvetbean caterpillar in soybean. Env.. Ent. 5:964–966.

Johnson, M. K. 1982. The target for the initiation of delayed neurotoxicity by organophosphorus esters: biochemical studies and toxicological applications. Rev. of Biochem. Tox. 4:141–212.

Jolly, D. A. 1991. Determinants of organic horticultural products consumption based on a sample of California consumers. Acta. Hort. 295:141–148.

Jolly, D. A., H. G. Schutz, K. V. Diaz-Knauf, and J. Johal. 1989. Organic foods: consumer attitudes and use. Food Tech. 43(11):60–66.

Jordan, N. R., and J. L. Jannink. 1997. Assessing the practical importance of weed evolution: a research agenda. Weed Res. 37:237–246.

Jussaume, R. A., and D. H. Judson. 1992. Public perceptions about food safety in the United States and Japan. Rural Soc. 57(2):235–249.

Kamel, F., T. Moreno, A. S. Rowland, L. Stallone, G. Ramirez-Garcia, and D. P. Sandler. 1998. Community participation and response rates in a study of farmworkers. J. Epid. 9:5117 (4 Suppl.).

Keifer, M. C., and R. K. Mahurin. 1997. Chronic neurologic effects of pesticide overexposure. Occ. Med.: State of the Art Reviews. 12(2):291–304.

Kimbrough, R. D., H. Falk, P. Stehr, and G. Fries. 1984. Health implications of 2,3,7,8-tetrachlorodibenzodioxin (TCDD) contamination of residential soil. J. Tox..Env. Health. 14:47–93.

Klassen, W. 1995. World food supply up to 2010 and the global pesticide situation. In Eighth International Congress of Pesticide Chemistry, Options 2000, N. N .Ragsdale, P .C. Kearney, and J .R. Plimmer, eds. Washington, DC: American Chemical Society.

Knutson, R. D., C. R. Hall, E. G. Smith, S. D. Cotner, and J. W. Miller. 1993. Economic Impacts of Reduced Pesticide Use on Fruits and Vegetables. Washington, DC: American Farm Bureau Federation.

Knutson, R. D., C. R. Taylor, J. B. Penson, and E. G. Smith. 1990. Economic Impacts of Reduced Chemical Use. College Station, Tex.: Knutson and Associates.

Kolpin, D. W., J. E. Barbash, and R. J. Gilliom. 1998. Occurrence of pesticides in shallow groundwater of the United States: Initial results from the national water-quality assessment program. Env. Sci. Tech. 32:558-566.

Ku, H. S. 1987. Potential industrial applications for allelochemicals and their problems. Pp. 449–454 in Allelochemicals: A Role in Agriculture and Forestry, G. Waller, ed. Symposia Series 330. Washington, DC: American Chemical Society.

Kurtz, D. A., ed. 1990. Long Range Transport of Pesticides. Chelsea, Mich.: Lewis Publishers.

Landres, P. B., J. V. Verner ,and J. W. Thomas. 1988. Ecological uses of vertebrate indicator species: a critique. Cons. Biol. 2:316–328.

Larson, S. J., P. D. Capel, and M. S. Majewski. 1997. Pesticides in Surface Waters: Distribution, Trends, and Governing Factors. Chelsea, Mich: Ann Arbor Press.

Leber, A. P. and T. J. Benya, 1994. Chlorinated Hydrocarbon Insecticides. Pp. 1503–1566 in Patty's Industrial Hygiene and Toxicology, 4th Ed., G.D. Clayton and F.E. Clayton, eds: Vol. II, Part B. New York: John Wiley & Sons.

Leung, H. W., and D. J. Paustenbach. 1989. Assessing health risks in the work place: a study of 2,3,7,8-tetrachlorodibenzo-p-dioxin. Pp. 689–710 in The Risk Assessment of Environmental and Human Health Hazards, D.J. Paustenbach, ed. New York: John Wiley & Sons.

Levine, E., and H. Oloumi-Sadeghi. 1991. Management of diabroticite rootworms in corn. Ann. Rev. Ent. 36:229–255.

Levine, R. 1991. Recognized and possible effects of pesticides in humans. Pp. 275–360 in Handbook of Pesticide Toxicology, Volume 1, W. J. Hayes, Jr. and E. R. Laws, Jr., eds. San Diego, Calif.: Academic Press.

Lichtenberg, E., and D. Zilberman. 1986. The econometrics of damage control: why specification matters. Am. J. of Ag. Ec. 68:261–273.

Lichtenberg, E., and D. Zilberman. 1997. Storage Technology and the Environment. Working Paper No. 90-05. College Park: University of Maryland, Department of Agricultural and Resource Economics,.

Lichtenberg, E., D. D. Parker, and D. Zilberman. 1988. Marginal analysis of welfare effects of environmental policies: the case of pesticide regulation. Am. J. of Ag. Ec. 70(4):867–874.

Lin, B-H., M. Padgitt, L. Bull, H. Delvo, D. Shank, and H. Taylor. 1995. Pesticide and Fertilzer Use and Trends in US Agriculture. Agricultural Economic Report Number 717. Washington, DC: US Department of Agriculture, Economic Research Service.

Linn, D.,M., F. H. Carski, M. L. Brusseau, and F. H. Chang, eds. 1993. Sorption and Degradation of Pesticides and Organic Chemicals in Soil. Madison, Wisc.: Soil Science of America; American Society of Agronomy.

Loague, K., D.L. Corwin, and T.R. Ellsworth. 1998. The challenge of predicting nonpoint source pollution. Env. Sci. Tech. 32(5):130A–133A.

Majewski, M. S., and P. D. Capel. 1995. Pesticides in the Atmosphere: Distribution, Trends, and Governing Factors. Chelsea, Mich.: Ann Arbor Press.

Mayer, F. L. Jr., and M. R. Ellersieck. 1986. Manual of Acute Toxicity: Interpretation and Data Base for 410 Chemicals and 66 Species of Freshwater Animals. Resource Publication No. 160. Washington, DC: US Fish and Wildlife Service.

McEwen, F. L., and G. R. Stephenson. 1979. The Use and Significance of Pesticides in the Environment. New York: John Wiley & Sons.

McKenna and Cuneo, L.L.P. 1996. Summary and Analysis of the Food Quality Protection Act of 1996, Public Law 104-170, August 3, 1996. Prepared for the American Crop Protection Association.

Mehler, L. N., M. A. O'Mally, and R. I. Krieger. 1992. Acute pesticide morbidity and mortality: California. Rev. Env. Contam. Tox. 129:51–66.

Melander, A. L. 1914. Can insects become resistant to sprays? J. Econ. Ent. 7:167.

Merck & Co., Inc. 2000. Venomous bites and stings (Chapter 287, section 24) in Merck Manual of Medical Information. Whitehouse Station, N.J.: Merck & Co.

Methyl Bromide Global Coalition. 1992. The Methyl Bromide Science Workshop Proceedings, M. W. Ko, and N. D. Sze, eds. Arlington, Va.: Methyl Bromide Global Coalition.

Mileson, B. E., J. E. Chambers, W. L. Chen, W. Dettbarn, M. Ehrich, A. T. Eldefrawi, D. W. Gaylor, K. Hamernik, E. Hodgson, A. G. Karczmar, S. Padilla, C. N. Pope, R. J. Richardson, D. R. Saunders, L. P. Sheets, L. G. Sultatos, and K. B. Wallace. 1998. Common mechanism of toxicity: a case study of organophosphorus pesticides. Tox. Sci. 41:8–20.

Misra, S. K., C. L. Hung, and S. L. Ott. 1991. Consumer willingness to pay for pesticide-free fresh produce. W. J. of Ag. Ec. 16:218–227.

Moody, R. P., C. A. Franklin, and L. Ritter. 1990. Dermal absorption of the phenoxy herbicides 2,4-D, 2,4-D amine, 2,4-D isooctyl, and 2,4,5-T in rabbits, rats, rhesus monkeys, and humans: a cross-species comparison. J. Tox. and Env. Hlth. 29:237–245.

Morgan, J., and B. Barbour. 1991. Marketing organic produce in New Jersey: obstacles and opportunities. Agribusiness 7:143–163.

NAPIAP (National Agricultural Pesticide Impact Assessment Program). 1997. Pesticides and the Bottom Line, D. R. Pike, F. Whitford, and S. Kamble, eds. Crop Science Special Report 1997-10. Urbana-Champaign: University of Illinois.

NASS. 1999. 1998 Childhood Agricultural Injuries. NASS Document SP CR8 (10-99) Washington DC. US Department of Agriculture.

National Opinion Research Center. 1994. General Social Surveys, 1972-1994: Cumulative Codebook. Chicago, Ill: National Opinion Research Center.

Natural Foods Merchandiser. 1997. Market Overview '96. June.

Noling, J. W. 1997. Role of soil fumigants in Florida agriculture. Pp. 14–24 in Fumigants: Environmental. Fate, Exposure, and Analysis, J. N. Seiber, J. A. Knuteson, J. E. Woodrow, N. L. Wolfe, M. V. Yates, and S. R. Yates, eds. ACS Symposium Series 652. Washington, D.C: American Chemical Society.

Norgaard, R.B. 1976. The economics of improving pesticide use. Ann. Rev. of Ent. 21:45–60.

NRC (National Research Council). 1986. Pp. 100–110 in Pesticide Resistance: Strategies and Tactics for Management. Washington, DC: National Academy Press

NRC (National Research Council). 1987. Regulating Pesticides in Food: The Delaney Paradox. Washington, DC: National Academy Press.

NRC (National Research Council). 1993. Pesticides in the Diets of Infants and Children. Washington, DC: National Academy Press.

NRC (National Research Council). 1994. Groundwater Vulnerability Assessment. Washington, DC: National Academy Press.

NRC (National Research Council). 2000. Genetically Modified Pest-Protected Plants: Science and Regulation. Washington, DC: National Academy Press

Oerke, E. C., H. W. Dehne, F. Schonbeck, and A. Weber. 1994. Crop Production and Crop Protection: Estimated Losses in Major Food and Cash Crops. Amsterdam: Elsevier.

Onstad, D. W., M. G. Joselyn, S. A. Isard, E. Levine, J. L. Spencer, L. W. Bledsoe, C. R. Edwards, C. D. DiFonzo, and H. Wilson. 1999. Modeling the spread of western corn rootworm (Coleoptera: Chrysomelidae) populations adapting to soybean-corn rotation. Env. Ent..28(2):188–194

Osteen, C. D., and P. I. Szmedra. 1989. Agricultural Pesticide Use Trends and Policy Issues. Agricultural Economic Report No. 622. Washington, DC: US Department of Agriculture, Economic Research Service,

OTA (US Congress, Office of Technology Assessment). 1988. Pesticides Residues in Food: Technologies for Detection (OTA-F-398). Washington, D.C: US Government Printing Office.

Ott, S. L. 1990. Supermarket shoppers' pesticide concerns and willingness to purchase certified pesticide residue-free produce. Agribusiness 6:593–602.

Padgitt, M., D. Newton, and C. L. Sandretto. 2000. Production Practices in US Agriculture, 1990-97, Statistical Bulletin. US Department of Agriculture, Economic Research Service. Washington, DC: US Government Printing Office. (in press).

Pannell, D. J. 1991. Pests and pesticides, risk and risk aversion. Ag. Ec. 5(4):361–383.

Parker, D. D., and D. Zilberman. 1993. Hedonic estimation of quality factors affecting the farm-retail margin. Am. J. of Ag. Ec. 75:458–466.

Pepper, I., C. Gerba, M. L. Brusseau, eds. 1997. Pollution Science. San Diego, Calif.: Academic Press.

Perrin, R. M. 1977. Pest management in multiple cropping systems. Agro-ecosystems. (Amsterdam). 3(2):93–118.

Peterle, T. J. 1991. Wildlife Toxicology. New York: Van Nostrand Reinhold.

Peterson, R.K.D. and B. A. Shurdut 1999. Human health risks from cockroaches and cockroach management; a risk analysis approach. Amer. Ent. 45:142–148.

Petty, R. E. 1995. Understanding public attitudes toward pesticides. Pp. 21-24 in The Public and Pesticides: Exploring The Interface, C. R. Curtis, ed. Washington, DC: US Dept. of Agriculture, National Agricultural Pest Impact Assessment Program

Pfadt, R.E. 1978. Fundamentals of Applied Entomology. New York: MacMillan Publishers Co.

Pignatello, J .J., C. R. Frink, P. A. Marin, and E. X. Droste. 1990. Field-observed ethylene dibromide in an aquifer after two decades. J. Contam. Hydrol., 5:195–210.

Pimentel, D., H. Acquay, M. Biltonen, P. Rice, M. Silva, J. Nelson, V. Lipner, S. Giordano, A. Horowitz, and M. D'Amore. 1992. Environmental and economic costs of pesticide use. Bioscience. 42(10):750–760.

Pimentel, D., J. Krummel, D. Gallahan, J. Hough, A. Merrill, I. Schreiner, P. Vittum, F. Koziol, E. Back, D. Yen, and S. Fiance. 1978. Benefits and costs of pesticide use in US food production. BioScience. 28 (772):778–784.

Pimentel, D., L. McLaughlin, A. Zepp, B. Lakitan, T. Kraus, P. Kleinman, F. Vancini, W. J. Roach, E. Graap, W.S. Keeton, and G. Selig. 1991. Environmental and economic impacts of reducing US agricultural pesticide use. In CRC Handbook of Pest Management in Agriculture, Volume I, 2nd Edition, David Pimentel, ed. Boca Raton, Fla.: CRC Press.

Pocchiari, F., A. Di Domenico, V. Silano, and G. Zapponi. 1979. Human health effects from accidental releases of tetrachlorodibenzo-p-dioxin (TCDD) at Seveso, Italy. Ann. N. Y. Acad. Sci. 320:311.

Porter Novelli. 1990. Pesticide Perceptions. Washington, DC: American Farm Bureau Federation.

Potter, M. F. and R. T. Bessin 1998. Pest control, pesticides, and the public: attitudes and implications. Am. Ent. 44:142–147.

Potts, G. R. 1986. The Partridge: Pesticides, Predation and Conservation. London: Collins Professional and Technical.

Powell, W. 1986. Enhancing parasitoid activity in crops. Pp. 319–340 in Insect Parasitoids: 13th Symposium of the Royal Entomological Society of London, J. Waage and D. Greathead, eds. London: Academic Press.

Rabb, R. L., R. E. Stinner, and R. van den Bosch. 1976. Conservation and augmentation of natural enemies. Pp. 233-54. In. Theory and Practice of Biological Control, C.B. Huffaker and P.S. Messenger, eds. New York: Academic Press.

Radosevich, S., J. Holt, and C. Ghersa. 1997. Weed Ecology, 2nd Edition. New York: John Wiley & Sons.

Rasmussen, P. E., K. W. T. Goulding, J. R. Brown, P. R. Grace, H. H. Janzen, and M. Korschens. 1998. Long-term agroecosystem experiments: assessing agricultural sustainability and global change. Science; 282:892–896.

Regev, U., H. Shalit, and A. P. Gutierrez. 1983. On the optimal allocation of pesticides with increasing resistance: The case of alfalfa weevil. J. of Env. Ec. and Mgmt. 10:86–100.

Regev, U., N. Gotsch, and P. Rieder. 1997. Are fungicides, nitrogen, and plant growth regulators risk reducing? J. of Ag. Ec. 48:167–178.

Risch, S .J., D. Andow, and M. A. Altieri. 1983. Agroecosystem diversity and pest control: data, tentative conclusions and new directions. Env. Ent. 12:625–629.

Robbins, C.S., D. Bystrak, and P.H. Geisler. 1986. The Breeding Bird Survey: Its First Fifteen Years 1965–1979. Resource Publication 157. Washington, DC: US Department of the Interior Fish and Wildlife Service.

Robinson, W. S., R. Nowogrodzki, and R. A. Morse. 1989. The value of honey bees as pollinators of US crops. Am. Bee Journal 129(7):477–487.

Roby, D. M., and W. W. Melichar. 1996. 1,3-Dichloropropene regulatory issues. Pp. 25–31 in Fumigants: Environmental Fate, Exposure, and Analysis, J. N. Seiber, J. A. Knuteson, J E. Woodrow, N. L. Wolfe, J. V. Yates, and S. R. Yates, eds. ACS Symposium Series 652. Washington, D.C: American Chemical Society.

Rosenheim, J. A., and B. E. Tabashnik. 1990. Evolution of pesticide resistance: interactions between generation time and genetic, ecological, and operational factors. J. Econ. Ent. 83:1184–1193.

Ryszkowski, L., J. Karg, G. Margalit, M. G. Paoletti, and R. Zlotin. 1993. Aboveground insect biomass in agricultural landscapes of Europe. In Landscape Ecology and Agroecosystems, R. G. H Bunce, L. Ryszkowski and M.G. Paoletti, eds. Ann Arbor, Mich.: Lewis Publications.

Sachs, C., D. Blair, and C. Richter. 1987. Consumer pesticide concerns: A 1965 and 1984 comparison. J. Cons. Aff. 21:96–107.

Saha, A., C. R. Shumway, and A. Havenner. 1997. The economics and econometrics of damage control. Am. J. Ag. Ec. 79:773–785.

Sanderson, W. T., V. Ringenburg, and R. Biagini. 1995. Exposure of commercial pesticide applicators to the herbicide alachlor. Am. Ind. Hyg. Assoc. J. 56(9):890–897.

Sawhney, B. L., and K. Brown, eds. 1989. Reactions and Movement of Organic Chemicals in Soils. Special Pub. No. 22. Madison, Wisc. Soil Science Society of America.

Scribner, S. L., T. R. Benzing, S. Sun, and S. A. Boyd. 1992. Desorption and bioavailability of aged simazine residues in soil from a continuous corn field. J. Env. Qual. 21:1115–1121.

Seiber, J. N., and J. E. Woodrow. 1995. Origin and fate of pesticides in air. Pp.157-172 in Eighth International Congress of Pesticide Chemistry: Options 2000, N.N. Ragsdale, P.C. Kearney and J.R. Plimmer, eds. Washington, DC: American Chemical Society

Seiber, J. N., and J. E. Woodrow. 1998. Air transport of pesticides. Rev. Tox. 2(14):287–294.

Senanayake, N. and H. Peiris. 1995. Mortality due to poisoning in a developing country: trends over 20 years. Hum. Exp. Toxicol. 14(10):808–811.

Smith, J. A., C. T. Chiou, J. A. Kammer, and D. E. Kile. 1990. Effect of soil moisture on the sorption of trichloroethene vapor to vadose-zone soil at Picatinny arsenal, New Jersey. Env. Sci. Tech. 24:676–682.

Sparks, T. C. and B. D. Hammock. 1983. Insect growth regulators: resistance and the future. Pp. 615–668 in Pest Resistance to Pesticides: Challenges and Prospects, G. P. Georghiou, T. Saito, eds. New York: Plenum Press.

Staiff, D. C., S. W. Comer, J. F. Armstrong, and H. R. Wolfe. 1975. Exposure to the herbicide, paraquat. Bull. Env. Contam. Tox. 14:334-340.

Steinberg, S. M., J. J. Pignatello, and B. L. Sawhney. 1987. Persistence of 1,2-dibromoethane in soils: Entrapment in intraparticle micropores. Env. Sci. Tech. 21:1201–1210.

Stone, W. B., and P. B. Gradoni. 1985. Wildlife mortality related to the use of the pesticide diazinon. Northeast. Env. Sci. 4:30–38.

Thomas, J. A., 1996. Toxic responses of the reproductive system.. Pp.547–582 in: Casarett & Doull's Toxicology, 5th ed.., C. D. Klaassen, ed. New York: McGraw-Hill..

Ulrich, A., W. H. Furtan, and A. Schmitz. 1987. The cost of a licensing system regulation: an example from Ccanadian prairie agriculture. J. of Pol. Ec. 95:160–178.

USDA (US Department of Agriculture). 1996. Agricultural Statistics 1995-96. Washington, DC: US Government Printing Office.

van den Bosch, R., and V. Stern. 1962. The integration of chemical and biological control of arthropod pests. Ann. Rev. Ent. 7.

van den Bosch, R., M. Brown, C. McGowan, A. Miller, M. Moran, D. Pelzer, and J. Swartz. 1975. Investigation of the Effects of Food Standards on Pesticide Use, Report to the US Environmental Protection Agency. Washington, DC:

Van Emden, H. F. 1965. The role of uncultivated land on the distribution of the cabbage aphid (*Brevicoryne brassicae*) on adjacent crop. J. Appl. Ecol. 2:171–196.

Van Emden, H. F. 1990. Plant diversity and natural enemy efficiency in agroecosystems. Pp. 63-77 in Critical Issues in Biological Control, M. Mackauer and L. E. Ehler, eds. Andover: Intercept LTD.

van Ravenswaay, E. O. 1995. Public Perceptions of Agrichemicals, Task Force Report No. 123. Ames, Iowa: Council on Agricultural Science and Technology.

van Ravenswaay, E. O. and J. P. Hoehn. 1991a. The impact of health risks on food demand: A case study of alar and apples. In The Economics of Food Safety, J. Caswell, ed. New York: Elsevier Science Publishing Co.

van Ravenswaay, E. O., and J. P. Hoehn. 1991b. Willingness to pay for reducing pesticide residues in food: Results of a nationwide survey. Staff paper no. 91-18. East Lansing: Michigan State University, Department of Agricultural Economics.

van Ravenswaay, E. O., and J. P. Hoehn. 1991c. Contigent valuation and food safety: The case of pesticide residues in food. Staff paper no. 91-18. . East Lansing: Department of Agricultural Economics, Michigan State University.

Wania, F., and D. Mackay. 1993. Global fractionation and cold condensation of low volatility organochlorine compounds in polar regions. Ambio 22:10-18.

Ware, G. W. 1983. Pesticides: Theory and Application. San Francisco, Calif.: W. H. Freeman and Company.

Warren, G. F. 1998. Spectacular increases in crop yields in the United States in the Twentieth Century. Weed Technology. 12(4):752–760.

Weaver, R. D., D. J. Evans, and A. E. Luloff. 1992. Pesticide use in tomato production: Consumer concerns and willingness-to-pay. Agribusiness 8:131–142.

White, D. H., C. A. Mitchell, L. D. Wynn, E. L. Flickinger, and E. J. Kolbe. 1982. Organophosphate insecticide poisoning of Canada geese in the Texas Panhandle. J. Field Ornithol. 53:22–27.

Whitmore, R. W., J. E. Kelly, P. L. Reading, E. Brandt, and T. Harris. 1993. National home and garden pesticide use survey. Pp. 18-36 in Pesticides in Urban Environment: Fate and Significance, K. D. Racke and A. R. Leslie, eds. Washington, DC: American Chemical Society,

Winter, D. K. 1992. Pesticide tolerances and their relevance as safety standards. Reg. Tox. and Pharm. 15:137–150.

Yarkin, C., D. Sunding, D. Zilberman, and J. Siebert. 1994. Cancelling methyl bromide for postharvest use to trigger mixed economic results. Calif. Ag. 48:16–21.

Zabik, J. M., and J. N. Seiber. 1993. Atmospheric transport of organophosphate pesticides from California's Central Valley to the Sierra Nevada mountains. J. Env. Qual. 22:80–90.

Zahm, S. H., and M. H. Ward. 1998. Pesticides and Childhood Cancer. Environmental Health Perspectives 106 (Supplement 3). Rockville, Md.: National Cancer Institute.

Zilberman, D., A. Schmitz, G. Casterline, E. Lichtenberg, and J.,B. Siebert. 1991. The economics of pesticide use and regulation. Science 253:518–22.

3

Economic and Regulatory Changes and the Future of Pest Management

The process of producing food and fiber is inherently a biological one, but it is conducted in an economic and sociological context. The socioeconomic environment in which US agriculture is placed has undergone considerable changes in the last few decades. The changes reflect trends that extend well beyond agriculture and include globalization of trade in general, increasing industrialization, and emergence of a knowledge economy. Reflecting the changing socioeconomic standards are changes in the regulatory environment in which the agriculture enterprise must function. This chapter examines recent economic, institutional, social, and regulatory developments in the United States and evaluates their impact on pesticide use in agriculture.

ECONOMIC AND INSTITUTIONAL DEVELOPMENTS AND THEIR IMPACTS ON PEST CONTROL

Over the last 2 decades, both the agricultural and general economies have undergone several major changes that affect and will continue to affect the way pesticides are used and pests are controlled. The changes include globalization, industrialization, decentralization and privatization, growth of the "knowledge economy", and growth of the organic food market.

Globalization of World Food Markets

Several policies have contributed to a reduction in trade barriers between nations and to expansion of export and import opportunities for producers and consumers in agriculture and manufacturing:

- Several rounds of General Agreement on Tariffs and Trade (GATT) negotiations led to a reduction in tariffs and a goal of a phaseout of subsidization and protection of many segments of agriculture.
- Regional "free trade" blocs have been established throughout the world and are constantly expanding. The European Union is the most long-lasting and successful example. The North American Free Trade Agreement (NAFTA) established another major trade bloc which includes the United States, Mexico, and Canada and is expected to be expanded to include Chile and other Latin, Central, and South American countries as well.
- The demise of communism in central and eastern Europe and the more open trade policies in China have substantially expanded the volume of trade among the previous communist world nations and the United States and other democracies.
- Developing countries in South America, Asia, and Africa have gradually abandoned protectionist strategies. They now tend to reduce tariffs and enable expanded foreign trade and investment in their economies.

Globalization and reduction of trade barriers increase competitive pressures and provide extra incentives to reduce costs and increase yields. They increase the demand for more effective and efficient pest protection and pest-control products. They also increase competitiveness in pest-control markets and can lead to expansion of facilities and markets for suppliers of superior pest-control products.

Recent developments have reduced but not eliminated barriers to international trade. Countries and regions have a wide array of legislative tools (including environmental and agricultural policies) that discriminate against foreign suppliers. The volume of international trade in agriculture depends on the capacity to recognize and meet the needs of foreign markets and on the economic well-being of the buyers. US exports to Japan might be hampered by product quality and design limitations. The recent economic slowdown in Asian markets led to a slowdown in US food exports to some of these countries.

Countries are still able to maintain their own separate environmental- and health-protection regulations even in the era of freer trade. Thus, Canada and Japan have had stricter pesticide-residue regulations than the United States, and food-quality requirements set by European nations present substantial obstacles for exporters from North Africa, the Middle East, and Eastern Europe (OECD 1997). Food-safety and other chemical-use regulations in target markets have affected chemical use in the United States by growers who export to these nations (Zilberman et al. 1994). Much the same applies to genetically modified organisms (GMOs) or their products.

The recognition that environmental and health regulations can be used as trade barriers has generated efforts to harmonize them and to reach a more uniform set of principles for pesticide use and environmental regulation in agriculture. Analyses of future roles of pesticides have to take a global perspective and be viewed within a context of increasingly harmonized sets of environmental regulations. Furthermore, the less advanced analytical capacity and expertise of some foreign countries have led to reliance on American regulatory and scientific decisions and knowledge in establishing pest-control policies abroad. Increased international coordination of regulations is likely to strengthen the interdependence of regulatory processes between the United States and other countries.

The widespread introduction of GMOs into North American agriculture has led to a change in attitudes and practice in this regard. Recent meetings, including the UN International Biosafety Conference held in Montreal, highlight the major policy disagreements among the 130 nations participating. The export and import of GMOs are at the center of these disagreements, and the precautionary approach—where the import of genetically modified products can be banned simply as a precaution, in the absence of scientific evidence that such products pose health or environmental risks—is the primary issue of contention (Pollack, 2000).

The globalization process affects the economic markets for agricultural chemicals. New chemicals and other pest-control technologies are now developed largely according to their global market potential. The pest-control solutions available to American farmers reflect the results of research, development, and production efforts that take place on a global basis. Thus, assessments of research strategies in the United States have to recognize efforts that are made elsewhere. In setting research priorities for the US Department of Agriculture (USDA) and other government agencies, how they fit with efforts in other countries and where the US national investment is more effective must be recognized, given a global perspective on research efforts and assessment of what is done elsewhere. Also there are growing tendencies to form research alliances between nations to take advantage of increasing returns to scale in areas of relative strengths. One noted example is the successful cooperation of the US-Israeli Binational Agricultural Research and Development Fund (BARD) (see Just et al. 1998). The net benefits of BARD research were estimated to be about twice the cost of the research program. The international food-research centers, such as the International Rice Research Initiative and the International Maize and Wheat Improvement Center are also major providers of knowledge; they generate technology useful mostly for developing countries, but there is still substantial spillover to the United States and developed nations (Alston and Pardey 1996). Thus, the perspective on research and development and on new products should be global and take into account all the collaboration and partnerships in research.

Industrialization of Agriculture and Food Processing

The structure of US agriculture has undergone drastic changes over the last 100 years. The industries that once employed more than 80% of the populace now rely on less than 2% of the people to feed the US economy and to export worldwide. Small family farms are increasingly replaced by larger organizations that rely heavily on purchased inputs, including labor inputs.

While the array of activities conducted on the farm has declined, the agribusiness and food sectors that provide input and process agricultural products have increased substantially. Tens of billions of dollars are spent every year globally on pest-control products. The agrochemical industry is closely linked to the petrochemical and pharmaceutical industries. It markets its products to a wide network of wholesalers and dealers. Over the years, the variety of products it provides has increased to include management information, pest scouting, and consulting in addition to raw materials. Modern agriculture has become knowledge-intensive, and many or most of the larger companies employ PhD scientists to manage their pest-control and irrigation activities, and a new set of professional crop-management consultants has sprouted throughout the United States (Wolf 1996).

The food-processing sector continues to increase the treatment and manipulation of food products beyond the farm gate, and its share in overall food revenues has increased. The industry has adjusted to changes in consumer preferences and lifestyles. It augments its revenue by providing value-added products that include prepared meals (for such institutions as restaurants and hospitals), ready-to-cook meals, and a wide variety of food products. Various companies—from small restaurants to international giants, such as Unilever, Nestle's, and Proctor and Gamble—have specialized in assessing consumer needs and producing and marketing products to meet them. The result is an immense set of differentiated food products, many of them with substantial brand recognition, and a food sector that is several times larger than agriculture.

Agriculture has begun to adopt some of the characteristics that typify the agribusiness and food sectors. Some producers of fruits and vegetables have begun to establish their own brand names, and producers of grains are being paid according to the quality of their products. More products are designed to meet specifications of retailers, and production and marketing efforts are coordinated with some of the giant chains (especially in fruits and vegetables). Industrialization is important in the livestock sector, especially in poultry. Such companies as Tyson Foods, Inc. have increased the efficiency of processing and the array of poultry products. Similar attempts are now being made to industrialize the production of swine.

Two arrangements that typify much of industrialized farming are vertical integration (in which one organization is responsible for a variety of activities such as farming, processing, shipping, and marketing) and contracting (in which the marketers and processors of agricultural commodities provide farmers with inputs, including genetic materials and guidance about production processes and specification of the final product). With contracting and vertical integration, some of the functions of traditional farmers are changing. People who market the final products are now making many of the decisions regarding input and chemical use. Processors and marketers may also be held liable for some of the environmental side effects of production.

Decentralization and Privatization

Globalization and the emergence of international governing organizations with decision-making power over nations are accompanied by processes that move in the opposite direction. National governments give much decision-making and economic power to regional and state governments, and the power that was concentrated in national governments is distributed among a wide array of organizations and infrastructure. The power can be specialized and reflect various degrees of geographic focus. Decentralization and privatization have many consequences, which can affect the research, development, and practice of pest control.

One such consequence is the transfer of technology from the public to the private sector. University knowledge has been transferred to the private sector for many years and has played an important role in the development of industries. For example, the Massachusetts Institute of Technology (MIT) has had an office of technology transfer for about 50 years, and graduates and professors at MIT (and many other universities) have developed important industries and economic projects. Over the last 20 years, the magnitude of technology transfer from the public to the private sector has changed drastically. Nearly 100 offices of technology transfer have emerged, and collectively they represent almost every major university. These offices and the processes that they manage have been crucial in the evolution of biotechnology and will play an important role in the future.

The impetus for establishing offices of technology transfer in many major universities was the awareness that private companies do not take full advantage of university innovations (Postlewait et al. 1993). This "waste" of knowledge reduced the social value of university research and led universities to develop institutions and arrangements to address the problem. Offices of technology transfer were established to identify university discoveries with commercial potential, to aid in issuing patents, and to search for private parties who would be interested in buying the rights to develop the patents. It was recognized that the private organization would develop the university patent only if it had exclusive rights. Thus, part of technology transfer is payment of royalties from revenues, income, or other benefits that accrue to the private companies from the patent.

Technology-transfer offices are responsible for

- Initiating the technology-transfer agreement.
- Collecting royalties.
- Protecting against patent violations.
- Ensuring that buyers of rights actually use them (many technology-transfer agreements impose penalties if buyers of a right do not put in the due effort).

Several thousand technology-transfer agreements between universities and private companies are in operation. These agreements generate revenue of about $150 million per year, but most of the revenue is generated by fewer than 10 of major breakthrough patents. The royalty rates change, but in most cases they are 1–5% of revenues generated by a patent. Practitioners agree that a multiplier of 40 reflects the direct contribution of such royalties to the gross national product. Obviously, the royalties do not capture much of the benefits, because it can take 8–10 years for an innovation to becom a commercial product, and the patent life is 17–20 years. The revenues from technology transfer, although substantial, are small relative to the amount of money spent on the research by the public sector directly ($9 billion) and constitute about 0.1% of the overall expenditures on education (Parker et al. 1998).

Scientific discoveries have become a necessary precursor and first step of technological innovation. That has been most apparent in the last 20 years, in which transfer of genetic-engineering techniques played a major role in establishing the biotechnology industry. Administrators in technology-transfer offices discovered that, in many cases, established private firms would not buy rights to develop new innovations that originated outside their own organizations. That was especially true with respect to new lines of biotechnology products. Technology-transfer offices were organized to connect university researchers with venture capitalists to establish startup companies that aim to develop university innovations. Most of the dominant firms in biotechnology (such as Genentech, Chiron, Amgen and Calgene) were established in this manner. Once the upstarts became solid and economically viable, some of them were taken over by the multinational pharmaceutical, chemical, and agribusiness firms. That move enabled the larger firms to augment their base of knowledge, research capacity, and product lines through increased returns to scale and registration, production, and marketing. Multinationals have an advantage in these stages of product development, and in many cases new products are integrated, distributed, and produced more efficiently once they are within their systems.

The process of commercializing university innovations has an important effect on industrial structure and productivity. University researchers engage in new lines of research and develop products and innovations that would not likely be developed by research and development activities in the private sector. As Parker et al. (1998) argue, universities are a dominant source of ideas for innovation and increased competitiveness. Because of perceived economic risks, some multinational firms might not have invested in a number of products that were developed in universities. However, venture capitalists and some multinational corporations, such as pharmaceutical and life-science firms, support research in and development of new technologies at universities. On transfer of these technologies to the private sector, the products are commercialized and introduced to the marketplace. Because these new products tend to increase market supply, the monopolistic power of industrial conglomerates decreases (Parker et al. 1998). Most university research focuses on basic problems, but sometimes it leads to practical products. University research has the potential to increase both productivity and competition in the marketplace.

The royalties of technology transfer are shared among university professors, universities, and departments. It has been a lucrative process for some university professors and changed the operation of university departments. Although some professor-entrepreneurs stop drawing salaries from the university or even donate funding, some continue to be on staff and use university resources. Thus, the border between public and private enterprises is somewhat vague. In addition to technology-transfer agreements and royalties, many companies become engaged in financing particular research lines with rights of first refusal of the research product. Thus far, it appears that technology-transfer agreements have not prevented publication of research results, but some scientists claim that the agreements do create publication delays (NRC 1997a). Such delays in publication could be especially problematic if they affect the careers of graduate students, postdoctoral associates, and other scientists beyond the principal investigator. In addition, it is not clear how much profit and commercial motives affect research agendas of university researchers. Those issues are being deliberated and the outcomes probably will affect the structure of university research.

Technology-transfer mechanisms evolve continuously, and this evolution should be studied and analyzed. Universities vary in the emphasis placed on technology transfer and education of their scientists on how to negotiate agreements with the private sector. Commercial firms believe that technology transfer offices generally overvalue inventions of their scientists and undervalue risks made by private firms (NRC 1997a). Increased sensitivity to differences in culture, mission, motives, and expectations among public and private research collaborators can increase the likelihood of success in these negotiations (NRC 1997a).

The USDA not only is required to transfer its knowledge to the public domain, but also is encouraged to transfer technologies that originate in its laboratories to the private sector for commercialization. The 1980 Stevenson Wydler Act and its amendment, the Federal Technology Transfer Act of 1986, set up Cooperative Research and Development Agreements as a mechanism for collaboration between government and private research laboratories (NRC 1997a). The USDA Agricultural Research Service (ARS) engages in collaborative alliances with a variety of companies, including large multinational firms (adapted from USDA-ARS Technology Transfer Information Systems databases) in a few recent cases the public has expressed some concern with the outcome of the partnerships between federal agencies and private corporations. This is an important topic for further research. The resolution of these issues could influence the design of public-sector research and the ability of the public and private sectors to use research results.

Privatization of Extension Services and Consulting

Agriculture has become more science-based and requires much more specific expertise to enhance productivity. As the support and funding of extension services decrease, new types of institutions and private consultants are emerging. It was stated earlier that some large farmers retain their own inhouse expertise. Private consultants serve some small farmers. Sometimes, they work independently; at other times, they work with agrochemical or irrigation companies. Transmission of knowledge in the past was mostly the responsibility of the public sector, but knowledge is increasingly privatized. In many cases, the clients of land-grant university extension services are now the consultants rather than farmers. In California, for example, extension offices work extensively with consultants and provide training for pesticide applicators and advisers.

The privatization of knowledge provision has changed how pesticide-use decisions are made and has introduced new ways to enforce and regulate chemical use. The professional conduct and responsibility of consultants might become more codified and scrutinized, and they will be liable for misuse of pest-control substances. The proliferation and expansion of consultants in pest control are closely related to the growing use of consultants for other agricultural activities, including irrigation and soil-fertility management. With the emergence of precision farming, consultants compile field data and analyze chemical-use information to develop more precise and productive chemical-input recommendations for their clients (NRC 1997b). This knowledge base could be very valuable in pesticide-use decisions and pest-management options. Effective training and continued education of these consultants will affect pesticide-application practices and the future of pest control in the United States and around the world.

Phaseout of Commodity Programs

A wide array of agricultural support programs that originated in the 1930s are gradually being phased out. The phaseout is consistent with the globalization process mentioned earlier and aims to improve efficiency and competition in the economy. The main reason that the commodity support programs were introduced was the tendency of agriculture to attain excess supplies and thus low prices and low income for farmers (Gardner 1978). However, in many cases, the commodity programs have backfired and actually provided an incentive to increase supply. Structural changes in agriculture are increasing the economic viability of agricultural businesses, and congressional mandates that provide price supports to farmers are expected to expire by the year 2002. Of course, reappearance of low commodity prices could lead to reenactment of some support programs. Farmers are increasingly encouraged to rely on insurance instruments provided by private firms and public-private partnerships to manage their production and revenue risks. Future markets and contracts are expected to play an increasing role in reduction of price risks. Insurance has been suggested as a tool to address productivity losses due to pests and to provide farmers some economic incentive to switch from chemical-pesticide use to alternatives. If such insurance instruments were instituted, farmers might forego the use of economically or environmentally expensive pesticides, knowing that they are insured against some types of risk associated with pests.

Some government support programs—such as the dairy, peanut, sugar, and tobacco programs—continue. Through these programs and the policy processed behind them, the government continuously monitors agriculture and, although it is unlikely, deregulation might be reversed (for example, if farm incomes become extremely low). The federal government's role in supporting farmer income has been de-emphasized, but the United States continues to be strongly committed to providing public goods with large social benefits, such as basic research, outreach, and protection of the environment (USDA, 2000).

Devolution

The US government and many other countries' goverments are shifting some major responsibilities to state and local governments. Increasingly, local governments are addressing natural-resource management and environmental-quality issues. Thus, we might be entering an era with global markets and reduced barriers to trade combined with a wide array of diversified local regulations and strategies to manage natural resources. This committee envisions pest-control strategies that could evolve to more regional approaches. A regional policy might, for example, incorporate the public view on the agricultural enterprise and ecological, economic, and social factors influencing agriculture in a particular region. Such regionalization suggests a need for flexible government policies and diverse pest-management strategies to address pest problems in varied circumstances.

Emergence of the Knowledge Economy

The growing importance of science in the development of technology and the proliferation of computers and information technology are gradually making knowledge and information dominant factors in production processes and major sources of wealth (Romer 1986). One manifestation of the increased value of knowledge is the growing importance of intellectual-property rights. The evolution of the legal systems of patents, plant-protection laws, and trade secrets enables owners of specific technological knowledge, which is critical for valuable production processes, to collect some return on their investments associated with the development and use of this knowledge. Major chemical companies and other entrepreneurs are racing to support research and buy rights to knowledge that will enable them to control valuable product lines. Ownership of the rights to process innovations could enable companies to control the fate and prices of many products that use their innovations. To some extent, the ability of a private firm to develop new products will depend on the availability and cost of new knowledge.

The life of patents is finite, and the price and profitability of pesticides and other substances decline after patents have expired. Policies and regulations that will restrict and constrain the use of substances after patent expiration might be valuable to industry, especially to firms that wish to introduce new products as substitutes for older products. This set of incentives should be considered in evaluations of environmental regulations and other legal instruments used to phase out old chemicals in favor of new ones. In many cases, environmental side effects will be used to justify phasing out of new materials. We have knowledge and experience about the side effects of existing chemicals, but their replacements could present some unknown risks.

Increasingly, refined measurement equipment and improved ability to analyze the composition of environments have led to more accurate identification of low-dose toxic materials in various environments. That can increase concerns with food safety and environmental side effects of chemical use, so more public education is needed. Improved ability to trace chemicals back to the source of their emission might result in stricter environmental regulation, especially because quantitative links between toxic concentration and risks to human health are in many cases ill-defined. The growing severity of environmental regulations might provide some justification for altering pest-control strategies by introducing new precision pesticide-application technology or for canceling some pest controls and setring the stage for introduction of new strategies.

Increased differentiation in management of environments and products is an emerging trend. The computer revolution has increased the ability to collect data and monitor the behavior of consumers, farmers, and ecosystems. The resulting information has the potential to yield more refined management strategies that adjust actions for specific spatial and temporal conditions. For example, the agricultural management system that is information-intensive enables producers to adjust inputs during the growing season in response to changes in the weather (NRC 1997b). Some major agribusiness firms have recognized the value of production-management services and are shifting their emphasis from providing inputs, such as seeds and chemicals, to selling production-management services. Similarly, agribusiness and food-marketing firms are providing farmers with detailed product specifications and, in some cases, production guidelines as part of contracts with farmers. Thus, agriculture is moving toward a more integrated system in which the information-intensive agribusiness has taken control of decisions traditionally made by the farmer.

The Organic-Food Market

Organic food is defined as food produced without the use of synthetic chemicals (pesticides, fertilizers, or other inputs) (Organic Trade Association 1998). The organic-food market is the most rapidly growing food segment, but it is still a very small portion of the total food produced in the United States. Consumers perceive organic food to be healthier and to have been produced without any pesticides at all. Health benefits of organic food over conventional food have not been conclusively proved scientifically (C. Winter, UC Davis, Food Safety Program, personal communication, April 8, 1998). There are some growers, packers and retailers that are trying an alternative to organic agriculture because organic agriculture focuses primarily on "no synthetics" but is silent on some environmentally friendly and socially responsible practices. For example, the Food Alliance of Portland, Oregon, requires its farmer-members to limit their use of chemicals, conserve soil and water, and provide fair and safe working conditions. The Food Alliance's aim is to develop a brand or label recognized by the consumer (see section on "eco-labelling" below) (Food Alliance 1998).

Market Size

The organic-food market in the United States was $178 million in 1980, $2.8 billion in 1995, $3.5 billion in 1996, and over $5.4 billion in 1998 (Organic Trade Association, 2000). The growth of organic foods is fueled by aging "baby boomers" who have large disposable incomes and who want healthy food. Organic food is predicted to grow by 10-20% per year for the next 10 years worldwide (Moore 1997). Land conversion to organic farms will limit the growth—land cannot be converted fast enough to match market demand. The traditiona1 food market is growing at 1% per year. At a growth rate of 20% per year, organic foods could constitute 20% of all food in the next 10 years. The rapid growth of the organic- and natural-foods market has sparked interest on Wall Street, with an unprecedented number of companies doing initial public offerings ($755 million raised from 1996 to 1997) and finding venture capital ($200 million raised in 1995) (Hoffman and Lampe 1998). But despite this growth, the percentage of the USDA research budget dedicated to organic farming is 0.1% (Lipson 1998).

There were 2,841 organic farms in 1991, 4,060 in 1994, and 12,000 in 1997. Acres of land dedicated to organic farms reached 550,267 in 1994 and 1,127,000 in 1997. Over 20% of US consumers purchased organic food during the first 6 months of 1997, and 97% of them were satisfied with their purchases. Tomatoes, lettuce, carrots, and apples are the four top-selling organic-food products. Across the country, 57% of upscale restaurants offer organic items (Business Editors/ Food and Wine Writers/ Consumer Reporters Advisory for September. Business Wire 9/17/97. Contact: Fineman Associates, Evette Davis). Organic dairy foods make up the fastest-growing organic-food class. Horizon Dairy is an organic dairy that has had triple-digit growth since 1992. Sales went from $446,000 in 1992 to $16 million in 1996 and $28 million in 1997 (Retzloff 1997). Organic wine, another growing product class, is 1% of the total premium wine market in the United States (Business Editors/ Ford and Wine Writers/ Consumer Reporters Advisory for September. Business Wire 9/17/97. Contact: Fineman Associates, Evette Davis 1998).

Consolidation of the industry is rapid; the number of organic-food companies is predicted to drop from 600 to 300 in the next 3 years (Moore 1997). Consolidation is necessary to drive the industry to a profitable business model. Only companies with "critical mass" will probably survive, and new entrants will find it difficult to establish themselves. Downward price pressure on organic foods also means increased pressure on farmers; the inefficient farmers probably will shut down and efficient organic farmers will expand acreage. Also influencing the trend is the fact that mainstream corporate farming operations are converting to organic production methods. For example, Shamrock/Disney's food company, Cascadia, will be sourced by a major corporate farming operation that is converting thousands of acres to organic farming systems.

California represents the leading edge of the changing organic enterprise. The market in California has doubled twice in 12 years, and 1,800 farmers are now in organic production. In 1998, there were 11,000 acres of organic certified grapes, 4,200 acres of rice, 3,500 acres of carrots, and 2,000 acres of lettuce. Organic acreage in the state totals 4% of all table grapes, 2% of wine grapes, 1% of lettuce and rice, and 5% of carrots. Organic farm revenues in the state increased from $75 million in 1993 to $95 million after 2 years and are expected to double by 2000 (Johnson 1998). Beyond California, the international market for organic products is also growing and discussed in more detail in Box 3-1.

Pavich Family Farms, in California started farming organically in 1972. It is in the San Joaquin Valley and in Arizona and successfully farms 3,700 acres of organic table grapes. Because of the growing demand, Pavich Family Farms embarked on marketing organic products from 50 other growers in Chile, Argentina, South Africa, and elsewhere in the United States. Pavich Farms markets 70 products, including vegetables, bananas, nuts, and apples. Steve Pavich, a company founder, sees his mission as dispelling rumors that large acreage cannot be farmed organically. He believes that the university system is set up to encourage "silver bullet" thinkers (linear thinkers), but organic agriculture is matrix thinking and a systems approach. Pavich feels that basic knowledge of soil microbiology is the critical scarce information and that such knowledge is the key to successful organic farming (Pavich 1998; Pavich Family Farms 1998).

USDA National Organic Standards Program

In December 1997, USDA announced its National Organic Standards Program proposed rule, which would end a patchwork of more than 3 dozen state and private certification standards for organic agricultural products. The rule addresses federal requirements for producing, handling, and labeling organic agricultural products. It provides requirements for certification of organic production, accreditation of state and private certifying agents, equivalence of foreign organic certification programs, approval of state organic programs, and user fees. The proposed rule, however, did not specifically eliminate the use of genetically engineered crops, sewage sludge, and irradiation on products designated as organic. USDA received a record 275,000 public comments against the inclusion of sludge, genetically engineered crops, and irradiation and only 19 in favor of their inclusion. USDA made fundamental revisions to the proposed rule in response to the public comments (USDA 1998), including banning the use of irradiation, sewer sludge, and genetic engineering in the production of any organic foods or ingredients (Glickman 2000).

Eco-labelling

Eco-labelling is a new technique used by retailers to provide consumers with the right to the information on and the production method of the origin of the food product they are purchasing. Eco-labelling is gaining attention in global trade, not only as it applies to pesticide residues, but also to address concerns about food produced from genetically modified seed. A growing trend in the US food industry is to use eco-labelling to inform the consumer of the specific agricultural practices used to make the product. The labels tout environmentally or eco-friendly production practices that conserve natural resources, protect the environment, and use low-chemical integrated pest management (IPM). This niche, which is broader than existing organic production, does not specifically exclude synthetic chemicals in crop production.

Efforts to increase revenue through "green" or "eco" labels are worldwide. In Europe integrated fruit production (IFP) represents economical and safer production with a goal of overall reduction in pesticide use. Growers have agreed to 13 guidelines for pome production in Europe,

BOX 3-1
International Organic-Food Market

The organic-food market is growing just as fast outside the United States as it is inside. Many governments—for example, those of Poland (Agrow, 1998a), Croatia (Agrow, 1998b), France (Agrow, 1998c), and China (Reuters, 1997)—are providing financial incentives to convert conventional production to organic production. Sales of organic products in Japan reached 150 billion yen in 1996 and sales are expected to be 250-300 billion yen in 1997 (Kyodo, 1997). The first all-organic supermarket opened there in 1997. Large mainstream retailers are lured by the potential rapid growth of demand for organic products. An official at the trading house Nissho Iwai Corporation predicted that 8-10% of Japan's 40 trillion yen ($333 billion) food market would turn to organic products; it is less than 1% now (Nakanishi, 1997). Nissho Iwai was one of Japan's first trading houses to strike deals with foreign farmers for organically grown food; it has contracts for grain, sugar, coffee, fresh and frozen fruits and vegetables. The trading house has teamed up with French organic-food maker La Vie S.A. to import 20 kinds of organic foods. In July 1997, the government started to work on domestic standards, and formed an independent controlling body to create national standards. Green Network Japan, founded by 2,000 farmers in November 1996, is creating a new market for small "green" producers in Japan (Nakanishi, 1997). A trading market for chemical-free rice and vegetables was established on January 1, 1998, on the Internet by the Nippon Field Science Association and IBM Japan. The market participants include about 1,000 farming households for organic produce, agricultural cooperatives, supermarkets, and food-servicing companies. The association predicted annual trade of 18 billion yen in the first year and 60 billion yen in 5 years with 10,000 farming households.

several of which result in lower pesticide use. The guidelines include a nonpermitted- and a permitted-pesticide list. Italy, Germany, Austria, and Belgium have high acceptance and grower participation in this region. In Tyrol, France, IFP started in 1989 with 10% of growers; by 1997, 90% of growers were participating. Controllers of the programs check at least 20% of the acreage each year and have very tight standards to maintain. They also require residue analysis on 10% of the crop. IFP has worked well on apples in humid areas to reduce insecticides (specifically organophosphates). IFP obtains cooperation from many groups involved in the process, and, notably, there is also high consumer demand for IFP fruit (Vickery 1998).

New Zealand

The New Zealand apple market is primarily a market of export to the United Kingdom (New Zealand markets 2% of the world's apples). Participants in the New Zealand IFP program are ENZA (sales arm of the

In Europe, organic farming is becoming "a mainstream option." The United Kingdom has seen a 20% annual increase in organic food sales since the middle 1980s, with demand outstripping supply. Nearly three-fourths of organic produce sold in the United Kingdom is imported. Tesco, a major British retailer, lowered prices of its organic produce and saw a massive increase in volume of sales. It set up a Center for Organic Agriculture with $8 million to develop efficient organic production systems and increase the national output of organic food (Judge, 1997). Farmers recognize the demand and have increased land conversion by 30%. In Austria, 200,000 hectares (494,000 acres) is organically farmed, compared with 49,000 hectares in the United Kingdom. In 1995, more than 10,000 Austrian farms were in conversion to organic. Sales of organic food were 40 million pounds ($67 million) in 1987 and 150 million pounds ($251 million) in 1994.

SwissAir has become the first airline to introduce organically grown products for in-flight meal service. Naturalgourmet™ is the name of the new concept. Introduction of organically grown food products into in-flight meals complements SwissAir's "environmental care" efforts. In a survey, passengers requested that organically grown products be used to as large an extent as possible but without lessening variety or taste. BioSuisse, the association of Swiss farmers who practice organic methods, and the Swiss consumer protection association will work with SwissAir on the Naturalgourmet project. The airline will serve organically grown baby food and will no longer serve genetically engineered products. The Swiss government pays a 20% subsidy to organic farmers but only about 2% to farmers who use only chemicals (SwissAir, 1997). In Switzerland, 7% of the total agricultural land is in organic production (75,000 hectares); this proportion is projected to increase to 13-15% by 2002 (Agrow, 1997).

apple-pear marketing board) and Horticultural Research (government-sponsored agricultural research). For New Zealanders to export apples, the commodity must go through ENZA. In 1996-1997, 88 growers were involved; by the 1997-1998 season, over 400 growers participated. It was estimated that for the 1998-1999 season there would be over 800 growers, over 50% of all growers. The program uses a pesticide rating system whereby pesticides are numerically rated according to their toxicity to various systems and each grower earns a particular number of points. Growers who use highly toxic compounds must reduce the total numbers of applications or they will not meet the standards set by ENZA for IFP fruit. Results after the first year showed a 40-60% decrease in spray; organophosphate-insecticide use decreased by 90% (one of the primary goals of the program), costs of production were slightly lower, and damage on fruit was slightly higher. Confirm(, an insect-growth regulator made by Rohm and Haas, figures prominently in the program as an organophosphate replacement (Dr. Scott Johnson, U.C. Davis, Dept. of Pomology, April 15, 1998, personal communication).

United States

In the United States, several IPM programs have identifiable logos, trademarks, and guidelines. They are based on a point system to measure attainment of certain goals. Programs are certified, and only participants can use the logo. Based in the University of Massachusetts, Partners with Nature is an information-based-program for vegetables (cole crops, peppers, potatoes, pumpkins, and sweet corn). In 1998, 50 growers were participating. The growers receive recognition for IPM-based efforts and attain points by engaging in various activities (the point system does not focus on reducing any specific products).

Wegman's Food Markets and growers—including the New York State Berry Growers Association, the Eden Valley Growers, and Comstock Michigan Fruit—have partnered with Cornell University in an educational IPM outreach program. The partners have developed IPM labeling programs for corn, beets, cabbage, carrots, and peas, mainly for the frozen-food markets. Gower compliance with IPM practices is verified jointly by Comstock and Wegman. The point system is not based on specific products used, but rather on pest monitoring and pest management. The requirements are minimal compared with California grower practices; most California growers are already using practices that exceed these requirements.

Some 250 apple, pear, and cherry growers in Washington and British Columbia participate in Responsible Choice (Kane 1999). The program ranks chemicals and includes mating disruption as a pest-management practice. Overall, participating growers use more environmentally friendly products than nonparticipating growers. The weakness in the program is that all growers in the program get the label "Responsible Choice" even if they have failed to meet point goals for pesticide reduction.

Wisconsin potato and vegetable growers have partnered with World Wildlife Fund in an eco-labelling effort. This program has an overall pesticide-reduction goal involving carcinogens, acutely toxic compounds, endocrine disrupters, and compounds affecting nontarget insects; it deals with water quality, soil quality, and resistance management. A 15% reduction in the target pesticides was achieved from 1995 to 1997; there has been a 30% reduction over the last 5 years (Mike Carter, director of government and grower relations, Wisconsin Potato and Vegetable Growers Association, July 1, 1999, personal communication).

In California, Mary Bianchi, farm adviser, works with wine-grape growers from central coast vineyards in Monterey and Santa Barbara counties. Team membership represents 40,000 of a total of 80,000 acres in Monterey County. The program uses a point system similar to that of the

Massachusetts program, Partners with Nature. Pest management represents 200 of the 1,000 total points. The objectives are to integrate best pest-management practices for quality wine production, to support education, and to enhance the environment. It is a grower-driven program, although the University of California and the California Department of Pesticide Regulation provide funding support. Pest-management guidelines include monitoring of pest and beneficial species, but no specific guidance regarding particular products is offered. It is a voluntary program with its own certification (Mary Bianchi, farm adviser, San Luis Obispo, Monterey and Santa Barbara counties, April 1, 1998, personal communication).

The Food Alliance is a nonprofit group in Oregon formed to promote environmentally friendly farm practices beyond organic agriculture. The Food Alliance requires its farmer-members to limit their use of chemicals, conserve soil and water, and provide fair and safe working conditions. Members of the Food Alliance who meet the requirements of the program can label their farm products with the group's eco-label (Food Alliance 1998).

Innovative Farming Systems to Reduce Pesticide Use

Biointegral Orchard Systems and Biointegral Farming Systems

Biointegral Orchard Systems (BIOS) and Biointegral Farming Systems (BIFS) are demonstration programs designed to help almond and walnut growers reduce the use of synthetic pesticides and fertilizers through the adoption of an IPM approach as expressed in a defined set of sustainable farming practices. The programs are coordinated by the Community Alliance with Family Farmers based in Davis and the University of California, Davis Sustainable Agriculture Research and Education Program. Through education field days, in-field technical support and small financial incentives, BIOS and BIFS helps growers to use a reduced pesticide approach. BIOS and BIFS personnel maintain good contact with growers and are viewed by growers as being supportive in addressing issues and problems. The Lodi Woodbridge BIFS has 40 growers, covering 20,000 acres, enrolled. Soil fertility, arthropod management, weather-based disease models, and cultural practices are the focus of the program. On the average, the growers reduced use of insecticides by 25% and of copper-based fungicides by 50%. Postemergent herbicide use (glyphosate) increased and pre-emergent herbicide use decreased. The West Side BIFS encompasses 12 farmers and 90,000 acres, with an average farm size of 7,500 acres. Cotton drives the system. In this BIFS, the number of sprays was not reduced successfully. BIOS and BIFS rely considerably on grower knowledge whereas Extension, which is research-based, relies on firmly anchored, replicated, statistically sound, objective knowledge generated by university research (University of California Sustainable Agriculture Research and Education Program Fact Sheet 1999).

Pesticide Environmental Stewardship Program

The Environmental Protection Agency (EPA) has developed the Pesticide Environmental Stewardship Program (PESP). This program provides funding to selected groups that work together to develop IPM programs that reduce pesticides. Growers, grower groups, utilities, and cities can become partners of PESP and then compete for grant money. Each PESP member must complete an EPA-approved IPM strategy within the first year of membership. (PESP 1999.)

REGULATORY CHANGES

Environmental Regulation

There has been growing concern about the environmental and health side effects of agricultural practices and other economic activities over the last 40 years. There is some empirical evidence, that as average income increases, concern about environmental protection also increases. The establishment of the EPA centralized and focused environmental-control activities. EPA became the major federal agency responsible for implementing the Federal Insecticide, Fungicide, and Rodenticide Act (FIFRA) and controlling environmental and human health effects associated with pesticide use. Over the years, similar agencies have sprung up in other countries, and some states have established their own environmental-protection organizations; for example, the California EPA is a large, powerful agency. In fact, the abatement of pesticide use is due to the activities of myriad agencies, including environmental-protection agencies, public-health agencies, boards of water- and health-quality protection, and departments of agriculture. Each agency has addressed a different aspect of pesticide use, and some have argued that these numerous agencies were responsible for a preponderance of requirements and overregulation.

During the 1980s and especially the 1990s, concern about global environmental problems increased, and several international agreements that had direct implications for pesticide use were signed. For example, concern about the depletion of stratospheric ozone led to the Montreal protocol that will ban the use of methyl bromide in developed nations by 2001 and in developing nations 10 years later. The Kyoto protocol in 1996 established limits on trading in the rights to emit carbon dioxide (this protocol must first be adopted by the carbon dioxide polluting countries).

The protocol will result in energy efficiency and possibly lead to increased pesticide use (for example, herbicides) as a substitute for energy. It will affect land-use patterns because countries will receive credit for sequestration of carbon dioxide and be fined for releasing carbon dioxide in forests and other land-use practices. The impact of such arrangements on pesticide use has to be studied further, but such agreements will likely lead to increased acreages of biomass that will be sold as sinks for carbon dioxide.

During the 1970s, EPA and similar agencies relied mostly on direct control of technology, liability rules, and litigation as the major tools to achieve environmental-policy objectives, especially in regulating production activities. Strict registration requirements have been and are still the major tools to address introduction of new substances for pest control and management. The command-and-control approach establishes environmental-quality targets, identifies practices to meet this target, and institutes a set of rules (mostly liability), incentives, and deadlines to adopt desired practices in reaching environmental-quality objectives. A notable example is the introduction and enforcement of scrubbers in power plants and of catalytic converters in cars to reduce air pollution. Some economists and others have suggested that the command-and-control approach is rigid and inefficient, curtails initiatives, and can have counteractive outcomes. Similarly, the confrontational litigious approach to enforcement of regulations led to a backlash against the environmental legislation among lawmakers and the public sector. Thus, there is a gradual shift toward a more flexible and cooperative approach to meet environmental-quality objectives (Ribaudo and Caswell 1999).

A more acceptable approach is the introduction of markets and trading in pollution rights. Government agencies can establish an aggregate target level if pollution is well below existing target levels. Firms then would establish pollution rights (often proportional to their share in overall pollution in the past) and then be allowed to trade in the rights. This approach has been used to address air-pollution problems and might be applicable to pesticide use.

The use of direct financial incentives (subsidies) to induce environmental protection is growing. Alternatively, some countries in Europe have used pesticide taxation. This approach is popular in Scandinavian nations and is being considered by other European countries (OECD 1995). To encourage environmental protection, governments have used subsidies to invest in technology or adopt cleaner application systems. For example, the 1985 Food Security Act made entitlement to certain government programs subject to environmentally friendly farming practices. The 1990 Farm Bill included a water-quality incentive program, which encouraged practices that reduced groundwater-contaminating chemicals.

Environmental-quality objectives can be achieved through government "rental" of lands and other natural resources (such as as water) whose use in agriculture can have adverse environmental effects. The best example of this approach is the USDA Conservation Reserve Program (CRP), in which the government rents land from farmers for a fixed term of 10–15 years and requires them to take out of production land that is identified as environmentally sensitive and that meets various criteria (SWCS 1992). In 1999, the US government allocated $1.3 billion for payments to farmers in the CRP (USDA 1999). Although the major criterion in the past for these rentals was soil erosion, the criteria expanded over the years to include wildlife and native-plant preservation and wetland protection. By targeting land rentals appropriately, governments can reduce some of the damage from pesticide emissions, such as groundwater runoff and contamination. Other funds are aimed at water rights that might be applied to pesticides. In principle, financial incentives are provided to farmers and other users to change their activities.

Another popular approach is voluntary agreements between firms and regulatory agencies to control particular types of pollution. Participating firms gain favorable publicity as environment-friendly companies, and it is argued (Segerson 1999) that this approach is especially effective if there are threats of strict environmental regulation (a threat of legislation behind the "stick"). The threat of EPA intervention in establishing water-quality standards in California was effective in motivating the different parties to work together to meet standards for the San Francisco Bay and Delta in 1993.

California has also introduced two regulations to control pesticide use. The first says that only certified individuals are permitted to make chemical-input recommendations and apply those chemicals. These professionals are responsible for environmentally hazardous activities and, in the long run, can be held liable for the application of chemicals and for environmental protection. California also requires documentation of chemical applications and storage of the information in a database. The reporting requirement becomes less expensive when the amounts of computation and documentation decrease and when the prescription application of chemicals is restricted to professionals. With more accurate and timely information, monitoring the use of pesticides and enforcement of regulations will become easier. Furthermore, public awareness of documentation activities encourages people to be more cautious in their pest-control activities.

In spite of the gradual shift toward financial incentives, the leading pesticide regulation remains direct control. A key regulation is the registration requirement, wherein chemicals and other compounds are subject to a wide array of tests to ensure effectiveness and mitigate risk. These registration requirements also increase the costs of introducing new pest controls and contribute to industrial concentration of the pest-control manufacturers. Another important feature of pest-control policies is the power to ban chemicals that have adverse effects. Over the years, the regulatory process has become more sophisticated, and the economic impacts of chemical bans have been taken into account. Regulators have recognized the heterogeneity and variability of impacts and now tend to rely on partial bans and use restrictions, taking benefit-risk criteria into account.

Until recently, most compounds have been regulated individually. However, most chemicals belong to a small number of "families", and there is a tendency, manifested in the Food Quality Protection Act (FQPA), to limit the aggregate use of and exposure to closely related compounds (such as the organophosphates). In the future, we could see more evaluation of pesticides in a more generalized context in which regulators consider the environmental and economic impacts of complementary and substitute compounds.

Food Quality Protection Act

FQPA was passed unanimously by Congress and signed into law on August 3, 1996, as Public Law 104-170 (FQPA 1996). FQPA amended in several major ways both laws regulating pesticide registration and use in the United States: FIFRA and the Federal Food, Drug, and Cosmetic Act (FFDCA). The passage of FQPA was the outcome of several factors, including the National Research Council (NRC) report *Pesticides in the Diets of Infants and Children* (NRC 1993). A major goal of FQPA is to protect children and infants from pesticide residues in their diets. The 1996 law also extended residue protection to include residue exposure aggregated from sources other than food (such as water, air, and dermal contact) and to account for cumulative exposure to multiple chemicals in those sources, at least for chemicals that act through a common mechanism of toxicity. FQPA was supported by the agricultural community to resolve the so-called "Delaney paradox", replacing it with a uniform health standard defined as "reasonable certainty of no harm". Under the "Delaney clause" of FFDCA, tolerances for pesticide residues on raw agricultural commodities were determined by a weighing of risks and benefits. The paradox arose from the fact that there was a zero tolerance for the same residues concentrating in processed foods. Thus, foods accounting for nearly half the total estimated dietary risk (all meat, milk, poultry, and pork products and many fruits and vegetables) fell ostensibly beyond the scope of the

Delaney clause, because under EPA guidelines they had no processed form. The Delaney clause was replaced with language to ensure uniform regulatory treatment of chemicals used in both raw and processed foods. Other provisions of FQPA were intended to encourage a more science-based and transparent regulatory process with focus on additional and adequate protections for subpopulations, particularly children. The over-arching goal of FQPA is to encourage development and use of pesticides with reduced risk to humans and the environment. That goal and the many provisions of the law will affect the number and types of pest-control agents that will be available for US agriculture in the future (EPA 1999).

As noted earlier, a major provision of FQPA was to address the inconsistencies that emerged from the Delaney Clause of FFDCA, primarily the dilemma caused by the zero tolerance standard for some pesticides in processed foods. Under the Delaney clause (FDCA Section 409), no finite tolerance was allowed for chemicals that were found to cause cancer (that is, to induce either malignant or benign tumors) in experimental animals and which concentrated during food processing (NRC 1987). Zero tolerance produced a variety of problems: as analytical methods improved, concentrations that were "zero" by older methods became measurable (Zweig 1970). That tie to the analytical method of detection led to an ever-shifting regulatory landscape as agencies and registrants attempted to keep up with improvements in analytical methods. In addition to improved analytical methods, extensive testing to determine oncogenicity occurred in the 1980s, partly as a result of reregistration requirements of FIFRA. The extent of testing in the 1980s indicated that 60% of all herbicides (on the basis of pounds applied) are oncogenic or potentially oncogenic. By volume of use, 90% of all fungicides and 30% of all insecticides fell into this category (NRC 1987). The combined effects of increased positive tests for oncogenicity and improved analytical capabilities threatened the registration status of many mainstream pesticides that were found both on foods that undergo processing (Section 409) and on raw agricultural commodities (Section 408). An additional problem with the Delaney clause is that zero tolerance for carcinogenicity but not other modes of death inadvertently biased test results in favor of the more toxic compounds because test animals succumbed before doses sufficient to induce cancer could be applied. FQPA eliminated the distinction between raw-food and processed-food tolerances so that all pesticide residues will be regulated under an amended FFDCA section 408. New section 408 requires all tolerances to be "safe", ensuring a "reasonable certainty of no harm" from pesticides. The law authorizes slightly higher residue concentrations on foods when pesticide use avoids greater health risks to consumers or substantial disruptions to domestic production of an adequate, wholesome, and economical food supply (Schierow, 1998).

The NRC report *Pesticides in the Diets of Infants and Children* (NRC 1993) addressed the question of whether regulations for controlling pesticide residues in foods adequately protected infants and children. There was concern that the exposure of infants and children to pesticides and their susceptibility to harmful effects would differ considerably from those of adults. Children consume more of some foods and water, on a body-weight basis, than adults. Many of those foods—such as apples, apple juice, applesauce, orange juice, pears, peaches, carrots, and peas—are treated with a variety of pesticides during production, and detectable residues are found on many of them. Furthermore, the foods most often sampled by the Food and Drug Administration (FDA) in its role of monitoring residues in the US food supply include just a few of those most often consumed by nursing and nonnursing infants. The NRC study committee concluded that infants and children are different from adults, both qualitatively and quantitatively, in their exposure to pesticide residues in food and that the differences were not taken into account in pesticide regulatory practices such as tolerance-setting and calculation of dietary exposure limits.

Evidence of quantitative and, occasionally, qualitative differences in the toxicity of pesticides between children and adults was also found. Although there was a general lack of data that addressed differential toxicity, examples did support differences in toxicity ranging as high as a factor of 10 for children vs. adults.

The combination of differences in exposure and in susceptibility led the committee to make several recommendations for a new approach to assessing risks for infants and children posed by pesticide residues, including dietary and nondietary exposures:

- Tolerances should be based more on health considerations than on agricultural practice.
- Toxicity-testing procedures that specifically evaluate the vulnerability of children and infants to pesticide chemicals should be devised.
- An uncertainty factor up to 10 should be considered when there is evidence of postnatal developmental toxicity and when data from toxicity testing relative to children are incomplete.
- Additional data should be collected on food-consumption patterns of infants and children within narrow age groups.
- Pesticide-residue data from different laboratories should be collected by comparable analytical and reporting procedures and made accessible through a national pesticide-residue database.

- All exposures to pesticides, dietary and nondietary, should be considered in evaluating risks to infants and children. Exposure distributions, rather than single-point data, should be used to characterize the likelihood of exposure to different concentrations of pesticide residues.

Virtually all of the recommendations of the committee were incorporated in some way in FQPA, primarily in its amendments to FFDCA. These included:

- Mandated consideration of the special sensitivity and exposure of children to pesticides
- Requirement of an explicit determination that tolerances are safe for children.
- Requirement of an additional safety factor of up to 10 based on available information and evidence, to account for the uncertainty of exposures and sensitivities of children.
- Requirement of consideration of exposure to all other pesticides with a common mechanism of toxicity in the setting of allowable residue concentrations.

The conventional method of calculating acceptable daily intake, or reference dose, (RfD) (Barnes and Dourson 1988) for pesticides included use of an experimentally determined no-observed-adverse-effect-level (NOAEL) derived from long-term animal toxicity tests. This is coupled with an uncertainty factor (or safety factor) for extrapolating from animals to humans and across the human population. The latter was generally 10 for intraspecies differences (other humans and animals) and 10 for interspecific or individual differences (U_i) Differences in susceptibility among species were assigned an additional uncertainty factor of 10 (U_s). A further uncertainty factor of 10 could be assigned for studies that encompassed less than a full life span (U_t), and a modifying factor (M, ranging from 0.1 to 10) could be used to account for issues related to the quality of the studies. Consequently $RfD = NOAEL \div (U_i \times U_s \times U_t \times M)$.

FQPA adjusted this "acceptable risk" for each pesticide by recommending an expanded uncertainty factor, up to 1,000. It also requires that all chemicals with a common mechanism of toxicity be included in calculating the RfD for a single chemical. For organophosphates, for example, calculating an RfD for one member must include the possibility that other organophosphates might co-occur in food or drinking water. The FQPA requirement that all exposure (dietary, home use, water, air) be considered in setting an RfD for a single chemical still further restricts the RfD for a given chemical.

The concept of a "risk cup" was developed to illustrate the new regulatory approach. A chemical would have, as before, an RfD-driven limit. But now the RfD would be lower to begin with (because of the use of the 1,000 rather than 100 as a safety factor). And the RfD could be reached more rapidly by the combined effect of including exposures to a single chemical from all sources and to other chemicals of similar mechanism. The risk cup, in effect, would shrink for a given chemical and thus more likely be filled or exceeded by pre-FQPA uses.

FQPA is having major impacts on pesticide development and registration. In part, this was the result of the smaller risk cup, a direct outgrowth of the single, health-based child-driven standard concept of FQPA. Registrants are required to confine RfDs of pesticides within a steadily diminishing level of acceptable risk. For example, EPA is considering a ban on all organophosphates because of the cumulative-risk concept and the common mechanism of action of this large group—some 1,800 tolerances are established. EPA's preliminary risk assessment of 28 organophosphates indicated that risks of some individual organophosphates (such as methyl parathion) exceeded acceptable levels even without the consideration of nonfood sources of pesticide exposure. EPA has determined that the risks to children are unacceptable in the case of methyl parathion and registrations of all fruit uses and many other food and nonfood uses will be canceled before the next growing season (Schierow 1999). Another overall result has been a trend of registrants to pursue registration of "reduced-risk" pesticides—that is, those with reduced inherent toxicity or reduced exposure potential (such as low-dose chemicals and low-persistence chemicals)—or of nonchemical alternatives to conventional pesticides (see figure 2-4 in chapter 2) .

To address risk cup-problems, EPA might propose and registrants might consider label changes that can include loss of selected uses, changes in or revocation of tolerances, changes in use rates, and other appropriate mitigation measures. Registrants are finding that it is important to work with the regulatory agencies and with the user community to ensure consideration of new data, in addition to the impact of these changes on the market. Such approaches will be essential as important pesticide groups such as organophosphates undergo evaluation by the participating regulatory agencies.

The workload imposed by FQPA has resulted in registration delays at EPA. Minor-use registrations and new registrations slowed at the same time when needs were most acute because of lost registrations, lost uses, and changes in tolerances. Farmers are concerned that if chemicals are withdrawn from use, few cost-effective pest-management tools will be available to replace chemicals lost through regulation. FQPA defines major crops as those grown on more than 300,000 acres. These crops include barley, canola, corn, hay, oats, rice, rye, sorghum, wheat, dry beans, popcorn, potatoes, snap beans, soy beans, sugar beets, sunflowers, tomatoes, sod, apples, and grapes. Other crops are considered minor crops, and pesticide usage on them thus constitutes "minor use". This NRC committee is using similar criteria to distinguish between major and minor uses.

FQPA requires review of all tolerances on a 10-year schedule, with 33% completed in 3 years (by August 1999), 66% within 6 years, and 100% within 10 years. EPA has met the initial 3-year goal, in part because of the substantial number of voluntary cancellations of potential reregistration candidates (EPA 1999). In this process, the priority for EPA review has been given to pesticides that pose the greatest risk to public health (organophosphates, carbamates, and B2 carcinogens)—a "worst-first" approach.

One of the most important aspects of FQPA is the requirement of tolerances for emergency exemptions. Section 18 of FIFRA authorizes EPA to allow state and federal agencies to permit the unregistered use of a pesticide for a limited time if EPA determines that emergency pest conditions exist and no registered pesticide would be effective. This means that FQPA required not only immediate implementation of all its new requirements, but also their application to emergency exemptions. Of all the changes in the act, this had the most profound effect on the ability of the agency to meet its deadlines, and it is responsible in part for the reduced number of new uses and new active ingredients. In its amendments to FIFRA, FQPA authorizes collection of fees from industry to complete the review of all current tolerances, thus ensuring that these chemicals meet current EPA standards; establishes minor-use programs in both EPA and USDA, including a USDA revolving-grant program to support data-collection requirements and other procedural provisions to assist minor-use pesticide applications; and reforms the antimicrobial-registration process to shorten regulatory review and decision times.

In addition to toxicity, human risks, cumulative effects, aggregate exposure, and variabilities among subgroups, FQPA requires that EPA consider endocrine effects for reregistration and re-evaluation of tolerances and assessment of risks. An environmental endocrine disrupter is defined as an exogenous agent that interferes with the synthesis, secretion, transport, binding, action, or elimination of natural hormones in the body that are responsible for maintenance of homeostasis, reproduction, development, or behavior (EPA 1997b). These agents can include chemicals with hormone-like effects that can mimic natural estrogens and chemicals with antihormone effects, which bind to androgen receptors and block testosterone-mediated cell responses. FQPA includes consideration of chemicals that might interact with pesticides to pose risks for humans and the environment. Both human effects (such as decreased sperm counts, testicular and prostatic abnormalities, skin cancer, breast cancer, endometriosis, and miscarriages) and wildlife effects (such as decreased fertility and abnormalities in gulls, fish, seals, and alligators) can result from exposure to endocrine-disrupting chemicals (NRC 1999). FQPA specifically mandated the development of a screening program for endocrine effects within 2 years of its passage, with implementation within three years of passage and a report to Congress within 4 years.

FQPA includes a provision to enhance enforcement of pesticide-residue standards. Specifically, the law enables EPA to impose civil penalties for some tolerance violations and authorizes increased funding for FDA residue monitoring. FQPA also contains a consumer "right to know" provision and a tolerance-uniformity section to harmonize tolerance-setting activities between state and federal regulatory agencies.

Several immediate issues have been raised as EPA, state agencies, pesticide registrants, and food producers have begun to grapple with the implementation of FQPA (Agrochemical Insider 1998). There are serious concerns among farmers and growers about the actions that were being considered by EPA under the aegis of FQPA, particularly a ban on all organophosphate insecticides. Speakers at a recent symposium examining FQPA and its impact on science policy and pesticide regulation confirmed that concern, with particular impact being felt in the production of fruits and vegetables, where the residues of organophosphates are highest. Minor crops are likely to bear the brunt of FQPA. Registrants of organophosphates are expected to sacrifice minor crops to maintain tolerances for major crops (Lichtenberg 1999). Proponents of the across-the-board ban had not recognized the widespread reliance on organophosphate pesticides for use on some crops (such as peas and lentils).

There is concern among some agricultural-commodity groups that alternatives to chemicals whose usage would be restricted or lost as a result of FQPA are not available. It might be necessary to have a bridge between the longer-standing pesticide laws and FQPA so that a smooth transition can occur. The bridge would include expedited review of emergency-use provisions (section 18) so that a critical pesticide use could legally continue until alternatives were available.

Loss of pesticides on minor crops might restrict US production and result in increased reliance on imports or in shortages in the marketplace. Those who need fresh fruits and vegetables most, including children in low income families, might suffer the consequences of reduced availability.

Some of the alternatives to existing chemicals could pose other problems, such as susceptibility to resistance, which will be undocumented if these products are rushed into service as substitutes.

EPA might rely increasingly on default assumptions to speed decision-making; this could result in unrealistic scenarios of risk, which might be readily overcome by systematically collecting real data. An example is provided by the safety factor of 10 and its potential use across the board.

In general, FQPA has set and will set priorities for science policies and regulatory decisions. It will both promote and complicate harmonization and further challenge those involved with pest management and those involved in recommending research priorities. A research agenda should help to define and refine the public-policy and science issues that FQPA advances. For example, software tools using sophisticated probabilistic models are being generated to address the aggregate- and cumulative-exposure assessments required by FQPA. Information relative to the temporal, spatial, and demographic exposure of various subpopulations to pesticides creates a demand for high quality current data that should be available to government officials in time to make regulatory decisions. The databases now available fall short of that need (CAST, 1999).

Important scientific issues have also been raised. A NRC study committee reviewed the broad issues of endocrine disrupting chemicals, some of which will directly affect FQPA's implementation steps for them. The NRC committee recommended new studies to assess the health and ecological effects of chemicals known as hormonally active agents. The committee also concluded that, although there is evidence of adverse effects of exposure to high concentrations of these substances, little is known about their impact at low concentrations, such as those that exist in the environment (NRC 1999).

The International Life Sciences Institute (ILSI) Subcommittee on Aggregate Exposure Assessment, Health and Environmental Sciences Institute is conducting a study on the development of methods for assessing exposures to pesticides from nonfood sources to measure aggregate exposure The study includes exposures in the home from the use of chemicals for in-home pest control or the intrusion of chemicals into the home from outdoor uses. Another ILSI working group issued a report on the scientific foundation of the "common mechanism of toxicity" cumulative-risk concept (Mileson et al. 1998).

Decreasing Worker Exposure to Pesticides

The 1992 EPA Worker Protection Standards

The 1992 EPA Worker Protection Standards (WPS), which are described in Chapter 2 included many provisions designed to decrease worker and handler exposure to pesticides. The major criticism of the 1992 WPS has been similar to the criticism voiced by EPA about the pre-1992 WPS: lack of compliance. Enforcement of the WPS is under the authority of the states (Arne 1997). The federal government at least partially finances enforcement efforts in the form of grants to the states. The states are required to report to EPA on enforcement efforts.

The current system of enforcement has a number of problems. First, it is generally more difficult to monitor activities on a large farm than in a factory or other circumscribed area. Second, most enforcement efforts are responses to complaints, so enforcement activities do not give a quantitative indication of general compliance. Third, the agencies responsible for compliance reporting can have political conflicts of interest. Fourth, it is difficult for workers and even medical personnel to diagnose pesticide poisonings (EPA 1998).

Each state determines which agency is responsible for enforcement. In most cases, states assign the role to a department of agriculture or a department of health and environment. There is concern that a state department of agriculture's mission to enforce the WPS is a conflict of interest with other missions of the department. Each state has its own official forms for reporting on inspection of agricultural operations. That has resulted in a lack of uniformity in the kinds of information collected. EPA requires the states to report on enforcement efforts and findings with a single form. Examination of the EPA form used for this reporting (form 5700-33H) shows that the states can give EPA only very general information on enforcement and compliance, even if the state forms have useful details.

Interviews with state and EPA officials indicate that there are political problems in having states enforce WPS and report on violations. Because states compete in terms of their compliance with WPS, state officials at many levels in the enforcement hierarchy have an incentive to underreport noncompliance.

Interviews with EPA officials indicate that EPA has not conducted any independent studies to gain an unbiased estimate of compliance with specific WPS regulations. The only information on compliance that was available to this committee came from one National Institute for Occupational Safety and Health (NIOSH) study, one university study, and two reports by farm-worker advocacy groups that have examined state investigations of worker poisonings and injuries.

In a 1997 NIOSH study (Bauer and Booker 1998), farm workers in Homestead, Florida, and in Kankakee, Illinois, were interviewed with the assistance of camp health aides under direction of a government contractor (Aguirre International) and with a questionnaire developed by NIOSH. A few examples of data collected in the interviews are revealing. The percentages of workers in Kankakee and Homestead that reported having been during the past 12 months in fields that were sprayed with chemicals (while the workers were present) in the preceding 30 days were 18% and 42%, respectively. The percentage that reported working in fields during the preceding 12 months that were "wet with chemicals" were 27% and 50% respectively. Water for washing in the field was reported to be available by 62% and 66% of workers. Soap was reported available by 36% and 59% of workers. According to the WPS soap and water must be provided in areas that have been sprayed within the last 30 days.

The workers were asked more specific questions about the last time they performed specific tasks. For example, during the last time they mixed or loaded pesticides 85% and 805% of the workers in Kankakee and Homestead were provided gloves, and 54% and 51% were provided respirators. No information was given about the type of pesticide that was being mixed or loaded.

A University of Minnesota study not directly aimed at studying compliance (Mandel et al. 1996) found that, of 502 farmers who used pesticides, 56% wore chemical-resistant gloves and 22% wore other protective clothing at least 75% of the time when using pesticides.

Two detailed reports that examined WPS state records from Florida and Washington were recently published by agricultural-worker advocacy groups (Davis and Schleifer 1998, Columbia Legal Services 1998). The groups accessed public documents on how state regulatory officials responded to reports of poisonings or noncompliance with WPS. They concluded that state officials were not vigilant in their investigations and that a number of the protocols for investigation were in need of substantial change. Evidence presented in both reports indicated that state regulatory officials were biased toward accepting the word of employers over employees.

Interviews with agricultural extension workers also indicate a lack of compliance with some WPS regulations. Easily observable actions required by the WPS are the clear posting of warning signs in fields where pesticides have been sprayed and removal of the warning signs within 3 days after the restricted-entry interval has expired. Agricultural extension agents noted that in some agricultural areas where they knew that pesticides had been applied to almost all fields (for example, southeastern cotton in August) there was less than 5% posting. In other regions, farmers sometimes left warning signs up long time after pesticides were applied. Leaving signs posted when they are not a signal of danger decreases the general warning value of the signs.

The studies and observations described in this section are all limited. It can be argued that they do not prove, conclusively, that there are major abuses of the WPS. However, without more detailed objective information on compliance, there is reasonable doubt that the 1992 WPS is accomplishing its goal.

Conducting an objective study of compliance with WPS will be difficult but important. It is critical that an independent organization and individuals with no personal, financial, or political conflicts of interest conduct an unbiased and objective study. Furthermore, the nature of the information to be collected requires sophisticated sociological, epidemiological, and statistical expertise. University personnel might be the most appropriate people to conductsuch studies.

Even if such studies were performed, evaluating the results in terms of what constitutes an acceptable health risk will be difficult. Before the studies are conducted, it is important to determine what percentage of compliance with each of the provisions of the WPS can be considered sufficient.

Additional Means of Decreasing Worker Exposure to Pesticides

Other approaches have been recommended for decreasing worker exposure to pesticides, described briefly below.

Ban on all or some uses of the most acutely and chronically toxic pesticides. Under the comprehensive pesticide reregistration engendered by FQPA, worker exposure and safety issues are under intensive direct scrutiny and analysis by EPA to decrease consumer exposure to pesticide residues on and in foods. One approach to reducing residues that is available through FQPA is the banning of specific uses of problematic pesticides. A large proportion of the residues encountered by consumers is on vegetables and fruits, which are the minor crops targeted by the 1992 WPS. Actions that decrease the use of toxic chemicals on minor crops to protect consumers should also decrease exposure of workers. The objectives of FQPA and some of the challenges inherent in banning specific uses are discussed in earlier in this chapter. The techology and monitoring required in enforcing FQPA could lead to the detection of illegal uses of chemicals that leave measurable residues on food products. Detection and prosecution of illegal use would be expected to provide deterrence in the future.

Prescription use of restricted pesticides. The adoption of a medical model for pesticide use has been discussed for many years (Dover and Croft 1984). The general idea is that some pesticides that are considered to be reasonably safe will be purchased "over the counter" and those considered more problematic will be "prescribed" only for specific uses by a professional trained in toxicology, IPM, and safe pesticide use. In the past, this approach was never considered seriously, because of the infrastructure that would be needed to support it, and because of the potential for abuse.

A recent report by the Council for Agricultural Science and Technology (CAST) (Coble et al. 1998) examined the pros and cons of prescription use. It discusses options for determining who would be considered a qualified prescriber, how the prescriber would actually function, and how potential legal problems could be handled. According to CAST, a professional prescriber would be required to have pesticide education and experience with local agricultural problems, including knowledge about nonpesticidal solutions to agricultural problems. The prescriber would have to be insulated from personal and political pressures. There would be an obvious problem in giving "provider licenses" to people who work for manufacturers or distributors, no matter how highly qualified they were. Federal agencies or state departments of agriculture could be a source of qualified personnel, but a potential problem of bias could arise there because these departments also have a goal of promoting profitable farming enterprises. Finally, independent consultants could fill the role of prescriber.

We lack sufficient qualified people from any source, so training programs would need to be put into place. A prescriber's function could range from writing local prescriptions and reporting to enhancing public and user knowledge of pesticide characteristics and IPM in general. Prescribers and companies that produce the "prescription pesticides" would have a higher liability exposure. The increased liability could be an important deterrent for both parties. Any increase in product labeling associated with prescription use will serve as a disincentive for manufacturers to take advantage of this approach. An alternative to prescription use would be "exceptions to labeled use" that could be permitted if a prescriber were involved, but this option would shift the legal burden from the manufacturer to the prescriber.

The CAST report warns that building the infrastructure needed for instituting prescription use would take time and money, and maintenance of the system would be expensive. The report suggests careful analysis before any steps in this direction are taken.

However, California has implemented a program similar to the pesticide-prescription approach through its Department of Pesticide Regulation when it began testing and licensing pest-control advisers (PCAs) in 1971. The department has since continuously raised the education and experience requirements for licensing PCAs and required annual course work for license renewal. The PCAs have the responsibilities of making determinations for pesticide use or alternatives such as biocontrol in agricultural and nonagruicultural settings. They are licensed to prescribe restricted use-pesticides, such as organophosphates and methyl bromide. They are required to report their recommendations to the state as part of a state pesticide-monitoring program (California Department of Pesticide Regulation, 2000).

Improvement of pesticide-application tools, packaging, and

formulations. Over the last 20 years, there have been important advances in pesticide-application technology. Many of the advances have been implemented, especially for class I pesticides. Entire engineering groups in agrichemical companies and in university departments are devoted to application technology. Ultralow volume, geographic information system applications, drip irrigation, chemigation, and prepackaged ready-to-use containers have all reduced the exposure of agricultural workers (see Table 3-1).

Some changes in formulations of pesticides that are used in household products are needed. The basic change would be to prohibit sale of highly concentrated formulations. That would lower the impact of accidental spills, ingestion by children, and spraying of higher-than-needed concentrations.

Addition of odorants or dyes to pesticides at concentrations that would match the risk of specific pesticides. A characteristic of pesticides that results in worker-safety problems is the lack of immediate feedback to exposed workers. When a worker is hurt because of malfunctioning or misused farm equipment, the cause of the injury is clear to the

TABLE 3-1 Application Technologies with Potential to Reduce Pesticide Risks

Problem	Application Technology
Drift reduction	Electrostatic sprayers
	Hooded sprayers
	Air-assisted sprayers
	Spray additives
	Low-pressure nozzles
More precise application	Baits
	Weed-identification sprayers
	Sprayer injection systems
	Speed and flow monitors
	Field mapping and GIS systems
	Variable-rate application
Applicator safety	Closed systems
	Direct-injection systems
	Prepackaged, ready-to-use containers

Source: Hall and Fox, 1997.

worker, and the worker can then take precautions to avoid a similar injury in the future. In contrast, a worker can be exposed to a harmful dose of a pesticide without knowing it. A worker who feels the symptoms of acute pesticide poisoning might not know whether the symptoms are due to a pesticide, hot weather, a virus, or food poisoning. Unless the worker has been told, or seen signs indicating, that he or she has been in a pesticide treated area before the restricted entry interval has elapsed, it is difficult to determine cause and effect. In the case of chronic or delayed-onset impacts of pesticides, it is always difficult for the worker to determine cause and effect.

If there were full compliance with EPA WPS, a worker would have a general idea of his or her level of exposure. Because evidence indicates that is impossible to get full compliance, it would be useful to develop a more foolproof mechanism to provide immediate feedback to workers who have been exposed to harmful levels of pesticides. In principle, a straightforward approach for decreasing the length of the feedback loop is to add odorants or dyes to pesticides, with the intensity of either signal being related to the danger of exposure.

The principle of using odorants to signal danger is not new. Odorants have been added to odorless natural gases for over 7 decades (Fieldner et al. 1931). The odorant is set at a concentration that a person could detect when the natural gas concentration reached one-fifth the minimal explosive concentration (Cain and Turk 1985). A substantial amount of research was conducted to determine what odorants in what concentrations would be adequate to warn citizens of natural gas leaks (Robertson 1980, Venstrom and Amoore 1968, Semb 1968). It demonstrated substantial variability among individuals in ability to perceive odors (Schemper et al. 1981, Cain et al. 1987, Cain and Turk 1985). Some of the variability was due to age, sex, and health. The sensitivity of a general group of people can vary by a factor of 16 for some odors (Amoore and Hautala 1983). A small percentage of people who are anosmic and cannot perceive odors, but they can perceive nasal irritants, such as allyl alcohol (Amoore and Hautala 1983).

Another component of odor sensitivity is level of awareness. People who are aware of the possible presence of an odor can sometimes detect it at a concentration that is only about 4 % of that needed by a person who is misinformed about the potential for presence of an odor (Amoore and Hautala 1983). People have been shown to quantify the concentration of some odors (such as pyridine) more easily than other odors (such as iso-3-hexene-1-ol). All those characteristics of odors are assessed in determining an optimal odorant for a given situation. In the case of agricultural workers, typically, a number of people work together, so there is little chance that an anosmic person will be isolated or unaware of the odorant. And anosmia can easily be determined with a scratch-and-sniff test (Cain and Turk 1985) during pesticide training sessions. Such tests could also be used in training because odor memory enhances detection. Choice of an odorant would depend on whether the goal is to have workers simply detect the odor or quantify its intensity.

Many commercial pesticide formulations have distinctive odors that come from the pesticidal compound, the formulation ingredients, or impurities. For example, workers can smell most formulations of chlorpyrifos. Unfortunately, the intensity of the odor associated with currently used pesticides is not correlated with the risks associated with exposure to them. Some cities (such as Phoenix, AZ) have ordinances that restrict the use of odoriferous pesticides on farms near residences and schools. That has sometimes resulted in farms using pesticides that are less odoriferous but more toxic than a pesticide with an unpleasant smell.

If the intensity or perceptibility of the odor associated with a pesticide were directly related to the potential harmfulness of the compound, workers would get clear indications of dangerous situations, whether or not their employers informed them of the situation. Because odoriferous compounds, by their nature, must be volatile, their intensity decreases with time. Formulations of odorants could be developed that would decrease in intensity in a manner that simulated the decreased danger associated with a specific pesticide application. It would probably be best to use a single general type of odor to signal pesticide danger so that workers and other people would get one clear signal. Keeping one basic odor associated with natural gas has worked well.

Tests of pesticide concentrations on the skin of pesticide applicators have shown a high degree of individual variation; for example, there was a range of a factor of 100 in residues of alochlor on the hands of 27 alochlor applicator and mixers because of differences in individual behavior (Sanderson et al. 1995). If applicators used formulations with odors, they would be aware of self-contamination rates. Because pesticides have differing rates of penetration of clothing, it might be difficult to develop formulations of one odorant that would perfectly mimic all compounds. Research on this subject is lacking.

Many problems could be associated with both development and licensing of pesticide formulations containing odorants. However EPA and industry have already demonstrated that such problems can be overcome. Because of special health risks associated with them, methyl parathion, paraquat, and methyl bromide now contain odorants (referred to as stenching agents).

In the case of methyl parathion, some household pest-control companies were illegally spraying inside houses with this compound (EPA

1997a), presumably because it is inexpensive. Many cases of severe illness resulted (EPA 2000). EPA worked with the manufacturer, Cheminova, to add a stenching agent to the formulation (packaging requirements were also changed). There have since been no reported infractions of the law.

Paraquat is one of the most toxic agricultural herbicides (Stevens and Sumner 1991). There is no effective antidote to paraquat once a person has been overexposed and it has been involved in many accidental deaths and suicides (Blondell 1996). In 1988, the manufacturer of paraquat, Zeneca, added a stenching agent, changed the color of the formulation, and added an emetic. Data sets from California and from the entire United States show about a 50% decrease in poisonings related to oral paraquat exposure since the formulation was changed. A Zeneca official indicated that some farmers do not like the stenching agent but that most continue to use the product.

The odor of the new formulation of methyl parathion can be smelled in recently sprayed fields, but it is not always easy to detect it (personal communication, P. Ellsworth, Univ. of Arizona, Nov 1, 1998). Poor detectability in field conditions might reflect the fact that the odorant was developed for closed buildings. The paraquat odor is difficult to detect after the compound is sprayed; this characteristic does not present a limitation, because paraquat is not dangerous after it is sprayed. The odor in paraquat is aimed at protecting mixers and applicators and intentional abusers of the pesticide. For most pesticide uses, it would be important to use an odorant concentration that could be detected in the field.

The identity of odorants in methyl parathion and paraquat is confidential business information, so detailed information on the composition of most of the odorants and on how the manufacturer adds odorants is not available. What is clear from these three cases is that the technical and regulatory problems associated with adding odors to pesticides are not difficult to solve.

In a number of informal studies, investigators have added dyes to pesticides as a means of training pesticide applicators (Paul C. Jepson, Oregon State University, November 1, 1998; Allan Hruska, Zamorano Univ., Honduras, Central America, November 2, 1998, personal communications). The training sessions are reported to have been highly successful. On the basis of short-term participant response, the participants were generally shocked by how much pesticide (dye) was deposited on their clothing and bodies. The demonstrations showed applicators who use ground equipment how important wind direction is in determining the amount of peticide deposited. In at least one case, a potato defoliant, Dinoseb—dinoseb2-(1-methylpropyl)-4-6-dinitrophenol—had a yellow color that dyed unprotected hands of applicators. Anecdotal information indicated that applicators handled this pesticide more carefully than other pesticides.

The incorporation of odors into pesticide formulations is feasible and effective. The use of dyes in formulations has received little attention but could work well. Those two related approaches for reducing health risks of agricultural workers require substantially less infrastructure to implement than the prescription-use approach. One of the major advantages of dyes and odorants is that workers are provided direct information on exposures. Assessing the best approaches and compounds for specific situations will require detailed research that could, at least in part, be conducted in the public sector.

The US farm-worker population is relatively small and well informed compared with agricultural workforces in poor countries. Any simple odor or dye materials and methods developed for shortening the feedback loop for US workers could be used by other nations where worker exposure is likely to be much worse.

Legislated Reductions in Pesticide Use

Several factors continue to provide an opportunity to develop safer products with the same efficacy as chemicals. Societal concerns about chemical pesticides—including worker safety, animal toxicity, and groundwater contamination—continue to increase in the United States and around the world. Northern European countries (Denmark, Sweden, and Netherlands) have legislated a 50% reduction by the year 2000 for farm uses of chemical pesticides. Switzerland pays farmers large subsidies to farm organically and provides minimal or no subsidies to farmers who use chemical pesticides in their crop-production systems. Many Asian countries have banned classes of toxic chemicals. The Clinton administration has stated that, by the year 2000, 75% of US farmers should be practicing IPM; and it has initiated programs to help farmers to reduce use of synthetic pesticides (Coble 1998).

REFERENCES

Agrochemical Insider. 1998. Special Edition, June 1998. Pasco, Wash.: Agricultural Development Group.

Agrow. 1997. Increase in Swiss Organic Farming. 292:9. PJB Publications Ltd.

Agrow. 1998a. Financial Support for Organic Farming in Poland. 304. PJB Publications Ltd.

Agrow. 1998b. Croatia Supports Organic Agriculture. 301. PJB Publications Ltd.

Agrow. 1998c. France to Invest in Organic Farming. 296:11. PJB Publications Ltd.

Alston, J. M, and P. G. Pardey. 1996. Making Science Pay: The Economics of Agricultural R&D Policy. Washington, D.C.: AEI Press.

Amoore, J. E., and E Hautala. 1983. Odors as an aid to chemical safety: Odor thresholds compared with threshold limit values and volatilities for 214 industrial chemicals in air and water dilution. J. of Appl. Toxicol. 3:272–290

Arne, K. H. 1997. State pesticide regulatory programs: themes and variations. Occ. Med. State of the Art Rev. 12(2):371–385.

Barnes, D.G., and M.L. Dourson. 1998. Reference dose (Rfd): Description and use in health risk assessment. Reg. Tox. Pharm. 8:471–486.

Bauer, S, and V. Booker. 1998. Worker to worker: collecting occupational health data on farmworkers through camp health aides. Presentation to the Midwestern Migrant Stream Forum, November 1998, San Antonio, Texas.

Blondell, J. 1996. Amended review of paraquat acute illness data. Correspondence, August 6, 1996. Washington, D.C.: US Environmental Protection Agency.

Cain, W. S., and A. Turk. 1985. Smell of danger: an analysis of LP-gas odorization. Am. Ind. Hyg. Assoc. 46(3):115–126.

Cain, W. S., B. P. Leaderer, L. Cannon, T. Tosun, and H. Ismail. 1987. Odorization of inert gas for occupational safety: psychophysical considerations. Am. Ind. Hyg. Assoc. J. 48(1):47–55.

California Department of Pesticide Regulation. 2000. Agricultural Pest Control Advisor License (PCA). [Online]. Available: http://www.cdpr.ca.gov/ docs/license/ apcainfo.htm

CAST (Council for Agricultural Science and Technology). 1999. The FQPA: A Challenge for Science Policy and Pesticide Regulation. [Online]. Available: HtmlResAnchor http:// www.cast-science.org/fqpa/fqpa-06.htm

Coble, H. D. 1998. A New Tool for Measuring the Resilience of IPM Systems – The PAMS Diversity Index. Paper presented at the IPM Measurement Systems Workshop, Chicago, Illinois, June 13, 1998. Ames, Iowa: Council for Agricultural Science and Technology (CAST).

Coble, H. D., A. R. Bonanno, B. McGaughey, G. A. Purvis, and F. G. Zalom. 1998. Feasibility of Prescription Pesticide Use in the United States. Ames, Iowa: Council for Agricultural Science and Technology (CAST).

Columbia Legal Services. 1998. Enforcement of farm worker pesticide protection in Washington State. Seattle, Wash.: Columbia Legal Services.

Davis, S., and R. Schliefer. 1998. Indifference to Safety: Florida's Investigation into Pesticide Poisoning of Farmworkers. Washington, DC: Farmworker Justice Fund.

Dover, M. J., and B. A. Croft. 1984. Getting tough: Policy issues in management of pesticide resistance. Study I. World Res. Inst. Policy Paper, Nov. 1984. .

EPA (US Environmental Protection Agency). 1997a. Questions and answers: methyl parathion. Washington, DC: EPA, Office of Pesticide Programs.

EPA (US Environmental Protection Agency). 1997b. Special Report on Environmental Endocrine Disruption: An Effects Assessment and Analysis. EPA/630/R- 96/012. Columbus, Ohio: EPA.

EPA (US Environmental Protection Agency). 1998. Proceedings of the EPA Workshop on Pesticides and National Strategies for Health Care Providers. EPA 735-R-98001. Columbus, Ohio: EPA.

EPA (US Environmental Protection Agency). 1999. Implementing the Food Quality Protection Act. EPA 735-R-99001. Washington, D.C.: EPA, Office of Pesticide Programs

Fieldner, A. C., R. R. Sayers, W. P. Yant, S. H. Katz, J. P. Shorhan, and R. D. Leitch. 1931. Warning Agents for Fuel Gases. American Gas Association Monogram. 4. Report of U.S. Bureau of Mines, Department of Commerce.

Food Alliance. 1998. TFA-Approved Farms. [Online]. Available: http:/www. thefoodalliance.org/farms.html. (9/21/98)

FQPA (Food Quality Protection Act). 1996. Public Law 104-170. August 3.

Glickman, D. 2000. National Organic Standards Remarks. US Department of Agriculture. [Online]. Available: http://www.ams.usda.gov/nop/glickman.htm.

Gardner, B. L. 1978. Public policy and the control of agricultural production. Am. J. of Ag. Ec. 60:836–843.

Hall, F.R., and R. D. Fox. 1997. The reduction of pesticide drift. Pp. 209-240 in Pesticide Formulation and Adjuvant Technology, C. L. Foy and D. W. Pritchard eds. Boca Raton, Fla.: CRC Press.

Hoffman, S., and F. Lampe. Natural Business Financial Market Overview. Presentation made at The Natural Business Financial and Investment Conference, 1998, Berkeley, California.

Johnson, B. 1998. Organic food industry sees double-digit growth. AgAlert, April 8th. California Farm Bureau Federation.

Judge, Elizabeth. 1997. Reuters. August 3rd.

Just, R. E., D. Zilberman, D. Parker, and M. Phillips. 1988. The Economic Impact of BARD Research on the US BARD Report, June.

Kane, D. 1999. Talk presented on The Food Alliance at Critical Elements in Transitioning to Biologically Based Pest Management Systems, June 30-July 2, 1999, Grand Rapids, Mich.

Kyodo New Service. 1997. July 2nd.

Lichtenberg, E. 1999. Panel discussion about agricultural impacts of FQPA. In Executive Summary, The FQPA: A Challenge for Science Policy and Pesticide Regulation. Ames, Iowa: Council for Agricultural Science and Technology(CAST).

Lipson, M. 1998. Searching For the O-Word: Analyzing the USDA Current Research Information System for Pertinence to Organic Farming. Santa Cruz, Calif.: Organic Farming Research Foundation.

Mandel, J. H., W. P. Carr, T. Hillmer, P. R. Leonard, J. U. Halberg, W. T. Sanderson, and J. S. Mandel. 1996. Factors associated with safe use of agricultural pesticides in Minnesota. J. Rural Health 12(4):301 310, Suppl. S.

Mileson, B. E., J. E. Chambers, W. L. Chen, W. Dettbarn, M. Enhich, A. T. Eldenfraw, D. W. Gaylor, K. Jamernik, E. Hodgson, A. G. Karezmar, S. Padilla, C. N. Pope, R. J. Richardson, D. R. Saunders, L. P. Sheets, L. G. Sultatoc and K .B. Wallace. 1998. Common mechanism of toxicity: A case study of organophosphorus pesticides. Tox.l. Sci. 41:8–20.

Moore, G. 1997. Investment Opportunities in Organic Food – Shamrock Capital Advisors. Presentation made at The Natural Business Financial and Investment Conference, August 3, Berkeley, California.

Nakanishi, N. 1997. Reuters News Service. Oct. 29th.

NRC (National Research Council). 1987. Regulating Pesticides in Foods: The Delaney Paradox. Washington, DC: National Academy Press.

NRC (National Research Council). 1993. Pesticides in the Diets of Infants and Children. Washington, DC: National Academy Press.

NRC (National Research Council). 1997a. Intellectual Property Rights and Plant Biotechnology. Washington, DC: National Academy Press.

NRC (National Research Council). 1997b. Precision Agriculture in the 21st Century: Geospatial and Information Technologies in Crop Management. Washington, DC: National Academy Press.

NRC (National Research Council). 1999. Hormonally Active Agents in the Environment. Washington, DC: National Academy Press.

OECD (Organization for Economic Co-operation and Development). 1995. Activities to Reduce Risks in OECD and Selected FAO Member Countries: Results of a Survey. Paris Cedex, France: OECD Publications.

OECD (Organization for Economic Co-operation and Development). 1997. Agriculture, Pesticides and the Environment: Policy Options. Paris Cedex, France: OECD Publications.

Organic Trade Association. 1998. Definition of Organic. [Online]. Available: http://www.ota.com/abouto.htm

Organic Trade Association. 2000. The Growth of Organic Sales. Greenfield, Mass.: Organic Trade Association.

Parker, D. D. Zilberman, and F. Castillo. 1998. Office of technology transfer, the privatization of university innovations, and agriculture. Choices first quarter, 1998:19–25.

Pavich Family Farms. 1998. Organic Industry Leaders – Pavich Family Farms and Cascadian Farm – Consider USDA Position on Organic Rule a Positive Move Forward. Press Release, May 8th. [Online]. Available: http://www.pavich.com/press/press10.htm.

Pavich, Steve. 1998. Presentation to the Committee on the Future Role of Pesticides, National Research Council. May 16, 1998, Irvine, California.

PESP (Pesticide Environment Stewardship Program). 1999. Background. [Online]. Available: http://www.pesp.org/pesp/background.htm.

Pollack, A. 2000. Talks on Biotech Food Day in Montreal Will See US Isolated. January 24th. The New York Times Company.

Postlewait, A. D., D. Parker, and D. Zilberman. 1993. The advent of biotechnology and technology transfer in agriculture. Tech. Forecasting and Social Change 43:271–287.

Retzloff, M. 1997. Building Successful Venture Capital Partnerships, Horizon Organic Dairy. Presentation made at The Natural Business Financial and Investment Conference, August 3, Berkeley, California.

Reuters. 1997. July 6th.

Ribaudo, M., and M. F. Caswell. 1999. Environmental regulation in agriculture and the adoption of environmental technology. In Environmental Regulation and Technology Adoption in Agriculture, F.Casey, A. Schmitz, S. Swinton, and D. Zilberman, eds. Boston, Mass.: Kluwer Academic Publishers.

Robertson, S. T. 1980. History of gas odorization. Pp. 1-5(a) in Odorization, F. H. Suchomel and J. W. Weatherly III , eds.. Chicago, Ill.: Institute of Gas Technologies.

Romer, P. 1986. Increasing returns and long-run growth. J. of Pol. Ec. 94:1002–1037.

Sanderson, W. T., V. Ringenburg, and R. Biagini. 1995. Exposure of commercial pesticide applicators to the herbicide alachlor. Am. Ind. Hyg. Assoc. J. 56(9):890–897.

Schemper, T., S. Voss, and W. S. Cain. 1981. Odor identification in young and elderly persons: sensory and cognitive limitations. J. Gerontol. 36:446–452.

Schierow, L. J. 1998. Pesticide Legislation: Food Quality Protection Act of 1996 (P.L. 104-170). Washington, DC: Congressional Research Service (The Library of Congress).

Schierow, L. J. 1999. Pesticide Residue Regulation: Analysis of Food Quality Protection Act Implementation (RS20043). Washington, DC: Congressional Research Service (The Library of Congress).

Segerson, K. 1999. Flexible incentives: a unifying framework for policy analysis. In Environmental Regulation and Technology Adoption in Agriculture, F. Casey, A. Schmitz, S. Swinton, and D. Zilberman, eds. Boston, Mass: Kluwer Academic Publishers.

Semb, G. 1968. The detectability of the odor of butanol. Percept. Psychophys. 4:335–340.

Stevens, J. T., and D. D. Sumner. 1991. Herbicides. Pp. 1356–1376 in Handbook of Pesticide Toxicology, vol. 3, W. J. Hayes and E. R. Law, eds. San Diego: Academic Press.

SWCS (Soil and Water Conservation Society). 1992. Pg. 5 in Implementing the Conservation Title of the Food Security Act. Ankeny, Iowa: Soil and Water Conservation Society.

Swiss Air. Press Release. 1997. June 3rd.

USDA (US Department of Agriculture). 1998. USDA to make fundamental changes in revised proposed rule on organic standards. Agricultural Marketing Service Press Release. [Online]. Available: http://www.ams.usda.gov/news/orgrule.htm.

USDA (US Department of Agriculture). 1999. USDA to issue more than $1.3 billion in CRP payments. Farm Service Agency News Release. [Online]. Available: htp://www.fsa.usda.gov/pas/news/releases/1999/10/0391.htm

USDA (US Department of Agriculture). 2000. USDA Mission. [Online]. Available: http://www.usda.gov/mission/miss-toc.htm.

University of California SAREC (Sustainable Agriculture Research and Education Program). 1999. Biologically Integrated Farming Systems Fact Sheet October 1999. [Online] Available: http://www.sarep.ucdavis.edu/bifs/factsheet.htm

Venstrom, D., and J. E. Amoore. 1968. Olfactory threshold in relation to age, sex or smoking. J. Food Sci. 33:264–265.

Vickery, P. J. 1988. Integrated fruit production (IFP): an overview of programs. Sandpoint, Idaho: Benbrook Consulting Services. [Online]. Available http://www.pmac.net/intefrt.htm.

Wolf, S. 1996. Privatization of Crop Production Service Market: Spatial Variation and Policy Implications. Unpublished Ph.D. thesis, University of Wisconsin-Madison, Institute for Environmental Studies.

Yarrow, J. 1998. Beyond Organic Agriculture. Growing Solutions, Summer 1998. Portland, Oreg.: The Food Alliance.

Zilberman , D., D. Sunding, M. Dobler, M. Campbell, and A. Manale. 1994. Who makes pesticide use decions: Implications for policymakers. Pp.23–39 in Pesticide Use and Produce Quality, W. Armbruster, ed. Oak Brook, Ill.: Farm Foundation.

Zweig, G. 1970. The vanishing zero: The evolution of pesticide analysis. In Essays of Toxicology, Vol. 2, New York: Academic Press.

4

Technological and Biological Changes and the Future of Pest Management

As a biological process, the agricultural enterprise is profoundly affected by the physiological, biochemical, ecological, and genetic attributes of the organisms involved. Anthropogenic activities within and beyond the agricultural enterprise have the potential to affect agriculture through their effects on the biology of organisms, including those in production (crops and livestock) and pests associated with them. In this chapter, we examine technological changes that have introduced new sources of mortality for pest organisms and evaluate their potential role in contributing to pest management in the future. We also examine how human activities have effected changes in the ecological milieu within which pest-crop interactions take place and the selection regime under which pests evolve, and we relate the changes to prospects for pesticide use in the future.

GLOBAL PESTICIDE MARKET TRENDS

Chemical-Pesticide Market

The global chemical-pesticide market is about $31 billion. It is a mature market with a growth of about 1-2% per year. Breakdowns by product category, crop, and by category for region for global and US markets are shown in Table 4-1 (Agrow 1998) and Table 4-2 (Aspelin and Grube, 1999).

TABLE 4-1 Global Chemical Pesticide Market (1997 Sales)

Product	Sales, billions of dollars	%
Herbicides	14.7	47.6
Insecticides	9.1	29.4
Fungicides	5.4	17.5
Others	1.7	5.5
Total	**30.9**	**100.0**

Crop	Sales, billions of dollars	%
Fruits, nuts, vegetables	6.5	21.0
Home and garden, turf, and ornamentals	5.25	17.0
Oil crops	1.75	5.7
Cotton	1.5	4.9
Cereals	4.0	12.9
Maize	2.5	8.1
Rice	2.5	8.1
Sugarbeet	1.0	3.2
Other	5.9	19.1
Total	**30.9**	**100.0**

Region	Sales, billions of dollars	%
North America	9.2	29.8
Western Europe	7.8	25.2
East Asia	7.1	23.0
Latin America	3.7	12.0
Rest of World	3.1	10.0
Worldwide Total	**30.9**	**100.0**

Source: Agrow, 1998.

Biopesticide Market

Biologically based pesticide products (also known as biorational products) generate sales of about $700 million per year (including transgenic crops) worldwide. The market for these products is expected to expand by over 20% over the next 5 years. Table 4-3 shows the sales figures for the global biopesticide market in the last few years and projected for 2004. These pesticides comprise living microorganisms or pheromones (animal-produced chemicals that serves as stimuli for behavioral responses in other individuals of the same species). The most successful biorational

TABLE 4-2 US Chemical Pesticide Market by Category (1997 Sales)

Product	Sales, billions of dollars	%
Herbicides	6.8	57.5
Insecticides	3.6	29.9
Fungicides & Other	1.5	12.6
Total	**11.9**[a]	**100.0**

[a]This estimate for the US pesticide market is larger than other estimates for 1997 sales because it includes expenditures for some nonagricultural pesticide applications. This includes applications by owner/operators and custom/commercial applicators to industry, commercial and governmental facilities, buildings, sites, and land; and homeowner applications to homes and gardens, including lawns.
Source: Aspelin and Grube, 1999.

TABLE 4-3 Global Biopesticide Market (in millions of dollars).

Market	Year				% change (1999-2004)
	1997	1998	1999	2004[a]	
Microbial	65	66	67	72	7.5
Transgenic plants	405	429	455	610	34.1
Miscellaneous	180	184	188	208	10.6
Total	650	679	710	890	25.4

[a]estimated
Source: Eppes, in press.

pesticides are the *Bacillus thuringiensis*-based (*Bt*-based) microbial pesticides with current sales of about $140 million. More than 40% of *Bt* sales are in the United States. Rapid growth of *Bt*-based biopesticides is occurring as replacements of competitive chemical products that are being banned or phased out in environmentally sensitive areas, in consumer and export markets in which concerns about food residue is high, and in organic food production.

The Industry

Agricultural-Chemical Companies

Of sales revenue of agricultural-chemical companies 7–13% is spent on pesticide research and development. The 1997 pesticide sales of leading agrichemical companies are ranked in Figure 4-1.

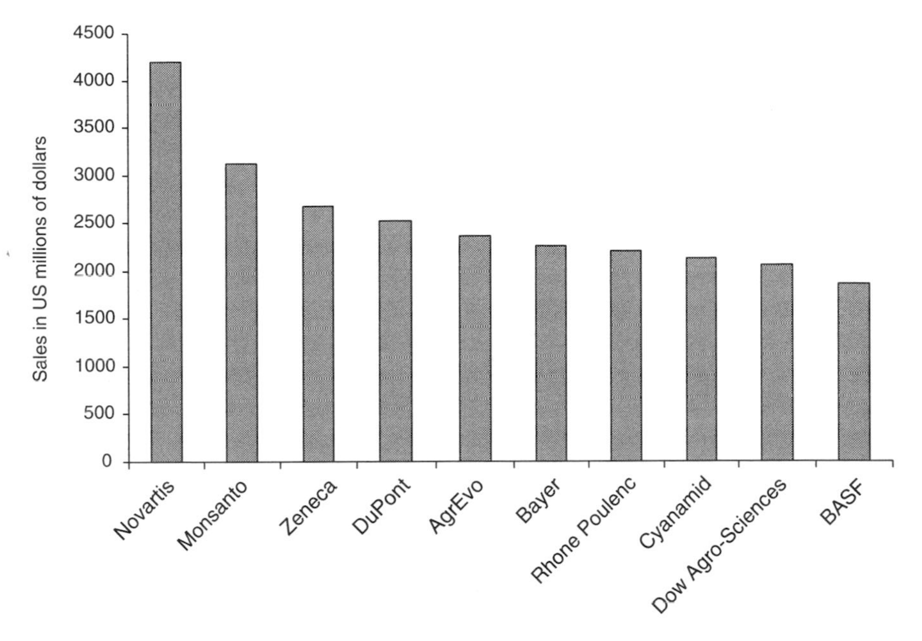

FIGURE 4-1 Pesticide sales of top 10 agrochemical companies, 1997.
Source: Panna 1998.

Major trends in the agrichemical industry today include

- Consolidation of multinational agrichemical companies.
- Rapid growth of transgenic-crop markets.
- Vertical integration of agrichemical firms with seed companies and food processors.
- Increase in generic pesticides (because chemicals are going off patent)
- Increase in consolidation and transformation of input distributors.

Despite those trends, the different companies are investing their resources differently in a wide variety of technologies. Table 4-4 shows which companies are pursuing the technologies. Table 4-5 highlights the foci of the companies' pesticide programs.

Major consolidation of multinational companies is under way. A wave of consolidations took place in the 1980s. In the 1990s, the merger of Sandoz and Ciba (forming Novartis) raised the bar even higher, creating an agrichemical company $2 billion larger than the next largest ($5 billion versus $3 billion). Business leaders continue to create empires that can rival Novartis in size. For example, AgrEvo has merged with Rhone-

TABLE 4-4 Comparison of Technologies Pursued by the Pesticide Industry

Technology	Companies	Products	Crops/Pest	Date of Entry
Transgenic seeds	Monsanto/D&PL	Bollgard (*Bt*)	Cotton / bollworm and budworm	1996
	Monsanto/Dekalb	Roundup Ready	Soybean / weeds	1996
	Calgene/Stoneville	BXN	Cotton / weeds	1996
	Monsanto/D&PL	Roundup Ready	Cotton / weeds	1997
	Ciba, Dekalb, N.King	Bt corn	Corn / borer	1996, 1997
	Mycogen, Pioneer	Bt corn	Corn / borer	1996, 1997
	AgrEvo	Liberty Link	Canola / weeds	1996
	Dekalb, Ciba, Pioneer	Liberty Link	Corn / weeds	1997
	Asgrow	Virus resistance	Vegetables and fruits / viruses	1996
Synthetic chemicals	Bayer	Admire/Provado	Multiple crops / sucking insects	1995
	AgrEvo	Applaud	Cotton / whitefly	1996
	Valent/Sumitomo	Knack	Cotton / whitefly	1996
	Rohm & Haas	Confirm/Intrepid	Cotton, vegetables and fruits / caterpillars	1996, 1997
	American Cyanamid	Pirate/Alert	Cotton and vegetables /caterpillars	1998
	Rhone Poulenc	Regent	Cotton, corn, rice, and vegetables / sucking insects	1998, 1995
	Zeneca	Abound, Heritage	Multiple crops / fungicide	1997
	BASF	Allegro	Multiple crops / powdery mildew	1997
	Novartis	Bion	Vegetables / viruses and fungi	1997

Natural products	Thermo Trilogy	Align	Vegetables /caterpillars	1995
	Thermo Trilogy	Neemix	Greenhouse / mildew	1995
	Dow	Tracer	Cotton and vegetables /caterpillars	1997
	Novartis	Proclaim	Cotton and vegetables /caterpillars	1997
Biocontrol agents	AgraQuest	LAGINEX	Rice and noncrops / mosquito	1997
	Mycotech	Mycotrol	Fruits and vegetables / whitefly	1996
	Thermo Trilogy	Spod-X	Cotton / caterpillars	1996
	Ecogen	AQ10	Grapes / powdery mildew	1996
	Thermo Trilogy	SoilGard	Greenhouse-potting mix /fungi	1996
	BioWorks	BioTrek	Field crops- / root rots	1996, 1997
	Eden Biosciences	Gray Gold	Greenhouse / gray mold	1996, 1997
	Liphatech/Gustafson	UW85	Cotton and alfalfa seed / fungi treatment	1997
	Abbott	DiTerra	Multiple crops / nematodes	1997
	Abbott	Trichodex	Grapes / gray mold	1997
	Abbott	Spherimos	Rice and noncrops / Mosquito larvae	1997

TABLE 4-5 Company Pesticide Programs[a]

	Pesticidal Natural Products	*Bacillus thuringiensis*	Living Fungi or Bacteria	Entomopathogenic Nematodes	Insect Viruses	Transgenic Crops	Pheromones	Synthetic Chemicals
Abbott	0	***	*	0	0	0	0	0
AgraQuest	***	0	*	0	0	0	0	0
AgrEvo	0	0	*	0	0	***	0	***
Am Cy	*	0	0	0	0	0	0	***
BASF	*	0	0	0	0	0	*	***
Bayer	0	0	0	0	0	*	0	***
BioWorks	0	0	***	0	0	0	0	0
Consep	0	0	0	0	0	0	***	0
Dominion	0	0	***	0	0	0	**	0
Dow	**	0	0	0	0	***	0	***
DuPont	0	0	0	0	*	***	0	***
Ecogen	0	***	**	**	0	0	***	0

EcoScience	0	0	**	0	0	0	0	0
EcoSoil Systems	0	**	***	0	0	0	0	0
Eden Biosciences	***	***	***	0	0	0	0	0
FMC	0	0	0	0	0	0	0	***
Monsanto	0	0	0	0	0	***	0	***
Mycogen	0	*	0	0	0	***	0	0
Mycotech	0	0	***	0	0	0	0	0
Novartis	**	0	0	0	0	***	0	***
Pioneer	0	0	0	0	0	***	0	0
RhonePoulenc	0	0	0	0	0	*	0	***
SafeScience	***	0	0	0	0	0	0	0
Rohm and Haas	0	0	0	0	0	0	0	***
Sumitomo	*	0	0	0	0	0	0	***
Thermo Trilogy	**	***	**	**	**	0	***	0
Zeneca	*	0	0	0	0	***	0	***

[a]=little or no activity
*=minor focus
**=major focus
***=main focus of program

TABLE 4-6 Sales of Transgenic Crops and Chemical Pesticides, 1995-1997

	Sales, millions of dollars			Change,
	1995	1996	1997	1996-1997, %
Herbicides	14,280	15,050	14,700	−2.33
Insecticides	8,750	8,745	9,100	4.06
Fungicides	5,855	5,895	5,400	−8.40
Plant-growth regulators and others	1,305	1,325	1,700	28.30
Total pesticides	30,190	31,015	30,900	−0.37
Total transgenic crops	75	235	650	176.60

Sources: Cultivar, 1997; Eppes, in press.

Poulenc to form Aventis (about the size of Novartis), and Novartis and AstraZeneca have merged to form Syngenta.

Another major trend in the market is the rapid growth of biotechnology products compared with chemical pesticides, stimulating the vertical integration of agrichemical companies with seed and food companies. These biotechnology products, chiefly transgenic-crop seed with pest-control attributes and herbicide tolerance, are dramatically changing market shares of agrichemical firms in soybeans (herbicide tolerance), cotton (insecticides for bollworm and budworm control), and corn (herbicide tolerance and insecticides for corn-borer control). Table 4-6 shows the total sales of transgenic crops relative to chemical pesticides in 1995–1997.

Agrichemical companies have invested billions of dollars to develop or access crop seed genetically engineered for caterpillar and disease control and tolerance to herbicide sprays. Monsanto is the industry leader, with reported investment that totals more than $8 billion to acquire Calgene, Delta PineLand, Cargill, Ecogen, DeKalb, Agracetus, Asgrow, Holden's, and PBI. Other examples are Dow's purchase of Mycogen and Pioneer's $30 million deal with Mycogen to obtain nonexclusive access to *Bt* genes for genetically engineered corn seed. Novartis and Mycogen partnered to launch a transgenic corn seed containing a *Bt* gene for corn borer control. A strategic alliance composed of Mycogen, Rhone Poulenc, and Dow was formed to develop crops with input and output traits. Other shifts in the industry include the purchase of Plant Genetic Systems for $750 million by AgrEvo (Hoechst-Schering-Roussel), DuPont's purchase of 20% of Pioneer for $1.7 billion, and the acquisition of Mogen by AstraZeneca to complement its share of seed companies (now called Advanta).

There is still heavy reliance on traditional chemicals for weed control, in part because of the widespread reliance on genetically engineered glyphosate-resistant crops. Other companies still depend on agrichemicals for primary income. These same multinationals are also entering into agreements with food companies to develop crops with value-added traits. Many agrichemical companies are shifting resources away from inputs (pesticides) and input traits (pesticidal genes) to output traits.

Examples of these traits are

- Improvements in feed value (for example, corn seed engineered with phytase enzyme or high levels of lysine, an essential amino acid).
- Higher-quality product (for example, fresh tomatoes with longer shelf-life or paste tomato with lower water content).
- Greater nutritional value (for example, rice with higher vitamin A content).

The industry is investing billions of dollars in genomics to characterize the genes of entire organisms. Industrial leaders expect that advances in genomics will lead researchers to the precise location and sequence of genes that contain valuable input and output traits.

A shift in R&D resources from input to output traits probably would have a large impact on the future of plant protection. Will the cycle of innovation on the input side continue? Because of the high investment required for development of chemical pesticides and transgenic crops, will large agrichemical and life science firms focus primarily on crops with large markets (such as row crops)? Whether companies will develop pesticides and input traits for minor use crops remains an open question.

A trend in agrichemical industry is the movement of many chemical pesticides off patent. As these chemicals become generic pesticides, manufacturers lose their monopolies on them. Large agrichemical companies are therefore aligning themselves with generic suppliers of chemical pesticides to reduce erosion in sales of the products that were formerly proprietary products. Several agrichemical companies have purchased outright or partially own generic companies. DuPont, for example, has entered into a joint venture and is now a 51% owner of Griffin. In May 1998, BASF purchased MicroFlo for the same reasons.

As more products become generic, profit margins erode for distributors, as well as for manufacturers. In response, distributors are consolidating and becoming "basic manufacturers" through acquisitions of proprietary products. For example, Gowan Corp., Griffin, and United AgriProducts (UAP) have taken this approach.

The consolidation of distributors can take two paths. In the first, a large distributor can acquire many smaller distributors. UAP, a practitioner of this approach, is now as large as or larger than many basic manufacturing agrichemical companies and has revenues of $2 billion. The second path is the formation of consortiums of smaller distributors into a group with more influence than individual firms. In California, for example, a group of distributors joined to form a consortium called Integrated Agricultural Producers (IAP). The consolidation of distribution and agrichemical companies has created some concern for farmers. They are worried that they will have fewer choices and that prices will go up. However, the advent of the Internet is tearing down old structures, and the entire distributor-manufacturer relationship is expected to change for the benefit of farmers, who will be able to order all their products directly over the Internet.

One of the most important trends for agrichemical companies is the growing shift towards the development and registration of reduced-risk pesticides. In 1993, the Environmental Protection Agency (EPA) began a program of expedited review of what were classified as reduced-risk pesticides. Expedited reviews can reduce the time to registration by more than half (EPA 1998). Since the introduction of this program, the number registered as reduced-risk pesticides has steadily increased. Table 4-7 lists almost 20 reduced-risk pesticides that have been registered since 1994.

For a pesticide to be considered of reduced risk, it must have at least one or more of the following characteristics (EPA 1997a):

- It must have a reduced impact on human health and very low mammalian toxicity.
- It must have toxicity lower than alternatives (0.01–0.1 as much).
- It displaces chemicals that pose potential human health concerns or reduces exposures to mixers, loaders, applicators, and re-entry workers.
- It reduces effects on non-target organism (such as birds, honey bees, and fish).
- It exhibits a lower potential for contaminating groundwater.
- It lowers use or entails fewer applications than alternatives.
- It has lower pest-resistance potential (that is, it has a new mode of action).
- It has a high compatibility with integrated pest management (IPM).
- It has increased efficacy.

The Food Quality Protection Act (FQPA) went further and mandated expedited registration of reduced-risk pesticides that could be expected to pose less risk to human health and the environment than other pesticides that meet existing safety standards. Since the enactment of FQPA, 62% of the 48 active ingredients registered have been considered "safer"

TABLE 4-7 Reduced-Risk Pesticides registered with US EPA since 1994

Chemical	Pesticide type	Use	Year registered
Acibenzolar-S-methyl	Plant activator	Tomato, lettuce, tobacco	Pending
Alpha-Metolachlor	Herbicide	All metolachlor uses	1997
Azoxystrobin	Fungicide	Turf	1997
Azoxystrobin	Fungicide	Almond, cucurbit, rice, wheat	1997
Azoxystrobin[a]	Fungicide	Grape, banana, peach, tomato, pecan, peanut	1997
Bifenazate	Insecticide	Ornamentals	1999
Cadre	Herbicide	Peanut	1996
Carfentrazone	Herbicide	Wheat, corn	1998
Cyprodinil	Fungicide	Stone fruit	1998
Diflubenzuron[a]	Insecticide	Below ground termite bait station	1998
Diflufenzopyr	Herbicide	Corn	1999
DPX-MP062	Insecticide	Cotton, tomato, pepper, cole crops, lettuce, sweet corn, apple, pear	Pending
Fenhexamid	Fungicide	Grape, strawberry, ornamental	1999
Fludioxonil	Fungicide	Corn	1995
Fludioxonil	Fungicide	Various seed treatments	pending
Fludioxonil[1]	Fungicide	Potato	1997
Hexaflumuron	Termiticide	Below-ground bait station	1994
Hexaflumuron[a]	Termiticide	Above-ground bait station	1997
Hymexazol	Fungicide	Sugar beets	1995
Imazamox	Herbicide	Soybeans	1997
Mefenoxam	Fungicide	Ornamentals, citrus, nuts, fruit trees	1996
Methyl anthranilate	Bird repellent	Food crops	1994
Pymetrozine	Insecticide	Potato, cucurbit, tomato, tobacco	1999
Pyriproxyfen	Insecticide	Cotton	pending
S-Dimethenamid	Herbicide	Corn, soybean, peanut	1999
Spinosad	Insecticide	Cotton	1997
Tebufenozide	Insecticide	Walnuts	1995
Tebufenozide	Insecticide	Pome fruit, cotton, leafy vegetables, cole crops, sugarcane, pecan, forestry, fruiting Vegetables, Ornamentals	pending
Trifloxystrobin	Fungicide	Pome fruit, grape, cucurbit, peanut, turf, banana	1999

[a]New uses of existing active ingredients.

Source: adapted from Aspelin and Grube, 1999

than conventional pesticides, and 77% of the new uses of pesticides have been considered "safer" (EPA 1999a.) In the year 2000, 10 of the 25 new chemicals seeking registration are classified as of reduced risk (EPA 2000). Those facts indicate that many of the new pesticides being registered meet one or more of the criteria stated above for reduced risk and that companies are responding to the expedited review of these pesticides by EPA. The consequence of this response is that future pesticides will pose even less risk to human health and the environment because they will be compared with pesticides that are now being registered as of reduced risk.

Biopesticide Companies

The most successful biopesticide is *Bacillus thuringiensis*. The market is dominated by Abbott (68% share), followed by Thermo Trilogy (26%), Ecogen (4%), and Mycogen (2%). Several small biopesticide companies (AgraQuest, BioWorks, Consep, Dominion Biosciences, Eden Biosciences and Mycotech) focus on biological pest control other than with *Bt* (table 4-5). New entrants in the last 5 years are AgBio Development, AgraQuest, BioWorks Inc., Dominion, and EcoSoils Systems. These companies focus primarily on minor-use crops, such as fruits, vegetables, wood products, and ornamentals. Some of the products that are getting a foothold in the marketplace are botanical extracts from the seed of the neem tree (NRC 1991), microorganism-based pesticides that include fungus-based insecticidees and fungicides, baculovirus, and beneficial nematodes. Many of the newer products are designed to prevent soil-based diseases from attacking crops and their roots (Dutky 1999). Because of the lower cost and shorter time to develop a biopesticide, as opposed to synthetic chemicals or transgenic crops, as seen in Figure 4-2, small biopesticide companies are able to enter this market. New startup firms develop alternative technologies for smaller markets that are neglected by larger agrichemical companies.

Use of Microbial Pesticides in Integrated Pest Management (IPM) Systems

The use of microorganisms as commercial pest-control agents was reviewed by Quarles (1996), Rodgers (1993), and Starnes et al. (1993). *Bt* by far has the longest history and is the most widespread microbial pesticide in use today. Strengths of microbial pesticides can include

- Safety with respect to nontarget organisms.
- Biodegradability.
- Low cost of development.

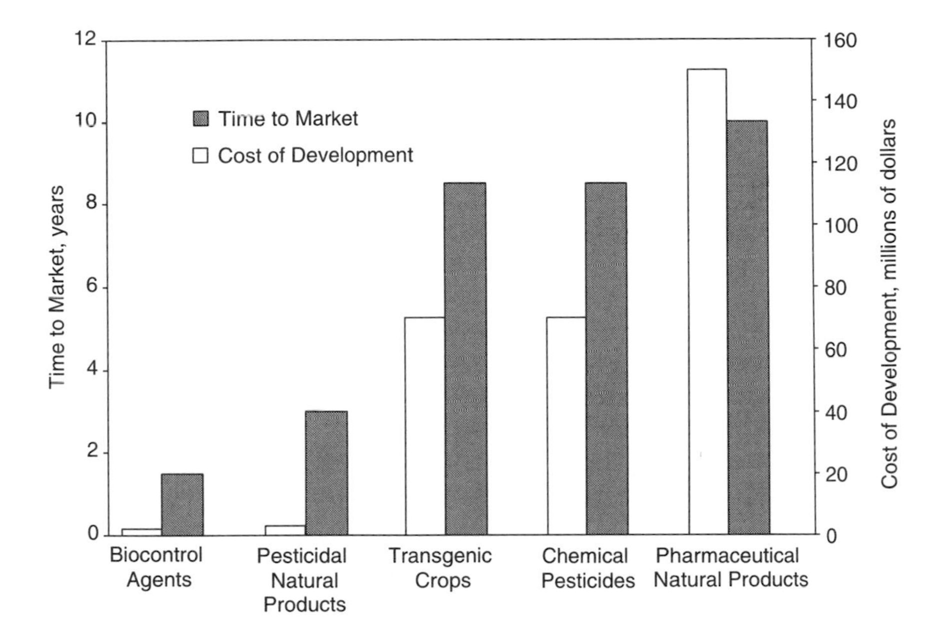

FIGURE 4-2 Cost to develop and time to market of various products.

- Suitability for IPM systems.

Drawbacks of microbial pesticides can include

- Short shelf-life of products.
- Inconsistency of efficacy within and between ecosystems.
- Short field residual life.
- Difficulty of use.

Bacillus thuringiensis

Bt-based microbial pesticides are used extensively in vegetable production in California because of concern about chemical residues on harvested products. These pesticides are also used when there is a need to decrease selection pressure on diamondback moth and cabbage looper pests that evolves primarily from heavy applications of conventional pesticides in production of cole, lettuce, and tomato crops (Zalom and Fry 1992).

John Trumble and his colleagues at the University of California, Riverside have done considerable work on *Bt* in IPM vegetable programs

(Moar and Trumble 1987, Trumble 1985, Trumble 1990). Trumble reported higher profits in IPM plots in production of celery and tomatoes relative to traditional chemical-spray programs (Trumble 1989, Trumble 1991, Trumble and Alvarado-Rodriguez 1993).

The Hunt-Wesson and Campbell Soup companies have introduced IPM programs in a large tomato-growing area of Sinaloa, Mexico. Results of the field tests indicate increases in yield per acre, decreased costs per ton, and improvements in tomato quality (Moore 1991). Campbell Soup has reduced the amount of pesticides used by incorporating *Bt* and other nonchemical approaches into most of their vegetable growing (W. Reinert, University of California, Davis, April 8, 1997, personal communication).

Bt has been successfully applied to cruciferous crops for many years (Sears et al. 1983, Ferro 1993). The practicality of *Bt* for controlling cotton caterpillars—such as the Egyptian cotton leafworm, *S. littoralis* Boisduval and *Helicoverpa* spp.—has also been demonstrated (Broza et al. 1984, Daly and McKenzie 1986). With the advent of transgenic crops, these IPM systems are no longer useful. The diamondback moth, *Plutella xylostella*, which has evolved resistance to all chemical classes and also to *Bt*, can be managed with IPM (Metcalf 1989). In Taiwan, two larval parasitoids, pheromone traps, and *Bt* reduced pest population densities on cauliflower and broccoli to less levels than in neighboring, conventionally sprayed plots (Asian Vegetable Research and Development Center 1991).

A survey of tree-fruit researchers throughout the United States indicated that the integrated use of pheromone mating disruptants, low doses (one-tenth of the recommended label rate) of pyrethroids, and the full rate (1 lb.) of *Bt* is a useful IPM program for leafroller management (Tette and Jacobsen 1992). The peach twig borer *(Anarsia lineatella* Zeller) is a major pest of almonds in California that has been controlled through an IPM program (Dr. Frank Zalom, University of California, Davis, April 15, 1998, personal communication). As the organophosphate insecticides are phased out because of adverse effects on hawks in the almond orchards, *Bt* reportedly is an economical replacement for organophosphates.

Bt has been used operationally for controlling forest and tree caterpillar pests for many years (Bowen 1991, Cunningham 1988, Elliott et al. 1993). *Bt kurstaki* (*Btk*) is the most widely used insecticide for forest defoliators, such as gypsy moth, spruce budworm, western spruce budworm, forest tent caterpillar, fall cankerworm, and hemlock looper. Improving the efficacy of *Bt* sprays continues to be a subject of research, but *Btk* is a successful stand-alone product and already has replaced established insecticides—such as carbamates, organophosphate, and pyrethroid pesticides—that were once considered essential to the system. The displacement occurred because of a combination of efficacy, economics, and system-specific attributes.

Bt tenebrionis (*Btt*), active on coleopteran pests, has been shown to control Colorado potato beetle under field conditions (Ferro and Gelernter 1989, Ferro and Lyon 1991, Jaques and Laing 1989) and is now used commercially in IPM systems. Roush and Tingey (1991) reported the development of an IPM system for Colorado potato beetle in New York. The program includes scouting, border sprays, or barriers to trap migrating overwintering adults and sprays of *Btt* on small larvae, cryolite on large larvae, and endosulfan on adults. The system enhances predators and manages pesticide resistance in potatoes through the use of "soft" chemicals and *Btt*, and rotation of pesticides in different chemical classes.

Although *Btt* has commercial promise for control of Colorado potato beetle, the advent of the pesticide imidacloprid has reduced the use of *Btt* substantially. Imidacloprid can provide season-long control when applied as a potato seed treatment, whereas *Btt* must be applied weekly when there are small larvae. Therefore, the chemical is easier and more effective and has become the product of choice for most potato growers. However, Colorado potato beetle has become resistant to all chemicals used against it and is expected to develop resistance to imidacloprid as well.

The discovery of the beetle-active *Btt* fueled the search for strains with novel activity. Companies in the *Bt* arena boast collections of several thousand *Bt* isolates. Many new crystal types have been discovered with activity against nematodes, mites, corn rootworm (*Diabrotica* spp.), adult flies, and ants. The potential for use of these new *Bt* strains in IPM is unknown. Most of the attention is on finding new genes that can be engineered into plants.

Ecogen and Novartis scientists have used various molecular techniques (including electroporation and transconjugation) to develop products that combine genes from *aizawai* and *kurstaki* strains to increase activity against key lepidopteran pests, such as armyworm (*Spodoptera* spp.) and cotton bollworm (*Helicoverpa* spp.). Novo Nordisk used classical mutation to improve *Btt*. This strain produces a larger crystal and crystal size correlates directly with field activity. In addition, fusion of genes from *Bt* into baculovirus is being used to expand the host range of *Bt*. Ecogen has introduced CryMax, which is a *Bt* strain engineered to contain multiple copies of a *Btk* protein, improving efficacy on target caterpillar pests.

It is well known that *Bt* remains active against the pest for only several hours on plant foliage under typical field conditions because of UV degradation, rainfall, and other environmental perturbations. In the early 1980s, Monsanto developed a recombinant plant-colonizing pseudomonad for delivery of *Bt* genes, with the objective of improving residual activity and efficacy of *Bt* proteins. This concept was developed by Mycogen into the products MVP™ and M-Trak™. The *Bt*-bearing pseudomonad is killed (to avoid regulatory hurdles for registering recombinant microorganisms) and sprayed on the crop as other *Bt* products are. The pseudomonad cell is reported to protect the *Bt* protein from environmental degradation, thus providing longer residual activity. These *Bt* products have had modest commercial success. A starch-encapsulation procedure for virus and *Bt* designed to improve the survival and efficacy of these microbial products in the field is under development by the US Department of Agriculture (USDA) Agricultural Research Service (ARS) in Peoria, Illinois.

Novo Nordisk discovered an enhancer of *Bt*, a natural substance produced in the *Bt* at a very low concentration (Manker et al. 1994, 1995). The same natural product was isolated previously by University of Wisconsin researchers as a fungicide, which they called zwittermicin A (He et al. 1994). When the compound is combined at higher concentrations with the *Bt* protein, efficacy against the most refractory caterpillars, such as *Helicoverpa zea* and *Spodoptera exigua,* is increased substantially in the field.

New genetically engineered and improved *Bt* products might provide more opportunities and choices for growers who use IPM programs. The most successful *Bt* products are ones that provide efficacy, ease of use, and consistency approaching traditional chemical pesticides. Improved armyworm (*Spodoptera*) and bollworm (*Helicoverpa*) products are the most important developments in the use of *Bt* microbials in agriculture. *Bt* products could capture a larger market share and replace some established chemical products. In any case, *Bt* strains probably will remain important replacements for chemical insecticides in fruit, vegetable, and forestry IPM systems.

Baculoviruses

Although they have not had the commercial success of *Bt*, baculoviruses could have important potential for use in IPM programs. They have a number of advantages. Baculoviruses are ideal for IPM because as far as is known, they are safe for nontarget insects, humans, and the environment. Baculoviruses might, in some cases, be the only effective biocontrol agents available for controlling insect species (Cunningham 1988) and they provide an avenue for overcoming specific problems, such as resistance. It is important to have a selection of control agents when designing pest-management strategies. Because viruses are not likely to elicit cross resistance to chemicals, they should receive more attention from university and industrial researchers (Cunningham 1988). The use of multiple biological products has the advantage of lowering the potential for evolution of pest resistance. Although practical applications of viruses are beginning to develop in IPM programs in US agriculture, the long-term value of viral approaches for widespread use has not been demonstrated.

Helicoverpa zea Boddie (cotton bollworm) nuclear polyhedrosis virus (NPV) was the first baculovirus to be marketed in the United States. It was developed by International Minerals and Chemical Corporation (IMC) but marketed by Sandoz under the tradename Elcar in 1976 after purchase of IMC's biological products division. Interest in Elcar declined with the introduction of pyrethroids, which are effective, inexpensive broad-range insecticides. The properties of viruses suitable for IPM systems have been well studied (Ignoffo and Garcia 1992).

In Europe, a number of companies—including Kemira Oy (Finland), Oxford Virology (United Kingdom), and Calliope (France)—have introduced viral products for the insecticide market or are developing them. Viral products include *Cydia pomonella* L. (codling moth) granulosis virus (GV), *Neodiprion sertifer* (Geoffrey)(European pine sawfly) NPV, *Spodoptera exigua* (Hübner)(beet armyworm) NPV, and *Autographa californica* (Speyer) (alfalfa looper) NPV. The largest use of baculoviruses is in Brazil, where *Anticarsia gemmatalis* Hübner (velvetbean caterpillar) NPV protects 5.9 million hectares of soybeans against the velvetbean caterpillar.

In North America, the effort with baculoviruses has been led mainly by government agencies (Cunningham 1988, Podgwaite et al. 1991, Otvos et al. 1989). The US Forest Service (USFS) has registered NPVs to control *Lymantria dispar* (L.) (gypsy moth), *Neodiprion sertifer*, and *Orgyia pseudotsugata* (McDunnough) (Douglas fir tussock moth) in forestry. The Canadian Forest Service holds registrations for *O. pseudotsugata* NPV and *Neodiprion lecontei* (Fitch) (redheaded pine sawfly) NPV. US companies actively involved in baculovirus research are American Cyanamid, Thermo Trilogy, and DuPont.

Louis Falcon (University of California, Berkeley) has demonstrated the successful use of codling moth granulosis virus in pear, apple, and walnut IPM systems in California and Washington (personal communication, May 12, 1991). Organic growers pay $30 per acre-treatment of virus and make applications five to 10 times per season, in contrast with conventional growers who spray monthly (three times per season) the organophosphate insecticide Guthion™ at $7.50-10.00 per acre-treatment. According to Falcon, the cost of total chemical inputs (insecticides, fungicides, acaricides, bactericides) is approximately $360 per season and the total cost of all pesticides (fungicides, miticides, insecticides), including the virus for codling moth is $316 per season. In the virus-treated orchards, natural enemies can survive to control mite pests, thus eliminating the need for miticides, which are required in Guthion-treated orchards. Although Dr. Falcon's program has been successful for organic growers, mainstream fruit producers have not switched to it, because

Guthion™ is an effective alternative. With the recent development of Guthion-resistant codling moth populations after 20 years of use and restriction of the preharvest interval for Guthion by the California Department of Pesticide Regulation, growers have more incentive to adopt the virus-pheromone IPM program.

Application of baculovirus for control of beet armyworm, *Spodoptera exigua* (Hübner) has been well studied in greenhouse systems (Smits et al. 1987a). Such characteristics as dose-response curves, larval feeding behavior, application techniques (Smits et al. 1988), timing, and strain (Smits et al. 1987b) have been integrated into recommendations for operational use of virus with other control methods. On head lettuce in California, a beet armyworm NPV was field-tested for 3 years and compared with chemical insecticides (Gelernter et al. 1986); results indicated comparable control with methomyl and permethrin.

At the present stage in the development of baculovirus products, several limitations are associated with the viruses' use as insecticides. A major limitation in agricultural systems is the slow rate of kill, which results in feeding damage. Kill rate, however, is not as crucial in forest systems, where cosmetic damage is not as important. Reduction of kill time will rely on improvements in formulation and application in the immediate future. However, this limitation can be managed in the short term by using baculoviruses in combination with other insecticides through IPM.

Lower production costs are essential for both recombinant and wild-type baculoviruses to compete with classical insecticides. There are active research programs in both in vivo and in vitro production (Bonning 1996). Although viruses are less expensive to produce in vivo than in vitro, the cost still exceeds that of *Bt.* Viruses are formulated to be applied in the same fashion as *Bt* strains. However, for extensive use in IPM, dramatic improvements in formulation and application technology are needed. In formulation, knowledge of stability and shelf-life is required to optimize storage and distribution. In application, droplet size, density, dosage, and components in the tank mix (for example, stickers, and UV protectants) need to be optimized.

Another limitation of baculoviruses is their host specificity, which can reduce their commercial potential. However, the host specificity is viewed positively from the environmental and IPM standpoints. Two viruses with relatively broad host ranges are *Autographa californica* (alfalfa looper) NPV and *Syngrapha falcifera* (Kirby)(celery looper) NPV, each of which kills over 30 insect species. The celery looper virus is reported to have commercial potential in cotton IPM systems (Wood 1992). Host range can be broadened through molecular means or by mixing two viruses.

In the long term, the development of recombinant baculoviruses that can kill rapidly will allow them to compete more effectively with classical pesticides. To increase the ability of baculoviruses to kill early, research to insert specific genes into the baculoviral genome is under way. These genes will serve as toxins or disrupters of larval development. Among the proteins being tested for exploitation are *Bt* endotoxin (which failed to improve the virus), juvenile hormone esterase, prothoracicotropic hormone (PTTH), melittin, trehalase, scorpion toxin, and mite toxin (Bruce Hammock, University of California, Davis, November 12, 1998, personal communication). The knowledge of the molecular biology of viruses has also promoted interest in modifying and improving baculoviruses with regard to host range and virulence.

The regulatory process that will be applied to recombinant baculoviruses is not yet clear. The recombinant virus system makes it possible to exploit a variety of proteins, including insect enzymes and hormones and proteins from other organisms. The recombinant viruses that will probably be commercialized first in the United States are the ones that carry genes expressing insect-selective nerve toxins, which are undergoing intensive safety and efficacy testing. In the United States public concern has not been voiced with regard to the safety of these viruses. However, objections have been raised in England as to the use of toxin genes in baculoviruses.

The growth and success of baculoviruses as commercial insecticides will depend on reducing production costs, developing practical and effective formulations, optimizing field performance, overcoming regulatory obstacles, and educating users and the public on their safety. To enable development of more economical and effective products, R&D efforts should focus on making improvements in baculovirus production, formulation, and application technologies in conjunction with genetic engineering of the viruses to enhance their kill rates and broaden their host range.

Entomopathogenic Fungi

Over 500 fungi are regularly associated with insects; some cause serious disease in their hosts, but few have been used commercially as control agents. Because of their dependence on specific environmental factors, such as relative humidity, fungi can be useful tools in IPM, especially as complements to other products. Fungi infect a broader range of insects than do other microorganisms, and infections of lepidopterans (moths and butterflies), homopterans (aphids and scale insects), hymenopterans (bees and wasps), coleopterans (beetles), and dipterans (flies and mosquitoes) are quite common. In fact, some fungi have very broad host ranges that encompass most of those insect groups. That is true of *Beauveria bassiana* (Balsamo) Vuillemin, *Metarhizium anisopliae* (Metschnikoff) Sorokin, *Verticillium lecanii* (Zimmerman) Viegas, and *Paecilomyces* spp., all of which have worldwide distributions; these are the most commonly used insect pathogens developed for commercial pest-management products (McCoy 1990).

B. bassiana has been identified in many insect species in temperate and tropical regions and is used for pest control on a moderate scale in eastern Europe and China. Mycogen produced a *B. bassiana*-based bioinsecticide, which has been shown to be highly pathogenic in coleopterans. The fungus also is amenable to mass production of conidia by semisolid fermentation. The product has been field-tested against citrus root weevil. Another *B. bassiana* strain, researched at the USDA ARS shows good control of rasping or sucking insects, such as thrips, whiteflies, and aphids.

M. anisopliae has been most extensively used in Brazil for control of spittlebugs on sugar cane. Using *Metarhizium* as the control agent, EcoScience Laboratories, Inc. has developed infection chambers in which insects (cockroaches and flies) brush against spores of the pathogen, which later germinate and infect the insect. However, this product has not been successfully commercialized.

V. lecanii is a pathogen that has demonstrated good control of greenhouse pests, such as *Myzus persicae* (Sulcer) aphids, on chrysanthemums. A distinct isolate of *V. lecanii* was obtained from whitefly and provided excellent control of greenhouse whitefly, *Trialeurodes vaporariorum* (Westwood), and of *Thrips tabaci* Lindeman on cucumber. *V. lecanii* was produced commercially as Vertalec for aphid control and Mycotal for control of whitefly from 1982 to 1986, and there is a resurgence of commercial interest in its use for control of aphids, whiteflies, and thrips because these greenhouse pests have developed resistance to chemical pesticides typically used for their control.

The fungal pathogen *Entomophaga maimaiga* has been recognized and used by USFS, states' departments of natural resources, and university personnel as a control for gypsy moth (Elkinton et al. 1991). Since its reappearance in the early 1990s, this pathogen largely has become self-perpetuating. It should be noted that public research agencies have played an important role in the development of gypsy moth pest-management strategies. In fact, the use of *Entomophaga maimaiga* with *Bt* and NPVs exemplifies the trend toward biologically based pest management in gypsy moth control in the last 10 years.

The effectiveness of fungi in controlling insect pests depends on the environmental conditions prevailing after application, particularly with respect to relative humidity. There is a need for research to develop moisture-retaining formulations that allow fungal growth at suboptimal relative humidity. Innovative biotechnology can also be used to engineer desirable traits into fungi and thus improve the effectiveness of some fungal pathogens. Transformation systems and recombinant-DNA techniques are now being used to study the mechanisms of pathogenicity and virulence at the molecular level (Carruthers and Hural 1990). Eventually, all this knowledge will enhance the ability to manipulate the genetics of these organisms.

Microbial and Natural-Product Fungicides

The use of microorganisms as fungicides was reviewed by Quarles (1996). There are many commercial microbial biopesticides for controlling plant pathogens, but they make up an insignificant portion of the fungicide market. Commercial adoption is hampered by the inconsistency of microbial fungicides in the field. However, as tools in IPM systems, microbial fungicides (Box 4-1) can be used in rotation with chemical products or stand-alone in conjunction with disease-forecasting models.

Natural products are organic compounds produced by microorganisms, plants, and other organisms. There are few natural-product fungicides for agriculture (Franco and Coutinho, 1991). One is validamycin, isolated from the bacterium *Streptomyces hygroscopicus* var. *limoneus* and used to control rice sheath blight (Yamamoto 1985). Extensive university research is focused on microbial control of plant pathogens (Adams et al. 1990, Taylor and Harman, 1990 Weller 1988). It has adressed microbial physiology or microorganism-plant interactions, especially for root patho-

BOX 4-1
Microbial Fungicides

- SoilGard (*Gliocladium virens*) for damping-off and root rots
- BioAg 22G, BioTrek 22-G, T-35 (*Trichoderma*), for damping-off and root rots
- Trichodex (*Trichoderma*) for many fungal diseases of vegetables and fruits
- Bio-Save 10 (*Pseudomonas syringae*) for postharvest *Botrytis* and *P enicillium*
- Epic/Kodiak (*Bacillus subtilis*) for cotton and legume seed treatment
- Mycostop (*Streptomyces griseoviridis*) for field, ornamental, and vegetable diseases
- AQ10 (*Ampelomyces quisqualis*) for powdery mildew
- Blightban (*Pseudomonas fluorescens*) for *Erwinia amylovora* fire blight

Source: Quarles, 1996.

gens. However, little is known of the identity of the plant pathogen-inhibiting natural products produced by these microorganisms and it is an important subject for future research.

The use of *Bacillus subtilis* as a fungicidal treatment has been demonstrated on a number of diseases, including cornstalk rot (*Fusarium roseum*) (Kommedahl and Mew 1975), onion white rot (*Sclerotium cepivorum*) (Utkehede and Rahe 1983), potato charcoal rot (*Macrophomina phaseolina*) (Thirumalachar and O'Brien 1977), bean rust (*Uromyces phaseoli*) (Baker et al. 1985), apple blue mold (*Penicillium expansum*) (Sholberg et al. 1995), and peach brown rot (*Monilinia fructicola*) (Gueldner et al. 1988). The vast majority of the work with *Bacillus subtilis* has concentrated on treatment of seeds or soil to control pathogens; in general, the use of biocontrol as a foliar treatment is much less developed than in the soil-rhizoplane environment (Blakeman and Fokkema 1982). Foliage has been sprayed with strains of *Bacillus* spp. for plant-pathogen control. For example, Sharga (1997) screened 270 bacilli and found two that provided good in vitro control of *Botrytis cinerea* and *B. fabae*. However, once these microorganisms were applied in the field, a rapid decline in the population of introduced bacilli was observed within 24 hours and was concomitant with loss of protection against *Botrytis*. Sharga concluded that the protection observed in greenhouse tests was due to intense competition for exogenous nutrients between *Bacillus* and *Botrytis*. In a second study, Baker et al. (1985) sprayed bean cultivars in the field with *Bacillus subtilis* and observed substantial control of rust (*Uromyces appendiculatus*) for plants sprayed three times a week.

Bacilli are known to produce antifungal and antibacterial secondary metabolites (Korzybski et al. 1978). University of Wisconsin and Cornell researchers have identified a novel fungicidal compound, zwittermicin A, produced by *Bacillus* spp. (Stabb et al. 1994, He et al. 1994). The compound substantially inhibits elongation of germ tubes from *Pythium medicaginis* cysts (Silo-Suh et al. 1994). A second fungicidal metabolite produced by the same strain was recently identified as the known amino sugar kanosamine (Milner et al., 1996). Wisconsin field data on control of below-ground root diseases for soybean with the *Bacillus* strain are promising (Osburn et al. 1995). However, only one report demonstrated its use as an above-ground treatment; Smith et al. (1993) described suppression of cottony leak on cucumbers using *B. cereus* strain UW85.

Another group of previously described metabolites of *Bacillus* is the cyclic lipopeptides of the iturin class, some of which are potent fungicidal agents. These compounds consist of a cyclic octapeptide with seven alpha-amino acids and one beta-amino acid with an aliphatic side chain. There are several groups of iturins that differ in order and content of the lipopeptide side chain. Generally, a suite of related molecules is produced with differences in the length and branching of the aliphatic amino acid residue. When tested against yeast, *Saccharomyces cerevisiae*, mycosubtilin was most active ($LC^{50} = 10\ \mu g/mL$), followed by iturin-A and bacillomycin L (both 30 µg/mL) (Besson et al. 1979). Iturin-C is inactive against fungi, including *Penicillium chrysogenum* (Peypoux et al. 1978).

Researchers at USDA ARS have investigated the structure-activity relationship of the iturins by synthesizing a number of analogues differing in amino acid chain length. They reported that the activity of the iturins increased with the length of the fatty acid side chain and the terminal branching in the order iso > normal > ante-iso (Bland et al. 1995) and that the "amounts of iturins obtained from natural production are inadequate to be commercially viable" on the basis of their work with a number of iturin-producing strains of *Bacillus*. However, only a small number of potential *Bacillus* strains have been screened for fungicidal activity. AgraQuest is developing a naturally occurring *Bacillus subtilis*, which has shown excellent activity for some foliar and fruit diseases. Investigators isolated it in a screening program directed at new strains with superior antifungal activity.

EcoSoils Systems, San Diego, California, has a unique approach to address the short-shelf-life problem of many microbials, such as *Pseudomonas*-based biofungicides. Industrial scientists ferment on-site continuously for control of turf diseases on golf courses. The company is applying the Bioject system to agricultural crops, especially for control of root rot diseases, such as *Phytophthora* root rot of avocado.

Natural products are another major source of new leads for pesticides. Most companies generate natural-product extracts either internally or from external sources and test them against test organisms or in mechanism-based screens. There are several advantages in using natural products directly as pesticides or as leads for new pesticides:

- Large diversity of chemical structures.
- High potential for finding new mechanisms of action.
- Lower risk to environment.
- Lower risk of toxicity to nontarget organisms.

Many of the new chemistries in fungicides, insecticides, and herbicides can trace their origins to natural products. For example, glufosinate is a derivative of bialophos, a natural product derived from *Streptomyces viridachromogens* (Leason et al. 1982). The methoxy-acrylate fungicides arose from strobilurin A, a natural product extracted from a mushroom (Clough et al. 1992). The active ingredient in the new spinosad insecticides used in cotton is isolated from the naturally occurring soil organism *Saccharopolyspora spinosa* (Thompson et al. 1994, Adan et al 1996, EPA

1997b). The new pyrrole insecticides originated in dioxypyrrolomycin, a natural product derived from an extract of *Streptomyces* spp. (Black et al. 1994).

Hypovirulence

Another potential use of microorganisms to protect plants is through hypovirulence—infection of a plant with benign strains that protect hosts from later infection by virulent strains. Hypovirulence is caused by a double-stranded RNA virus that infects the fungus, and reduces its virulence relative to the crop plant. The hypovirulent state can be induced through infection with viruses and related materials. Hypovirulence has achieved its widest success in combating chestnut blight, a disease that radically altered the eastern forest landscape after introduction of an invasive fungus (NRC 1996).

Transgenic Crops

Choice of crop variety has always been a cornerstone of crop protection, especially for disease and insect control. There are numerous review articles on plant resistance to insects (Harris 1980, Kogan 1982, Hedin 1983). The sources of resistance to pests in crops have been classified as nonpreferred, antibiosis, and tolerance (Pfadt 1971). Insect preference for a host plant is related to physical structure of the surface (hairs, wax, and so on), color, taste, odor, and light reflection. A resistant variety might be nonpreferred by virtue of lacking one or more of the preferred factors or characteristics of the host plant. Antibiosis is the adverse effect of the plant on the pest, which can be due to a deleterious chemical or the lack of specific nutrients in the plant. Tolerance is the ability of certain plants to withstand pest attack by virtue of general vigor or ability to repair tissue damage caused by a pest attack.

Plant resistance as a pest-management factor has achieved some outstanding results (for example, against grape phylloxera, woolly aphid, Hessian fly, and wheat stem sawfly) (Kogan 1982). Desirable features of pest-resistant plants are specificity, cumulative effectiveness (with effect on the pest compounded in successive generations), and persistence and harmony with the environment, ease of adoption, and compatibility with other IPM tactics (Kogan 1982). Disadvantages of using pest-resistant plants for pest management include the long development time (3-15 years), genetic limitations (due to lack of available resistance genes), and evolution of pests that have overcome and are no longer controlled by the bred-in resistance. To overcome some of the problems associated with traditional resistance-breeding approaches, transgenic technologies have been adopted. Inserting resistance genes from other genomes into a crop plant can reduce the problems associated with a lack of available resistance genes.

At present, resistance is the predominant defense against several plant diseases, such as rust diseases, that would otherwise reduce cereal-crop production in much of the world. In the case of rust diseases, a plant cultivar generally is resistant to only one race of a pathogen. Other races of the pathogen can infect the plant, and the shift in the distribution of the pathogen races leads to a boom-and-bust syndrome of the rust diseases. Strategies of resistance-gene deployment in which multiple cultivars (with different race-specific resistance genes) or single cultivars with multiple race-specific genes are planted in fields have been found to be effective against the syndrome. Race-specific resistance genes deployed in this manner can be quite successful in controlling plant diseases (NRC 1996).

Crop seed engineered with a single gene for virus or insect control or herbicide tolerance is commercially successful (Table 4-3). In 1998, more than 5 million acres were planted with cotton seed engineered to contain a *Btk* gene for caterpillar control; in the same year, 1 million acres were planted with dual-trait cotton seed (engineered to contain *Bt* and a gene expressing herbicide tolerance). Potato containing a *Btt* gene for controlling Colorado potato beetle is also on the market. Cotton, corn, and soybean crops engineered to resist the herbicide Roundup(were launched during the last 2 years; Roundup Ready(soybean seed was planted on 9 million acres in 1997. Also, Liberty Link(canola and corn seed engineered to resist glufosinate are on the market. Industrial leaders anticipate that plants containing chitinase genes and other genes for fungal and bacterial plant-pathogen control will be on the market in the near future; many of these newer genetically engineered plant varieties are already undergoing field tests (Figure 4-3).

Transgenic plants, whether engineered to contain an insecticidal protein, such as an endotoxin protein from *Bt* or a chitinase gene to control root rot pathogens, appear to have the same advantages as traditionally bred pest-resistant crop varieties. For example, cotton engineered with a *Bt* Cry IA(c) or (b) protein

- Is selective to lepidopterans.
- Increases the persistence of *Bt* for season-long control.
- Is compatible with the environment (reducing the use of more toxic chemical pesticides).
- Is compatible with other pest-management tactics, such as use of natural enemies or chemical pesticides, aimed at other pests.

In general, transgenic approaches also have the same disadvantages

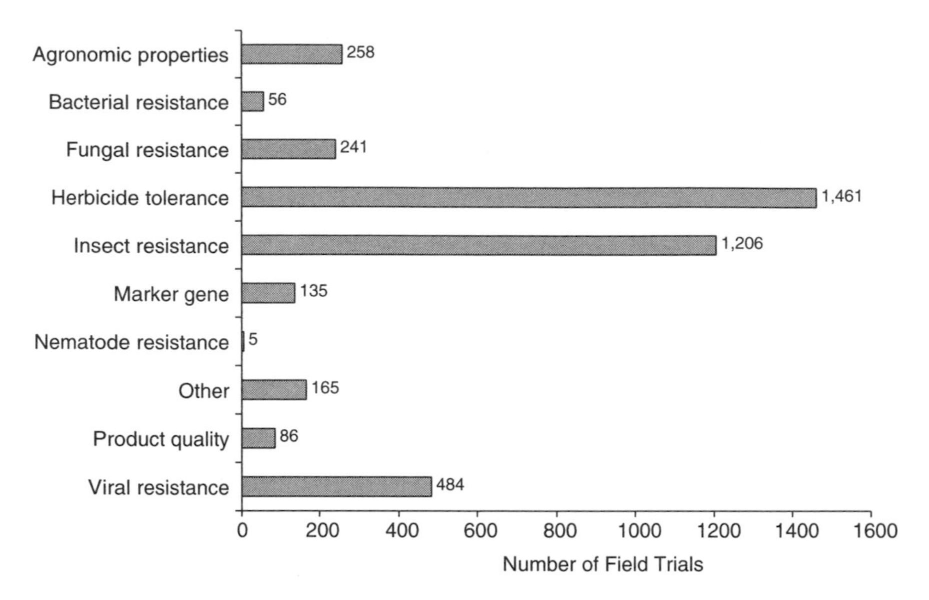

FIGURE 4-3 Agricultural-environmental biotechnology modifications, 1987-1998.

as traditionally bred crops. First, the period needed for development of transgenic plants—for such steps as finding appropriate genes, tissue-culture selection process, and backcrossing—is as long as traditional plant-breeding methods (John Callahan, AgraQuest, Inc., personal communication, May 22, 1998). Second, although much progress has been made in the discovery of new genes for introduction into plants, the ability to introduce the genetic material has surpassed the ability to discover new genes to engineer. Very few, if any, genes have been found for control of nematodes, sucking insects, and mites. In cotton, for example, transgenic plants with *Bt* genes will provide growers with another alternative for bollworm, but the need for products that control whiteflies, mites, and lygus bugs remains. Cholesterol oxidase, an enzyme from the microorganism *Streptomyces* and discovered by Monsanto Company, is toxic to cotton boll weevils and has been engineered into cotton plants; the enzyme is not effective on other insect pests.

The current choices of transgenic plants that contain single *Bt* genes are "first-generation" plants and will be followed by more sophisticated "second-generation" and "third-generation" plants with greater flexibility for use in IPM systems. These advances could include plants with inducible and tissue-specific expression systems whereby expression of the protein is "turned on" in response to insect feeding or some other stimulus. Also on the horizon are crops engineered with multiple genes ("stacking"). This new advance has been undergoing extensive field trials over the last few years. Table 4-8 shows the number of field tests of genetically engineered crops containing one or more genes in 1987–1998.

For example, companies have already introduced crops stacked with *Bt* and herbicide tolerance. Crops with multiple genes for proteins with different modes of action (for example, gene 1 containing *Bt* is combined with gene 2 containing a protease inhibitor) will be useful to prevent or delay development of resistance by a pest. Some protease inhibitor and lectin genes have been engineered for caterpillar control. Macintosh et al. (1990) discovered that some protease inhibitors can be combined with *Bt* proteins for enhanced caterpillar kill.

A great deal of research has focused on the search for new genes for plant-pathogen control. In the next 10 years, agriculture will see the introduction of plants engineered to inhibit fungal and bacterial plant pathogens. Virus-resistant crops are already on the market.

Another limitation of both traditionally bred and transgenic crops is the evolution of pests that are unaffected by the bred-in resistance (the pest develops resistance to the plant resistance traits). This is a key issue for crops engineered with single-gene pest-control traits, such as *Bt*. Concerns about adoption of transgenic plants without IPM was expressed by Gould (1988), who wrote that, "the successful engineering of highly resistant crops could lead to the elimination of IPM techniques that aim at using intense, pest suppressive measures only when pests are likely to cause economically important damage." Considering the advantages and the cost and time needed to develop new transgenic crops, it is important to sustain the life of these crops for a long period. Strategies designed to limit the development of resistant pests on the basis of understanding of pest population dynamics (such as studies on pest population establishment and growth, genetics, movement, behavior, number of generations required to develop resistance, and generation time) are critical for optimal and sustained use of transgenic plants in IPM systems.

Much research is still needed to develop effective resistance-management strategies and IPM systems incorporating *Bt* plants. There is a need for more research on pest genetics, resistance mechanisms, cross resistance, pest behavior and biology, and practical aspects of the implementation of refuges (for maintenance and development of susceptible insects) and areawide research on pest population levels and on detection and monitoring of resistance. The absence of USDA funding and coordination of public and private research on *Bt*-resistance management limits progress on resistance management. McGaughey et al. (1998) state that, "the challenge to US agriculture is to find ways to design and implement resistant management programs that will allow a realization of the benefits of this new agricultural biotechnology delivery system, while at the

TABLE 4-8 Number of Field Tests of Genetically Engineered Crops Containing Single or Multiple Genes

	No. of Field Tests No. of Genetic Interventions								
Year	One	Two	Three	Four	Five	Six	Seven	Confidential Business Information	Total
1987	5	—	—	—	—	—	—	—	5
1988	16	—	—	—	—	—	—	—	16
1989	27	3	—	—	—	—	—	—	30
1990	36	12	1	1	—	—	—	—	50
1991	77	10	2	1	—	—	—	—	90
1992	131	20	3	5	—	—	—	—	159
1993	254	38	6	4	—	—	—	—	302
1994	505	49	16	—	2	1	—	—	573
1995	629	63	14	4	4	—	—	—	714
1996	501	88	10	1	8		1	2	611
1997	588	146	9	10	6	1	—	2	762
1998	764	250	15	24	5	3	—	2	1,063
Total	3,533	679	76	50	25	5	1	6	4,375

Source: Stewart, in press.

same time not jeopardizing the long-term benefits of this environmentally useful insecticide." *Bt* has been used for more than 30 years and is the most successful biopesticide ever developed. It is important to sustain its use as a biopesticide spray.

Genetic Engineering of Pests

Most research aimed at pest management based on genetic engineering has focused on genetic manipulation of crop plants, economically important tree species (Carozzi and Koziel 1997; Raffa 1989), and biocontrol agents (Hoy et al., 1997, Bonning and Hammock 1996). Less emphasis has been placed on engineering of weeds, pathogens, and insect pest species in ways that would decrease economic damage. However, the recent successes in stable transformation of mosquitoes based on the use of *Hermes* and *Mariner* transposons (Coates et al. 1998) has provided more impetus for genetic engineering of pests (Kidwell and Wattam 1998). The concept of modifying pests by use of classical genetic manipulations dates back at least to the 1940s, when Serebovsky (1940) and Vanderplank (1944) suggested that chromosomal abnormalities and hybrid sterility could be used for insect control. The most heralded success in using genetics for pest control is the eradication of the screwworm from areas of the United States (Baumhover 1966), Mexico (Krafsur et al. 1987), and Libya (Vargas-Teran et al. 1994) through release of irradiated, sterile insects. The genetically damaged males released in these programs mated successfully with native females, which laid eggs that could not develop properly. Repeated releases finally resulted in population extinction.

A considerable theoretical and empirical knowledge base about sterile-release techniques developed in the 1970s and 1980s (Asman et al. 1981, Whitten and Foster 1975; Davidson 1974, Prout 1978, Foster et al. 1988). More sophisticated genetic manipulation approaches, such as the use of conditionally lethal genes (Davidson 1974) and chromosomal translocations (Asman et al. 1981), also received attention during that time. Conditional lethals could spread into populations during favorable times but could induce a genetic load if they inhibited proper diapause initiation or hindered survival at high temperatures. Release of strains with single or double translocations could impose a genetic load on a native population while replacing native genes with those of the released strains. At least theoretically, a population of malaria-vector mosquitoes or plant-pathogen-vector planthoppers could be replaced with an artificially developed nonvector strain (Curtis 1968). Because of the lack of enough simple success stories, support for this general kind of pest management declined in the late 1980s and early 1990s (but see Kerremans and Franz

1995). Nevertheless, the early theoretical and empirical studies determined the limits and potential of this species-specific control technology.

In the 1990s, entomologists and molecular biologists recognized that transposable elements, such as the P element in *Drosophila*, could be used in a manner similar to, but more efficient than, the use of translocations. That spurred new theoretical and empirical studies (for example, Curtis 1992, Ribeiro and Kidwell 1994, Meister and Grigliatti 1994). Although replacement of a native population with insects that had translocations required release of huge numbers of laboratory-reared insects (at least as many as in the native population if there was one translocation and no relative fitness reduction), it seemed theoretically possible to release a relatively small number of insects that had transposable elements (about 10% of the native population) to replace the native insect strain (Kidwell and Ribeiro 1992; Ribeiro and Kidwell 1994). A number of laboratory experiments with *Drosophila* have shown that a new P element introduced into a population spreads rapidly and is usually fixed in the population within 10 generations (Good et al. 1989). An important experiment by Meister and Grigliatti (1994) demonstrated that, when an alcohol dehydrogenase gene was artificially added to a P element, the P element was still able to invade a *Drosophila* population. Furthermore, the dehydrogenase gene was active in a number of cases.

If a transposon "loaded" with a gene for vector incompetence (inability to transmit malaria) (Olson et al., 1996) could be used to transform *Aedes aegypti*, the transformed strain could be released into the wild, and presumably the native malaria-transmitting strains of the mosquito would become refractory (Crampton et al. 1990, Hastings 1994, Curtis 1994, Pfeifer and Grigliatti 1996). The same approach could be taken for insect-transmitted plant diseases. This approach has not been without its critics, especially in the case of malaria control (Spielman, 1994, Pettigrew and O'Neill 1997). Some of the major concerns are about getting enough transmission reduction to substantially decrease disease severity (Spielman 1994); about the fact that a transposon that has invaded a native population cannot be used again; and about the potential for horizontal transmission of transposons between pest and nonpest species of animals (Robertson and Lampe 1995).

Another newly proposed approach for driving desirable genes into native populations involves use of rickettsial symbionts in the genus *Wolbachia* (Werren 1997). When a male insect carrying *Wolbachia* mates with a female that does not carry *Wolbachia*, no offspring are produced because of cytoplasmic incompatibility. The reciprocal cross is fertile, so the insects bearing the microorganism spread through the population (Sinkins et al. 1995). Because the *Wolbachia* is maternally inherited, any other maternally inherited trait present in the initial *Wolbachia*-carrying insects will "hitch-hike" to fixation. Strategies have been developed that involve use of strains with *Wolbachia* and a virus that carries a gene for disease refractoriness (Sinkins et al. 1997). Other strategies involve isolation of the *Wolbachia* genes that code for cytoplasmic incompatibility; these genes are used to transform insect strains directly (Sinkins et al. 1997).

In addition to its use to develop novel control strategies based on transposons and microorganism-based incompatibility, genetic engineering could be used to improve the efficiency of classical genetic control strategies, such as introduction of conditional lethal genes or sex-ratio distortion genes (Gould and Schliekleman, in prep.). Researchers using classical genetic techniques have been severely limited in their ability to find genes that confer conditional lethality or sex-ratio distortion. Each time a project is developed for a specific pest, a new search must be undertaken to find useful genes or translocations. With the tools of genetic engineering, promoters and useful genes could be found that would be effective in many species. Once transformation of diverse insect species becomes more routine, it will be possible to transform a variety of species with appropriate genes.

Whether genetic engineering can be used for the direct control of pest species is an open question. In many cases, it seems easier to engineer crop plants for insect and disease resistance than it is to engineer the pest or vector. Most genetic control strategies to be used on pest species (such as the insertion of dominant lethal genes) work only for obligately sexual, outcrossing species (Atkinson and O'Brachta 1999, Thomas et al. 2000). Because many weeds are self-fertilized, they might not be amenable to this approach, and other weeds that currently outcross could become resistant to genetic control by increasing their frequency of inbreeding. Many pathogens and insect pests are parthenogenetic; these species would not be candidates for the new genetic control strategies.

Genetic control strategies are exceptionally specific in their action, so they are likely to be environmentally friendly. Once a genetic control strategy is put into place, it can maintain itself and not require continued expenditure of funds. Many obligately outcrossing pests (such as weeds and household and veterinary pests) could be difficult to control through plant engineering. There is a need to assess carefully what the niche might be for direct genetic control techniques. The niche and the environmental benefits might be larger if substantial funding is used immediately to support research. If research support is insufficient or too late, the niche for this environmentally benign control technology could become occupied by other more expensive and less benign techniques. Public concern about the use of genetically engineered insects could, however, overshadow any of the potential environmental benefits of this technology and could block its implementation.

Targets of Chemical Pesticides

In the last 50 years, there has been an intensive search for compounds that will control pests without damaging crops or the environment. During this time, the way the search for new pesticides has evolved with respect to how it is done and the criteria by which discoveries are judged. The current methods used to find and develop new pesticides are sophisticated and complex. A new compound has to be not only highly efficacious, but also environmentally acceptable. Thus, testing on the environmental behavior and toxicology of a candidate compound comes much earlier in the discovery process than it did in the past. If a candidate cannot meet the new criteria for environmental safety, as well as efficacy, it is not advanced.

The number of compounds that have to be tested before a viable candidate is found has increased from 2,000 to 40,000–50,000 (Stetter 1998). There are many reasons for the increase, including environmental regulation. For example, an upcoming material needs to "beat the standard". That is, for a herbicide to be truly promising, it has to act at extremely low rates to outperform the materials alredy being used. Thus, many more compounds must be screened than would be needed simply to identify a compound with equivalent efficacy. Another problem is resistance. And, because many of the known target sites are no longer available, materials that contain novel modes of action are sought.

The current method of screening for new pesticides starts with high-throughput screens that can handle hundreds of thousands of compounds per year. These screens include miniaturized systems in which whole organisms are grown on microtiter plates or other systems but require microgram or nanogram quantities of material for tests, which can be run in a minimum of space and time (Jansson et al. 1997). There is also increased use of mechanism-based screens in either modified microorganisms or in vitro enzyme assays, where a particular molecular or enzymatic target is targeted. In most cases the molecular targets are ones that are peculiar to the pest and are not found in nontarget organisms (for example, essential amino acid biosynthetic pathways) (Abel 1996). The advantage of mechanism-based screens is that a search can be restricted to compounds that will selectively control the pest with less chance of nontarget toxicity. In addition, with the development of resistance to many current pesticides, compounds with new mechanisms of action have become more valuable. Mechanism-based screens potentially can find compounds that hit new target sites where there is no known resistance. A limitation is that activity detected with in vitro assays might not show activity on the whole organism.

Another change in the discovery process for new pesticides is the time when research on the mechanism of action begins. In the 1950s to 1970s, research on determining the mechanism of action of new leads would not begin until after a compound that could lead to a product had been found or even registered for use. Now this research begins as soon as a new chemical is discovered to be biologically active. One reason for the change is to eliminate compounds with high potential for mammalian toxicity. Companies now start testing for mammalian (and other nontarget) toxicity early in the screening process. This has involved the development of high-throughput screens to test for nontarget toxicity. Another reason is to determine whether a new mechanism of action has been discovered. This information is used to develop new mechanism-based screens and to conduct computer modeling to guide the synthesis of analogues to locate better, more potent compounds. It is also used to determine the potential for resistance.

Combinatorial Chemistry

Combinatorial chemistry encompasses systems in which a large number of chemical compounds are made via multiple parallel synthesis. In this system, a set of building blocks (N) is coupled through repetitive steps (x) to synthesize a number (N^x) of different compounds. The process results in a chemical library of a few hundred to millions of compounds, depending on the number of building blocks and the number of repetitive steps (Lyttle 1995.) The technique was first used to make peptides and oligonucleotides, but more recently it has been adapted to make nonpolymeric small molecules (Warr 1997a). In a recent publication, it was reported that one person working 7 months was able to generate a 400-member library of heterocyclic carboxamides; this resulted in identification of a herbicidal compound with 4 times the activity of the original lead compound (Parlow et al. 1997.)

Combinatorial chemistry is often done on a solid support phase, which greatly simplifies automation, product isolation, and purification. The disadvantage of this approach is the limited size of the sample, the functionalization needed for attachment to the solid phase, and the chemistries needed for linkers and capping to prevent unwanted reactions (Warr 1997a). Another problem with combinatorial chemistry is the need for sufficient diversity in the chemicals being synthesized to optimize the chances of discovering useful compounds. The latter problem has been addressed by making "virtual" libraries via specialized software programs (Warr 1997b) and then synthesizing a small number of compounds with the highest diversity. This procedure was used successfully to find an isozyme-specific inhibitor of glutathione-*S*-transferase (Lyttle et al. 1994).

Combinatorial chemistry has been widely adopted by many pharmaceutical and agrichemical companies to increase the potential for finding new, profitable products. It has been estimated that a company needs to synthesize and screen 40,000–50,000 compounds to find a compound that can lead to a new product. Using traditional methods of synthesis, one chemist can produce only 50–100 compounds per year. Using combinatorial and parallel-synthesis methods, one chemist has the potential of making several thousand compounds per year (Warr 1997a). Combinatorial synthesis is used not only to find new leads, but also to optimize leads.

The compounds generated via combinatorial chemistry are processed by high-throughput screening techniques, described previously. The advantage of these techniques is the diversity of structures that can be synthesized and tested. One disadvantage is the size of the sample that is initially synthesized. The small sample limits the amount of initial testing that can be done, and in many cases mechanism-based screens or some simple in vivo initial screens are used (Lyttle 1995). Another disadvantage is that, by their very nature, compounds within a particular combinatorial library are related. Depending on the measure of diversity used, combinatorial chemistry might or might not be a good source of broadly diverse chemistry for lead generation.

Development and Commercialization of New Chemicals

Functional genomics serves as a source of bioassays, monitoring, and diagnostics. The efficacy of the pesticide is only one of many criteria used by industrial researchers to determine whether it will be developed. After finding a new lead area of chemistry that lends itself well to synthetic modification, additional screens (beyond whole organism screens) are put into place to determine ecological, toxicological, and environmental characteristics of compounds. New screens have been developed to determine the toxicity of compounds (Yamanaka et al. 1990, Swanson et al. 1997). Tests are also conducted early in the discovery process to assess the toxicity of potential leads to nontarget organisms, such as fish or aquatic invertebrates. This information is used to guide the synthesis of analogues to try to eliminate nontarget toxicity, if present, while maintaining activity against the target pest. If nontarget toxicity cannot be eliminated, the chemical structure is not pursued further.

Screens are commonly used for quickly determining the soil binding and degradation characteristics of compounds (Lamoreaux 1990; Daniels and McTernan 1994). These screens determine the leaching potential and persistence of compounds with the goal of finding compounds that have the lowest leaching potential and minimal persistence but that still control the target pest. Compounds that have very high leaching potential or long soil persistence are eliminated early in the discovery process unless there are other applications for which the compounds can be useful (for example, household pests).

Once a compound with the efficacy needed to control the pest with minimal impact on the environment and nontarget organisms is found, the development phase can begin. This is the most expensive part of bringing a new product to market. Development, testing, and registration can typically take 8-12 years and cost over $50 million for each pesticide (IANR 1994). Registration involves at least 142 tests, most of which are aimed at determining the environmental and toxicological characteristics of new compounds. EPA will register only those compounds that meet strict criteria for human and environmental safety. Industrial scientists must use this same battery of tests to maintain the registration of existing compounds.

Pesticide research and development have come a long way from the procedures used to find earlier compounds, such as DDT and related compounds. There is now a much better understanding of the desirable and undesirable characteristics of pesticides. Companies have set up systems to eliminate, early in the discovery phase, compounds that do not meet the new standards, to ensure that time and resources are not wasted on compounds that will not be acceptable as pest-control agents.

Application Technology

Pesticides effectively control many insects, diseases, and weeds. However, to be effective, pesticides have to target the crop or animal of interest. Pimentel (1995) estimated that only 0.00001-1% of the pesticides sprayed actually reach the target pests. The reasons for the low rate of interception are many, including the distribution of the pest, the size of the pest population at the time of application, uncertainty as to how and when the pest will encounter the pesticide, the use of broadcast sprays, and drift. One way to reduce the risk posed by pesticides is to increase the amount of pesticide that reaches the target pest while decreasing the total amount of pesticide applied to the field. That can be accomplished only through improvements in application technology (see table 3-1).

Spray drift is one of the biggest concerns regarding the movement of pesticides to nontarget organisms. Off-target losses can range from 50 to 70% of the applied pesticide because of evaporation and drift (Hall and Fox 1997). Drift from aerial applications is greatest and that from ground applications is least (Hall and Fox 1997).

There are several ways to reduce drift. One way is to use spray additives that affect the drop size of sprays by increasing the number of large droplets and decreasing the number of small droplets (Hall et al. 1993).

However, those agents can also affect the spray pattern, and lead to inefficient placement of the pesticide. Another method to decrease the number of fine droplets during spraying is to use new nozzles that are designed to decrease the number of fine droplets. The nozzles work by increasing droplet size through a reduction in the velocity of the liquid just before it is discharged (Ozkan 1997).

Spray additives can also be used to decrease the rate of evaporation of droplets. It has been shown that the most rapid evaporation occurs during the first 20 seconds of flight (Hall et al. 1993). Additives can reduce the rate. For example, a conventional formulation of permethrin lost 55% of its original volume after falling 2 m at terminal velocity at 40°C and 17% at 25°C. The loss could be reduced to 40% and 9%, respectively, with the incorporation of an antievaporant in the formulation (Hall et al. 1993).

The use of aerial electrostatic sprays has the potential to increase the amount of pesticide that reaches the crop while reducing drift. Carlton et al. (1995) showed that using a bipolar-charge protocol (having an opposite charge on each end of the spray boom of an aircraft) increased deposits by a factor of 4.3 on cotton leaves. Penetration into the canopy and wraparound leaf effects also increased. However, the technology is limited by the lack of charging nozzles suitable for the commercial market (Carlton et al., 1995). In addition, the effectiveness of electrostatic sprayers depends on environmental factors, such as temperature and relative humidity, height of the spray boom over the canopy, and distribution of the pesticide spray throughout the canopy (electrostatic sprayers tend to deposit most of the pesticide material in the top of the canopy).

Another method of increasing pesticide efficacy while decreasing drift potential is through the use of air-assisted sprayers, which allow the use of ultra-low-volume applications of pesticides in oil-based diluents. Mulrooney et al. (1997) used such a system to increase the effectiveness of malathion while reducing the total amount of pesticide applied. Bohannan and Jordan (1995) were able to increase the effectiveness of postemergence applications of various grass herbicides (sethoxydim and clethodim) by applying them in oil diluents. Sethoxydim and clethodim applied at 25 and 16% of their recommended dose rates gave over 90% control of yellow foxtail (*Setaria glauca*). However, that was not the case with two other herbicides, imazethapyr and bentazon, possibly because of the lack of a suitable formulation that was compatible with oil.

Hooded sprayers, spray towers, and spray tunnels are examples of mechanical methods for increasing the amount of pesticide reaching the target while reducing off-target movement (Hall and Fox 1997). However, the effectiveness of these physical methods varies with environmental conditions, crop canopy, and other such factors. Research needs to continue to determine the most efficacious way to reduce drift using physical barriers.

Researchers are examining other methods for selective application of a pesticide that targets the pest while minimizing application to nontarget organisms. Such methods should decrease the volume of pesticide spray while maintaining efficacy. Weed-sensing sprayers that use red and near-infrared sensors to detect the presence of green tissue have been successfully used in chemical fallow and hooded sprayers to reduce herbicide use by 14-90% (Ahrens1994, Hanks and Beck 1998, Blackshaw et al. 1998b.) However, these sensors and accompanying analytical equipment are relatively expensive, and the sensors are limited by the size of the weed that can be detected and are affected by environmental conditions, such as soil type, and wind speed (Blackshaw et al. 1998a).

The use of baits in combination with a pesticide has the potential of reducing the amount of a pesticide applied while maintaining efficacy. For example, the amount of malathion insecticide needed for control of Mediterranean fruit fly (*Ceratitis capitata*) in Israel was reduced from 1 kg/hectare to 150 g/hectare through the use of baits (Grinstein and Matthews 1997).

One of the biggest reasons for overapplication of pesticides is variation in the speed of the application equipment. New technology that constantly monitors ground speed and adjusts the application of the pesticide accordingly is being developed. The primary means of adjusting application rate is to alter the pressure of the spray. However, because the performance of many spray nozzles is pressure-dependent, applicators are limited in how much they can alter the pressure without affecting the distribution pattern of the spray (Paice et al. 1996). Other systems under development include use of a direct-injection system in which the spray volume is kept constant but the amount of active ingredient is varied (Ghate and Perry 1994).

Precision-agriculture techniques are new technologies under development that can vary the amount of herbicide application. In the future, users might map a field to identify locations of various weed populations and then select the amount and type of herbicide spray that match the weed density and population for that area of the field (Paice et al. 1996). One of the main limitations in adopting these mapping and sensor technologies is the need for sprayer technologies with rapid response and rate controllers that can automatically adjust the amount of active ingredient. Injection-metering devices, twin fluid nozzles, pulsed-nozzle systems, and rotary-spray generators with prediluted herbicides can decrease response time to less than 1 second.

Finally, one of the major risks in pesticide application is exposure of the applicator during the mixing procedures and disposal of used containers. Closed systems are being developed for many pesticides through minibulk systems and innovative water-dissolvable bags, which virtually eliminate exposure of the applicator to the pesticide during mixing. Furthermore, many companies are providing returnable containers, such as the "minibulks" (refillable tanks that holds typically 50 to 100 gallons of pesticide) that greatly reduce disposal costs and risks associated with the disposable smaller 2.5 gallon pesticide containers (Frei and Schmid 1997). A closed system greatly reduces exposure to the pesticide because the pesticide flows from the storage tank directly into the application device (the sprayer) and only does so if the container is properly connected to the sprayer. Additional work needs to be done in these technologies to reduce the risks even further.

Precision Agriculture

Precision agriculture can be defined as a bundle of technologies designed to adjust input use to variations in environmental and climatic situations over space and time so as to reduce residues associated with input use. Many of these technologies rely on space-age communication and incorporate the use of global positioning systems. Modern irrigation technologies that adjust input use according to variability in soil and weather conditions and rely on weather stations and moisture-monitoring equipment are also examples of precision technologies. Precision technologies have the potential to increase input-use efficiency, increase yields, and reduce residues of chemicals that can contaminate the environment. In many cases they can lead to input saving; but in others, the yield effect can also entail increased input use (NRC 1997).

There has been substantial variation in rates of adoption of what can be generically defined as precision technologies. Some modern irrigation technologies have high rates of adoption in high-value crops. Some components of precision agriculture, such as yield monitors, are gaining acceptance among grain producers. But overall rates of adoption of many components of precision agriculture have not been high (NRC 1997). Although some of the technologies can be obtained through consultants, adoption can be hampered by the cost of investment. Furthermore, the management software needed to take advantage of this highly technical information has not been fully developed. Assuming that management software does become available, adoption of precision farming technologies is likely to accelerate as their costs decline and productivity increases are demonstrated.

Precision technologies will allow farmers to take advantage of a wider arsenal of pest-control tools, to adjust application rates, and to target areas where pest problems are most severe. Because agricultural pollution can result from pesticides that do not reach their intended targets, precision technologies have the potential to reduce pesticide emissions into the environment and to reduce the negative side effects of pesticide use. However, precision agriculture is early in its adoption phase, and environmental benefits are not proved. Future research should focus on environmental impacts of precision technologies in the field and watershed scales (NRC 1997). The emergence of precision technologies is contributing to the increased value of information in agriculture, and it could also contribute to the value and use of agricultural management consultants who have access to data and analytical capability and can provide effective management strategies that take advantage of large bodies of information.

Remote Sensing and Pest Management

Satellite and airborne remote sensing data, coupled with geographic information systems and global positioning satellites, are potentially powerful tools for monitoring pest infestations and their crop impacts. Remote sensing has been effective in identifying new colonies of weeds in low-access areas and monitoring the spread of metapopulations (Lass and Callihan 1997, Lass et al. 1996, Everitt et al. 1993, Carson et al. 1995; Anderson et al., 1993). Thus, this technology can help to set priorities for herbicide application and to direct treatment to the locations where it can have the greatest impact on weed populations. In addition, airborne sensors that detect multispectral reflectance differences between crop and other vegetation canopies can identify crop-vigor variation in response to insect pests and plant pathogens (UC Davis, 1997; Baldy et al., 1996; Hill et al., 1996; Kline et al., 1996; Liedtke, 1997; Vickery et al.; 1997; Ustin et al., 1997). As this technology is improved, it is likely that pesticide use will decrease with the change from whole-field to pest-location-specific applications. Remote sensing for pest management in forest ecosystems has been widely used for several decades. Proposed improvements in application of remote sensing for pest management include increased availability of sensors, improvements in quality of images, and increased resolution of the spectral information (UC Davis, 1997).

A large number of earth-observing satellites scheduled for launch in the next few years will enable specific monitoring of environmental conditions and natural-resource management. Some commercial multispectral satellites will be collecting hyperspatial data (with spatial resolution of 1-5 m), and many others will be in resoulution range of 5-100 m. Other satellites will collect multispectral data in hundreds of bands (NASA 1998). Remote sensing is likely to become an increasingly important tool for identification of pests and their impacts and later for site-specific application of pesticides.

Increasing Knowledge of Pest Ecology

Population dynamics, the field of ecology most relevant to pest management, has become a mature field of inquiry in the last 20 years. Many of the broad generalizations made in the past have been reassessed because they failed to explain important patterns in the populations of plants and animals (Denno et al. 1995, Kareiva 1993). Recent advances in population dynamics, at both theoretical and empirical levels, are applicable to agricultural systems (Kareiva 1996, Turchin 1999).

Insects and early-successional plant species have figured prominently in many basic studies of population dynamics because they are prominent parts of the landscape and because their population fluctuations can be monitored and manipulated within a reasonable time and spatial scales. Scientists with basic and agricultural job responsibilities have cooperated in some major studies because the goals in both are similar: the prediction of future population densities. An intellectual synergism has often resulted from such cooperative studies (Murdoch et al. 1984).

Biological-control researchers can boast of many successes, but the factors that lead to success or failure are rarely understood. Work by basic and applied scientists led to the careful dissection of the population dynamics involved in successful biological control of the citrus red scale (Murdoch et al. 1995, Murdoch and Briggs 1996). Insights gained from that work could help in designing future biological-control efforts aimed at this and other pests. Recent work on interactions among predators of herbivores (Rosenheim 1998) has shown that two species of predators are not always better than one in decreasing population densities of herbivores. That is an important issue in designing biological-control systems. Finally, the finding that plants, in response to damage by a herbivore, produce specific compounds that then attract parasites and predators of that herbivore also offers new insights into patterns of insect population densities (Tumlinson et al. 1998). These and other new findings could be used in breeding crop cultivars that are more compatible with biological control (NRC 1996).

Herbicide costs are an important component of agricultural production in some crop systems. Knowledge that allows for decrease in herbicide use can improve agricultural efficiency in those systems. Without an understanding of weed population dynamics, the decision to forgo the use of a herbicide can be risky. Research on weed population dynamics (Cousens and Mortimer 1995) indicates that long-term predictions about weed densities might not be possible, but that shorter-term, more specific predictions are. New crop-specific models (White and Coble 1997) have been developed and have been proved useful in predicting whether herbicide treatment is economically justifiable. In other systems, research is under way to determine whether certain crop cultivars or combinations of plantings can be developed that will regulate weed populations at low densities with little input of herbicide and cultivation (Jordan 1993).

One of the major drawbacks in using ecological knowledge in agriculture is the information-processing and communication barrier. It will take time before most farmers are routinely using on-line extension information and maintaining their own computer-based records, but the transformation is under way. As ecological knowledge of pest population dynamics improves and becomes more accessible to farmers, the risks inherent in decreasing (or tailoring) pesticide can be diminished. In the last decade, there has been a tremendous increase in Internet-accessible information on pest identification and pest biology. Web pages are also being developed to help farmers with identification of the beneficial organisms in their fields. It is challenging to link the basic Web pages to interactive programs that can offer alternative solutions to problems associated with an identified pest. As more information is available on the impact of specific pesticides on specific beneficial organisms, it should be possible for farmers to choose pesticides according to which beneficial organisms are in their crops. Studies of population dynamics of plant pathogens have been hampered by our inability to monitor fungi, bacteria, and viruses when they are at low densities.

Decision-Support Systems for Pest Management

A wide array of issues—including crop productivity, environmental protection, and profitability—creates a complex background against which agricultural producers or professional managers are expected to make decisions about optimal pest management. Decision support systems (DSS) can condense the knowledge base required to make decisions through the use of models and databases of expert knowledge (Wagner 1993). DSS are defined as knowledge-based systems that help producers and managers to make strategic and tactical management decisions on pest-control treatments, time of planting and harvesting, cultivar selection, and marketing options (Plant and Stone 1991). Moulin (1996) described two types of DSS. The first consists of models based on production functions calculated with empirical or mechanistic relationships between variables, such as soil properties, temperature precipitation, and yield. The models can be used to improve understanding of biotic and abiotic factor relationships with production or as management tools for decisions related to production. Models of the second type are used strategically to provide supportive information for broad policy decisions or tactically to provide specific recommendations. Most pest-management DSSs are of the second type and are based on rules derived from expert knowledge; they are often focused on pesticide selection rather than a complete set of management practices for managing a particular pest. The low representation of alternative tactics in DSSs is a result of inability to predict the outcome of these tactics. Cultural and biological pest-management tactics rely on more subtle influences on pest populations; so more environmental factors play a role in determining population dynamics and there is greater variability in success rates. DSSs have not been extensively used by producers because of the sparseness of specific information input requirements (NRC 1997).

ECOLOGICAL CHANGES AFFECTING AGRICULTURE

The distribution and abundance of organisms on the planet have long been influenced by global environmental change. For the most part, such changes have taken place on geological time scales. In recent years, however, it has become apparent that anthropogenic changes in global environments can affect the distribution and abundance of organisms on ecological time scales. Over the last 2 centuries, industrialization, urbanization, and agricultural development have contributed to alterations in global environmental patterns that are historically unprecedented in their magnitude. Among the forms of environmental change most likely to affect American agriculture and forestry, several stand out: an increase in atmospheric carbon dioxide resulting from fossil-fuel combustion, and a concomitant increase in temperature; stratospheric ozone depletion linked to chlorofluorocarbon emissions, and a consequent increase in exposure to short-wave ultraviolet-B radiation; increases in the frequency of biological invasions associated with internationalization and globalization of trade; and an increase in extinction rates due to expansion of development, particularly in tropical regions. Each of those has the potential to alter the relationships among agricultural and forest pests, their hosts, and growers or producers.

Carbon Dioxide and Global Warming

Increases in rates of fossil-fuel consumption worldwide have steadily increased atmospheric concentrations of carbon dioxide over the last 100 years. Increased carbon dioxide has two potential impacts on agricultural productivity. First, as a critical resource for photosynthesis, carbon dioxide availability can affect growth rates, yields, and chemical composition of both crop plants and weeds. Generally, increases in carbon dioxide concentrations are associated with increases in productivity, particularly for annual crops. In the case of annual crop plants exhibiting C3 photosynthesis, such productivity increases average about 30% with a doubling of contemporary carbon dioxide (Cure and Adcock 1986,Rogers and Dahlman 1993). Individual species vary, however, in their responsiveness (C4 plants—such as corn, sorghum, sugar cane, and millet—exhibit a considerably lower response). The available evidence suggests that some direct growth and physiological responses to increased levels of carbon dioxide would be less dramatic in woody plants than in annual crops (Watson ey al, 1996). Conversely, woody plants show greater alterations in secondary chemicals than do annual plants and thereby could show the greatest alterations in resistance to insects and pathogens (Penuelas and Estiarte 1998).

Differential impacts on weedy species and crops could affect the economic impacts of weeds. Outcomes are difficult to predict. C3 plants tend to benefit to a greater extent from the "carbon dioxide fertilization effect;" Watson et al. (1996) estimate that, although 80 of 86 of the plant species that contribute 90% of per capita supplies of food are C3 plants, only four of 18 of the worst weed species are C3 plants. One possible outcome of increased carbon dioxide, then, might be a change in the composition of the principal weed flora.

Changes in carbon dioxide can alter plant composition, as well as plant growth rates. Under conditions of increased carbon dioxide, the production of nonstructural carbohydrates in many species is increased and protein and mineral-nutrient concentrations decline (Mooney and Koch 1994). Exposure to increased carbon dioxide can also lead to an increase in root:shoot ratios. Such changes in tissue composition and biomass allocation can have important affects on the extent and distribution pattern of herbivore or pathogen damage. In particular, dilution of protein and mineral nutrients by increases in nonstructural carbohydrates can lead to compensatory increases in food intake by herbivores (Slansky 1992) and concomitant increases in tissue losses. Increases in starch content of weed species and increases in allocation of biomass to roots or rhizomes could potentially reduce herbicide efficacy, particularly in C3 weed species.

Increased carbon dioxide also has the ability to affect plant-pest relationships through by affecting global temperatures. Through the "greenhouse effect", increased carbon dioxide is predicted to increase average global temperatures by 1.5–4.5°C. Such an increase in temperature can have profound effects on both crops and pests. Higher temperatures can lead to a variety of problems in crop plants. Many critical developmental processes have relatively narrow optimal temperature ranges and are sensitive to excesses. Increased temperatures can affect the quality of various cool-season crops, which have winter chilling requirements for bud set (for example, apple, cherries, and other tree fruits) or require lower night-time temperatures to maximize quality (for example, spinach and broccoli). Excessively high temperatures can also affect pollen viability, as in maize (Schoper et al. 1987) or tuber initiation with potato (Kooman et al. 1996). Another manifestation of the greenhouse effect is increased variability in temperatures; extremes in weather conditions can cause losses in a wide array of crop types.

Global warming can increase pest problems in agriculture and forestry as a result of expansion of insect, weed, and pathogen ranges into hitherto inhospitable regions. Temperature can influence surrounding developmental rates, fecundity, and activity of individual pests, as well as the size, distribution, and continuity of populations of pests (Drake 1994). Cold temperatures limit the distribution of many insect pests, particularly those of tropical origins. At present, for example, the red imported fire ant (*Solenopsis invicta)*, accidentally introduced into North America from Brazil, is effectively restricted to southern states, where they cause considerable damage (Callcott and Collins 1996). This distribution parallels that of the introduced *Apis mellifera scutellata*, the so-called African ("killer") bee, initially brought to Brazil from Africa and now reported to occur in the US South and Southwest. Porter et al. (1991) predicted that, in Europe, a 3–6°C increase in European temperature by 2025-2070 could expand the range of the European corn borer *Ostrinia nubilalis* by 1,220 km northward; Mochida (1992) predicted that a 3°C increase in temperature in Japan would lead to expansion of the ranges of eight major pests of rice or soybean. Damage inflicted by agricultural and forest pests could potentially increase as a result of decreased development time and increased voltinism. An increase in the number of breeding cycles per season raises the possibility that resistance to insecticides could evolve more rapidly.

A reduction in winter temperatures could increase the severity of plant-disease outbreaks; disease problems tend to be exacerbated under such conditions. Presumably, pathogens could also experience range expansion as a result of warming trends. Disease incidence and severity are subject to change in animal agriculture. Increases in average temperatures are predicted to increase the activity of vector species, such as culicid mosquitoes, and microorganisms of tropical origin could make inroads at higher latitudes.

Constraining any definitive predictions of impacts of increased carbon dioxide is the acknowledged buffering capacity of genetic adaptation and technological change. There is considerable evidence of genetic variation in temperature tolerance among crop cultivars and it is well documented for such important crops as wheat, rice, maize, and soybeans

(Watson et al. 1996). Existing agronomic practices have the potential to ameliorate adverse effects of increased global temperatures; the greatest uncertainty lies in whether economic systems have the flexibility to incorporate the necessary technological innovations in a timely way.

Increased Ultraviolet-B Radiation

As a result of stratospheric ozone depletion, due principally to emissions of chlorofluorocarbons, levels of ultraviolet-B-radiation (UV-B) (280-315 nm) reaching the earth's surface have increased in all temperate latitudes (Madronich 1993). Increased UV-B radiation is an environmental stress factor that can alter the physiology and development of terrestrial plants in several ways (Caldwell et al. 1989, Tevini 1993), ranging from such cellular and subcellular effects as DNA damage and degradation of photosystem II proteins to such whole-plant effects as stunting, shortening of internodes, alterations in root:shoot biomass ratios (Tevini and Teramura 1989), and alterations in plant primary and secondary chemistry (Ros and Tevini 1995,;McCloud and Berenbaum 1994). At the ecological level, these developmental effects can manifest themselves as yield reductions in several crop species (Tevini 1993). Increased UV-B can also affect interspecific competition in plants (Gold and Caldwell 1983), reduce plant resistance to pathogens (Orth et al. 1990, Panagopoulous et al. 1991), affect pollinator-plant relationships (Collins et al. 1997, Feldheim and Connor 1996), and alter host-plant suitability for herbivores (McCloud and Berenbaum 1994).

Plants have the capacity to reduce damaging effects of UV-B in several ways. Among these is the production of compounds that absorb damaging wavelengths of UV radiation. In several plant species, increased UV-B exposure induces production of flavonoids and other phenolics (reviewed in Caldwell et al. 1989.). Plants can also ameliorate the oxidative damage brought about by UV exposure by enhanced synthesis of enzymatic and nonenzymatic antioxidants (Larson 1988). Alpha-tocopherol, ascorbic acid, and glutathione are all effective scavengers of reactive oxygen species and in several temperate crop species increase in response to UV exposure (Foyer 1993, Grace and Logan 1996).

Inasmuch as antioxidant defensive compounds also serve as essential nutrients for insects, UV-B exposure can potentially increase plant suitability for herbivores. Indeed, Hatcher and Paul (1994) found that elevated UV-B increased nitrogen content of *Pisum sativum* and digestibility of its foliage by *Autographa gamma*, a caterpillar. This enhanced nutritional suitability, however, appears to be counterbalanced in other plants by increases in phenolics, flavonoids, and other UV-screening compounds, many of which are toxic or deterrent to insects.

Predicting overall effects of increased UV-B levels on US agriculture is difficult because of the scarcity of appropriate studies. The impact might be minimal. Genetic variability in resistance to UV-B has been documented in crop species (Teramura 1983, Teramura and Sullivan 1991, D'Surney et al. 1993) and weed species (such as *Plantago lanceolata*; McCloud and Berenbaum 1994); this allows for the possibility of adaptation in the case of weeds and selective cultivar use in the case of crop plants. Moreover, the nearly universal adoption of the Montreal protocols for global reductions in chlorofluorocarbon production suggests that ozone depletion could diminish, rather than accelerate, in the future.

Increased Frequency of Biological Invasions

Biological invasions constitute a form of global change as devastating ecologically and economically as more widely recognized global climatic changes (Vitousek et al. 1996). The growth of agriculture, industry, and development over the last 200 years has led to redistribution of species in all major taxa. In the United States, for example, over 5,000 species of alien plants have been introduced and established, infesting over 100 million acres of terrestrial and aquatic habitats; the acreage occupied by these plants is increasing 8–20% per year. The proportion of the flora that is made up of alien weeds varies by state from less than 10% to, in the case of Hawaii, almost 50% (Pimentel et al. 2000).

The impacts of alien species on agriculture and forestry are enormous (Niemela and Mattson 1996). Many of the introduced species are themselves pests; life-history attributes that are conducive to colonizing new habitats—including exploitative reproductive traits, broad resource requirements, small bodies, and high vagility—predispose these species to pest status. That problem is compounded by the fact that many species that are not pests in their native range become important pests in their new, acquired range because factors that historically regulated their population, such as natural enemies, are not present. Moreover, introduced species can act as disease vectors and facilitate the entry of new pathogens into nonnative agroecosystems.

Introduced species can reduce yields by removing crop biomass directly (as in the case of alien insects) or by competing with crop plants for water, minerals, or light for photosynthesis. Introduced species can also cause problems by altering trophic dynamics and ecosystem structure.

Forest ecosystems in North America have been radically altered, and dominant species driven to near extinction, by such invasive pests as chestnut blight, gypsy moth, and balsam woolly adelgid (Niemela and Mattson 1996). On rangeland, toxic species can poison livestock (even, on occasion, pollinator species); in agricultural and forest ecosystems, introduced species can alter drainage patterns and accumulation of biomass and create fire hazards. Cheatgrass, a fire-adapted annual, has invaded over 5 million hectares of shrub-steppe habitat in the Great Basin and, by eliminating native perennials, contributed to the frequency of fires, which increased by a factor of 20 (Vitousek et al. 1996).

Introduced species, via introgression and hybridization with native species, can also interfere with evolutionary processes (Darmency 1997), with many possible outcomes, including extinction of native species, introduction of pesticide-resistance genes into susceptible genomes, and transfer of agriculturally undesirable traits into native species. All told, the management of alien weeds alone in the United States exacts a cost of $3.6–4.5 billion per year (Westbrooks 1998).

As for the impact of biological invasions on the future of pesticide use in the United States, suffice it to say that some of the most controversial examples of pesticide use in the last half-century have involved attempts to eradicate introduced species. An ill-fated campaign launched in 1957 to eradicate *Solenopsis invicta*, the red imported fire ant, culminated in expensive failure, public concern, and political concern about the safety of the chemical control agents used (Kaiser 1978, Hinkle 1982). Efforts to contain Mediterranean fruit flies (*Ceratitis capitata*) in California in 1989-1990 with malathion-laced baits have also led to a public outcry (Myers et al. 1998). Pesticides have often been used initially to eradicate localized populations of introduced species because they kill more quickly than do most biological control methods. Such use, while often highly desirable in agricultural contexts, might be less appropriate for eradicating nonnative species in preserves, refuges, and wildlands, where use of chemicals is inconsistent with management practices and can pose unacceptable risks to nontarget species, and in urban green spaces, where pesticide use can present unacceptable risks of exposure to large numbers of people.

In view of the increasing globalization of trade, accidental introductions will probably increase in frequency in the near term. Possible consequences of this accelerated pace of invasion are increased frequency of pesticide failure due to resistance acquisition and increased demand for narrow-spectrum chemicals or biologically based approaches.

Loss of Biodiversity

Whether accelerated biodiversity loss contributes a separate form of global change or is the consequence of the combined effects of many human-caused forms of global change is debatable. What appears to be clear, however, is that extinction rates globally are higher than rates recorded in the recent past (Pimm et al., 1995; Purvis et al., 2000; Wilson 1993). Effects of biodiversity loss on agriculture vary with site and system. In general, how-

TABLE 4-9 Number of Papers Published in 1996 that Report on Biologically Active Natural Substances

Activity	No. of Publications
Anthelmintic	13
Antibacterial	176
Antifeedant	19
Antifungal	91
Insecticidal	10
Molluscicidal	7
Pheromonal	52
Plant-growth modulating	17

Source: Claeson and Bohlin, 1997.

ever, potential effects include loss of critical germplasm for crop improvement, breakdown of such mutualistic associations as those between pollinator and plant and between plant and mycorrhizae, and decrease in availability of natural enemies for use in biological-control programs. With respect to direct effects of extinctions on pesticide use, in that natural products have often provided leads for development of chemical control agents (or have themselves served as control agents), biodiversity loss could eliminate potential sources of new chemicals that have novel modes of action or novel sites. In a review of all articles appearing in 1996 in over 130 major journals dealing with natural products, Claeson and Bohlin (1997) tabulated the number of publications that reported a wide range of types of biological activity. In that single year, over 150 publications documented pesticidal activity of some sort (insecticidal, antifungal, pheromonal) (Table 4-9). Many random-screening efforts focus on floras and faunas in regions of the world at greatest risk of biodiversity loss through extinction (such as tropical rain forest) and could be compromised as well.

EVOLUTIONARY CHANGES IN PESTS

Evolutionary interactions between pests and farmers predate conventional pesticides by thousands of years (Day 1974). Evidence of such interaction is best preserved in weed characteristics that enable specific genotypes to avoid effects of traditional cultural control practices. As a result of changes in their timing of reproduction and seed size, some weed populations display resistance to winnowing, a process used by prehistoric and modern farmers to eliminate weed seeds from grain seeds, including those being saved for the next season's planting (Gould 1991). Other genotypes produce seedlings that physically resemble crop seedlings and thereby decrease the efficiency of hand and hoe weeding procedures (Barrett 1983).

The historical record of insect and microorganisms adaptation to traditional agricultural practices is harder to trace than that of weeds, but there is evidence that prehistoric crop breeding indirectly produced cultivars resistant to pest attack. In turn, pest genotypes that are capable of surviving on these traditional cultivars have evolved (for example, sheath blight fungal genotypes found on traditional Tibetan cultivars of rice). In recent history, a major corn pest, *Diabrotica barberi* (northern corn rootworm), has adapted to the yearly rotation of soybean and corn by maintaining a quiescent egg (diapause) for an additional year (Krysan et al. 1986, Fisher et al. 1994); then eggs hatch in the season that corn is planted. New data indicate that a closely related corn pest, the western corn rootworm *Diabrotica virgifera*, has evolved in a different way to avoid the effect of corn-soybean rotation (Sammons et al. 1997). It feeds on corn, but late in the season it lays many of its eggs in soybean fields, which with crop rotation, will become cornfields in the following year.

Even biological control is not immune to the effects of pest evolution. A number of studies show that pests can evolve resistance to their parasitoids (e.g. Petersen 1978, Henter and Via 1995) and pathogens (e.g., David and Gardiner 1960, Alexander et al. 1984). But one unique advantage possessed by self-perpetuating biocontrol agents is their ability to respond evolutionarily to changes in the pests (Sasaki and Godfray 1999, Fellowes et al. 1998).

As explained in chapter 2, there has recently been much well-deserved attention to the evolution of pest resistance to synthetic pesticides. Carefully maintained records show that over 500 insects, 200 weeds, and 160 pathogens have evolved resistance to at least one pesticide (Delp 1988; Bills and Whalon 1991; Ffrench-Constant 1994; Georghiou and Lagunes 1988; Heap 2000; Holt et al. 1993; NRC 1986; Warwick 1991). Many pests are documented to have become resistant to new pesticides in less than 3 years (Forgash 1984, Shelton and Wyman 1991, Sun et al. 1992, Schwinn and Morton 1990). Unlike biological control agents, synthetic pesticides cannot respond evolutionarily to pest adaptation. That response is the responsibility of modern agriculture. A major factor in the "pesticide treadmill" involves two responses to pesticide resistance. The first is to increase the dose and frequency of use of the less effective pesticide (Georghiou 1986); this typically results in higher levels of pest resistance and damage to natural enemies and the environment. The second response is to develop and commercialize a new pesticide. The treadmill concept assumes that this two-step process will continue until the pest meets a resistance-proof pesticide or until the supply of effective new pesticides is exhausted.

There are certainly cases in which specific pests seem to lack genetic variation for adapting to a particular pesticide (such as the boll weevil and the organophosphate Guthion). But there are also cases in which, in the absence of new pesticides, a resistant pest has caused the collapse of a cropping system (such as Canyete Valley and Nicaraguan cotton industries, Barducci 1972). A difficult question for the future is whether products and approaches can be developed to ensure efficacy against specific pests.

One generalization that emerges from agricultural history and evolutionary theory is that the greater the impact of control measures on pest populations, the more extreme are their evolutionary responses. For example, traditional winnowing procedures were not very efficient, so weeds could find their way into seed bags without great resemblance to the crop seed. Modern winnowing procedures are highly efficient, so only the weed genotypes that precisely mimic the crop escape detection (Gould 1991). It is the hope of modern agriculturists that they can make the requirements for pest adaptation so stringent that pest evolution by natural selection will fail. That was certainly the hope of some early advocates of the use of insect hormones and pheromones in control; some even went so far as to suggest that counteradaptation to hormones was unlikely or impossible (Williams 1967). Early on, that proved not to be the case, however and multiple studies have demonstrated that this group of control substances rapidly selects for resistance. Within a few years, simple laboratory experiments resulted in resistant insects (e.g., Shemshedini and Wilson, 1990).

In recent years, many synthetic pesticides have been developed that are more toxic to target pests or more selective in the spectrum of pests that they affect than older classes of compounds. For example, about 1 lb. of an organophosphate insecticide per acre was needed to control *Heliothis virescens*, a major cotton pest. The pyrethroids that replaced organophosphate insecticides provided control at 0.01–0.10 lb/acre. The newly registered insecticide can control this pest with only 0.04 lb/acre. The same trend has been seen with some fungicides and herbicides. Environmental and health benefits are often associated with these two trends, but the trends could lead to problems with evolutionary sustainability. The heightened target toxicity has often been the result of new compounds that affect specific biochemical targets in the pest. Whereas traditional fungicides affected a number of target sites (NRC 1986), some of the new herbicides affect only a single site. Organophosphates and pyrethroids attack components of the nervous system common to all insects, but some of the newer insecticides affect target sites that are only found in specific groups of insects. For example, the toxins from *Bt* are effective at very low concentrations when a target insect has a midgut receptor to which the toxin can bind. However, the presence of these receptors is often limited to a few insect genera (Frankenhuyzen 1993).

When a target site, such as acetylcholinesterase in the nervous system is so highly conserved evolutionarily that it is used by insects and vertebrates, it can be relatively difficult for pests to evolve alterations in the molecule. When a fungicide affects many biochemical targets, it is unlikely that a single population of the pathogen will contain genetic variation needed to alter each of the targets. However, rapid resistance to fungicides that act at a single target site has arisen. In contrast, when a target site such as the midgut receptor for the *Bt* toxin CryIIIA is present in one beetle genus but absent in related genera, there is an expectation that the structure of these receptors is evolutionary labile and could easily change in response to heavy selection pressure. That expectation is not always met.

Higher levels of toxicity mean that each pest encounters a relatively small number of toxin molecules because the farmer uses less insecticide. Detailed evidence is lacking, and it will be difficult to predict the speed of resistance development. It might be easier for a pest to evolve alterations to metabolize the lower concentrations of the new synthetic compounds than larger concentration of older compounds. Alternatively, protein sequestration and metabolism of low concentrations of pesticides might emphasize evolution of improved affinity of the detoxifying protein for the pesticide compared with the maximal turnover rate or concentration of the protein. If the trend for developing more specific, highly toxic synthetic pesticides continues, it will be necessary to determe whether such pesticides are predisposed to rapid pest adaptation.

The new field of pesticide-resistance management, which has interwoven the disciplines of IPM and population genetics, concedes that pests have the potential to adapt to almost all control tactics. If future synthetic compounds are more prone to pest adaptation than older compounds, resistance management could become more critical. The premise of resistance management is that, through an understanding of pest genetics and ecology, approaches can be developed for using pest-control tools in ways that will diminish the rate of pest adaptation. One such approach involves presenting pests with a combination of a stringent control and a refuge from that control. EPA, agricultural companies, USDA, and university scientists are testing this approach on a large scale with transgenic cotton, corn, and potato cultivars that produce the insecticidal protein derived from *Bt*. Another tactic for avoidance of pest resistance is managing for the susceptible gene. It is the susceptible gene in a pest population that makes a pesticide useful as a control agent. Conversely, it is the absence or loss of the susceptible gene that makes a pesticide ineffective. With careful monitoring of susceptible types, location of refugia, and relaxation of selection pressure, resistance can be avoided (Roush et al. 1990, Maxwell et al. 1990). If those approaches succeed, they and other resistance management approaches might gain broader acceptance and use.

Metcalf (1989) analyzed rates of insect adaptation to succeeding waves of new insecticide classes and concluded that the pace of insect adaptation to pesticides was increasing. That could be due, in part, to the factors discussed here, but Metcalf proposed an additional mechanism. He presented the hypothesis that, in addition to evolving mechanisms enabling them to cope with a specific toxin, pests are evolving a general type of genetic flexibility that "preadapts" them to dealing with future compounds. Metcalf did not offer a detailed explanation of the form of this preadaptation, but a number of scenarios are feasible (Gould 1995). When the ability to control a pest with a specific compound is lost, the organism typically only has 20–1,000 times lower sensitivity to the compound. One or a few genes typically account for most of the resistance. It is generally found that these genes also confer some resistance to related pesticides, and sometimes to chemically unrelated pesticides (for example, DDT and pyrethroids). We would be likely to study gene A if it conferred 100-fold resistance to the selecting pesticide Z and conferred 2- to 10-fold resistance to other compounds. We would be much less likely to study gene B, which might have evolved in response to selecting pesticide Z, if it conferred only 2-fold resistance to pesticide Z, because 2-fold resistance seems like a minor problem and is difficult to study experimentally. By ignoring genes that are difficult to study, such as gene B, we could miss important aspects of pest and pathogen evolution. There might be many genes, like gene B, that confer only minor resistance to the selecting toxin but confer an overall heightened resistance to most toxins. It is even possible that gene B would some day be found to confer 100-fold resistance to a new compound.

Another mechanism that could lead to preadapted pests would be alterations in a pest's metabolic flexibility. It is possible that some pest genotypes incur more physiological disruption than other genotypes because of alterations in metabolic pathways. As pests continue to be selected for altered metabolism, shifts in the general physiology of the organism that decrease the degree of disruption might be selected for. Work with the sheep blow fly (*Lucilia cuprina*) has shown that resistant populations have genes for altered target sites but also have a gene for ameliorating the fitness cost incurred by having the resistance-conferring form of the gene (McKenzie and Game 1987). If genes that decrease such fitness costs have general buffering impacts on pests with other resistance genes, pest populations might accumulate traits that allow them to withstand metabolic changes caused by major resistance genes.

Very few data are available for testing Metcalf's general hypothesis, but, given the potential consequences, it seems prudent to pay more attention to this possibility. Beyond the preadaptation hypothesis, there is a reason to expect future pests to have greater ability to respond to selection: the globalized economy. With globalization comes increased movement of pests. For example, molecular studies of mosquito resistance to organophosphates suggest that resistance arose in one generally isolated population in Africa and then was spread to Europe and the Western Hemisphere by airplanes (Raymond et al. 1991). Increased air travel and increased global shipping of agricultural products provide pathways for the spread of resistance genes. It can also increase the raw genetic variability possessed by the average pest and pathogen population via the continual influx of new genes from distant locations.

Environmental-protection groups have been raising another issue related to pest adaptation. They contend that as genetically engineered crops are released into the environment, genes from these crops will spread into weed and pathogen populations and will cause increased crop losses. Rigorously derived data prove that genes can be transferred between some crops and their weedy wild relatives, for example, sorghum and Johnsongrass, cultivated rice and wild rice, wild radish and cultivated radish (Klinger et al. 1992, Krimsky and Wrubel 1996). In such cases it is likely that a herbicide-resistance gene will move from a crop to a weed and could then make the weed impossible to control with the specific herbicide and related herbicides.

REFERENCES

Abel, L.M. 1996. Biochemical approaches to herbicide discovery: Advances in enzyme target identification and inhibitor design. Weed Sci. 44:734–742.

Adams, R. M., C. Rosenzweig, R. M. Peart, J. T. Ritchie, B. A. McCarl, J.D. Glyer, R. B. Curry, J. W. Jones, K.J. Boote, and L. H. Allen, Jr. 1990. Global climate change and US agriculture. Nature 345:219–224.

Adan, A., P. Del Estal, F. Budia, M. Gonzalez, and E. Vinuela. 1996 Laboratory evaluation of the novel naturally derived compound spinosad against *Ceratitis capitata*. Pestic. Sci. 48:261–268.

Agrow. 1998. 299. Surrey, UK: PJB Publications Ltd.

Ahrens, W.H. 1994. Relative costs of a weed-activated versus conventional sprayer in northern Great Plains fallow. Weed Tech. 8:50–57.

Alexander, H. M., J. Antonovics and M. D. Rausher. 1984. Relationship of phenotypic and genetic variation in *Plantago lanceolata* to disease caused by *Fusarium moniliforme* var. *subglutinans*. Oecologia 65:89–93.

Anderson, G. L., J. H. Everitt, A. J. Richardson, and D. E. Escobar. 1993. Using satellite data to map false broomweed (*Ericameria austrotexana*) infestations on south Texas rangelands. Weed Technol. 7:865–871.

Asian Vegetable Research and Development Center. 1991. Demonstration of IPM of diamondback moth on farmers' fields in the lowlands. Progress Report, Asian Vegetable Research and Development Center.

Asman, S. M., P. T. McDonald, and T. Prout. 1981. Field studies of genetic control systems for mosquitoes. Ann. Rev. Ent. 26:289–343.

Aspelin, A. L., and A. H. Grube. 1999. Pesticide Industry Sales and Usage. 733-R-99-001. Washington, DC: US Environmental Protection Agency, Office of Pesticide Programs.

Atkinson, P. W., and D. A. O'Brachta. 1999.Genetic transformation of non-drosophilid insects by transposable elements. Ann. Ent. Soc. Am: 92:930–936.

Baker, C. J., Stavely, J., Mock, N. 1985. Biocontrol of bean rust by *Bacillus subtilis* under field conditions. Plant Dis. 69:770–772.

Baldy, R., J. DeBenedictis, L. Johnson, E. Weber, M. Baldy, B. Osborn, and J. Burleigh. 1996. Leaf color and vine size are related to yield in a phylloxera-infested vineyard. Vitis 35(4):201–205.

Barducci, T. B. 1972. Ecological consequences of pesticides used for the control of cotton insects in Canyete Valley, Peru. Pp. 423–438 in The Careless Technology: Ecology and International Development, M. T. Farvar and J. P. Milton, eds. Garden City, N.Y.: The Natural History Press.

Barrett, S. C. H. 1983. Crop mimicry in weeds. Econ. Bot. 37:255–282.

Baumhover, A. H. 1966. Eradication of the screwworm fly as agent of myiasis. J. Am. Med. Assoc. 196:240–248.

Besson, F., F. Peypoux, G. Michel 1979 Antifungal activity upon *Saccharomyces cerevisiae* of iturin A , mycosubtilin, bacillomycin L and of their derivatives; inhibition of this antifungal activity by lipid antagonists. J. Antibiotics 32 (8):828–833.

Bills, P., and M. Whalon. 1991. Weeds: 216 species of weeds are resistant worldwide. In The Occurrence of Resistance to Pesticides in Arthropods. Rome: Food and Agriculture Organization.

Black, B.C., R. M. Hollingworth, K.I. Ahammadsahib, C. D. Kukel, and S. Donovan. 1994. Insecticidal activity and mitochondrial uncoupling activity of AC 303,630 and related halogenated pyrroles. Pest. Biochem Physiol. 50:115–128.

Blackshaw, R.E., L. J. Molnar, and C.W. Lindwall. 1998b. Merits of a weed-sensing sprayer to control weeds in conservation fallow and cropping systems. Weed Sci. 46:120–126.

Blackshaw, R.E., L.J. Molnar, D.F. Chevalier, and C.W. Lindwall. 1998a. Factors affecting the operation of the weed-sensing Detectspray system. Weed Sci. 46:127–131.

Blakeman, J.P., N. J. Fokkema. 1982. Potential for biological control of plant disease on the phylloplane. Ann. Rev. Phyt. 20:167–192.

Bland, J. M., A. R. Lax, M. A. Klich. 1995. Iturin-A, an antifungal peptide produced by *Bacillus subtilis*. Pp. 105-107 in Proceedings of Plant Growth Regulatory Society of America, 22nd.

Bohannan, D.R., and T.N. Jordan. 1995. Effects of ultra-low volume applications on herbicide efficacy using oil diluents as carriers. Weed Tech. 9:682–688.

Bonning, B. C. 1996. Development of recombinant baculoviruses for insect control. Ann. Rev. Ent. 41:191–210

Bonning, B. C. and B. D. Hammock. 1996. Development of recombinant baculoviruses for insect control. Ann. Rev. Ent. 41:191–210.

Bowen, A.T. 1991. Innovative IPM application technology. J. Arboculture.17:138–140.

Broza, M., B. Sneh, A. Yawetz, U. Oron, and A. Honigman. 1984. Commercial application of *Bacillus thuringiensis* var. *entomocidus* to cotton fields for the control of *Spodoptera littoralis* Boisduval (Lepidoptera: Noctuidae), J. Econ. Ent. 77:1530–1533.

Caldwell, M. M., A. H. Teramura, and M. Tevini. 1989. The changing solar ultraviolet climate and the ecological consequences for higher plants. Trends Ecol. Evol. 4:363–367.

Callcott, A. M. and H. Collins. 1996. Invasion and range of imported fire ants (Hymenoptera Formicidae) in North America from 1918-1995. Fla. Ent. 79:240–251.

Carlton, J.B., L.F. Bouse, and I.W. Kirk. 1995. Electrostatic charging of aerial spray over cotton. Trans. Of ASAE. 38:1641–1645.

Carozzi, N., and M. Koziel, eds. 1997. Advances in insect control: the role of transgenic plants. London: Taylor & Francis.

Carruthers, R. I., and K. Hural. 1990. Fungi as naturally occurring entomopathogens. Pp. 115-138 in New Directions In Biological Control: Alternatives For Suppressing Agricultural Pests And Diseases , R. R. Basker and P. E. Dunn, eds. Proceedings of a UCLA symposia on molecular and cellular biology, series v. 112. Los Angeles, Calif.: University of Califonria, Los Angeles.

Carson, H. W., L. W. Lass, and R. H. Callihan. 1995. Detection of yellow hawkweed with high resolution digital images. Weed Technol. 9:477-483.

Claeson, P., and L. Bohlin. 1997. Some aspects of bioassay methods in natural-product research aimed at drug lead discovery. Trends in Biotech. 15:245–248.

Clough, J.M., P. J. De Fraine, T. E. M. Fraser, and C. R. Godfrey. 1992. Fungicidal beta-methoxyacrylates: from natural products to novel synthetic agricultural fungicides. Pp. 372-383 in Synthesis and Chemistry of Agrochemicals III, D. R. Baker, J. G. Fenyes, and J. J. Steffens, eds. Symposium series 504. Washington, D.C.: American Chemical Society.

Coates, C. J., N. Jasinskiene, L. Miyashiro, and A. A. James. 1998. Mariner transposition and transformation of the yellow fever mosquito, *Aedes aegypti*. Proc. Natl. Acad. Sci. USA 95:3748-3751.

Collins, S., J. K. Conner, and G. E. Robinson. 1997. Foraging behavior of honey bees (Hymenoptera: Apidae) on *Brassica nigra* and *B. rapa* grown under simulated ambient and enhanced UV-B radiation. Ann. Ent.. Soc. Am. 90:102–106.

Cousens, M. and M. Mortimer. 1995. Dynamics of Weed Populations. England: Cambridge University Press.

Crampton, J. M., A. Morris, G. Lycett, A. Warren, and P. Eggleston. 1990. Transgenic mosquitoes: future vector control strategy? Parasitol. Today 6:31–36.

Crowe, A.S. and J.P. Mutch. 1994. An expert systems approach for assessing groundwater contamination by pesticides. Ground Water 32(3):487–498.

Cultivar 1997. The Impact of Transgenic Plants in the Markets. November 18-25.

Cunningham, J.C. 1988. Baculoviruses: their status compared to *Bacillus thuringiensis* as microbial insecticides. Outlook on Ag. 17:10–17.

Cure, J. D., and B. Adcock. 1986. Crop responses to carbon dioxide doubling: a literature survey. Ag. Forest Meterol.38:127–45.

Curtis, C. F. 1968. Introducing vector incompetence. Nature 218:368–69.

Curtis, C. F. 1992. Selfish genes in mosquitoes. Nature 357:450.

Curtis, C. F. 1994. The case for malaria control by genetic manipulation of its vectors. Parasitol. Today 10:371–74.

Daly, J.C., and J. A. McKenzie. 1986. Resistance management strategies in Australia: the *Heliothis* and 'Wormkill' programmes. Pp. 951-959 v. 3 of Proceedings of the British Crop Protection Conference - Pests and Diseases, November 17-20, 1986, Brighton, England.

Daniels, B.T., and W. F. McTernan. 1994. A screening method to identify the probabilities of pesticide leaching. Ag. Water Manag. 25:23–4

Darmency, H. 1997. Movement of resistance genes among plants. Pp 209-220 in T. M. Brown, ed. Molecular Genetics and Evolution of Pesticide Resistance, M. T. Farvar and J. P. Milton, eds. Symposium Series 645. Washington, D.C.: American Chemical Society.

David, W. A. L. and B. O. C. Gardiner. 1960. A *Pieris brassicae* (Linnaeus) culture resistant to a granulosis. J. Insect Pathol. 2:106–14.

Davidson, G. 1974. The Genetic Control of Insect Pests. London; New York: Academic Press.

Day, P. R. 1974. Genetics of Host-Parasite Interaction. San Francisco: W. H. Freeman.

Delp, C. J. 1988. Fungicide resistance problems in perspective. Pp. 4–5 in Fungicide Resistance in North America, C. J. Delp, ed. St. Paul, Minn.: APS Press.

Denno, R. F., M. S. McClure, and J. R. Ott. 1995. Interspecific interactions in phytophagous insects – competition reexamined and resurrected. Ann. Rev. Ent. 40:297–331.

Drake, V. A. 1994. The influence of weather and climate on agriculturally important insects – an Australian perspective. Australian J. of Ag. Res. 45(3):487–509.

D'surney, S.J., J. T. Tschaplinski, N. T. Edwards, and L. R. Shugart. 1993. Biological responses of two soybean cultivars exposed to enhanced UVB radiation. Env. Exp. Bot. 33:347–356.

Dutky, E. M. 1999. Plant Pathology Update. College Park: University of Maryland.

Elkinton, J. S., A. E. Haje, G. H. Boettner, and E. E. Simmons. 1991. Distribution and apparent spread of *Entomophaga maimaiga* (Zygomycetes: Entomophthorales) in gypsy moth (Lepidoptera: Lymantriidae) populations in North America. Env. Ent. 20:1601–1605.

Elliott, H.J., R. Bashford, A. Greener, and S. G. Candy. 1993. Integrated pest management of the Tasmanian eucalyptus leaf beetle, *Chrysophtharta bimaculata* (Olivier) (Coleoptera: Chrysomelidae). For. Ecol. Man. 53:29–38.

EPA (US Environmental Protection Agency). 1997a. Pesticide Regulation (PR) Notice 97-3. Available. [Online]: http://www.epa.gov/opppmsd1/pr_Notices/pr97-3.html.

EPA (US Environmental Protection Agency). 1997b. Pesticide Fact Sheet, Spinosad. Available. [Online]: http://www.epa.gov/opprd001/factsheets/spinsoad.htm

EPA (US Environmental Protection Agency). 1998. Staff Background Paper #2.4 – General Overview: Reduced Risk Pesticide Program. Available. [Online]: http://www.epa.gov/oppfead1/trac/safero.htm

EPA (US Environmental Protection Agency). 1999a. Implementing the Food Quality Protection Act. EPA 735-R-99001. Washington, D.C.: EPA, Office of Pesticide Programs.

EPA (US Environmental Protection Agency). 1999b. Fact Sheets on New Active Ingredients. [Online]. Available: http://www.epa.gov/opprd001/factsheets.

EPA (US Environmental Protection Agency). 2000. Interim Fiscal Year 2000 Work Plan. Office of Pesticide Programs. [Online]. Available: http://www.epa.gov/opprd001/workplan/fy2000.htm

Eppes, J. In press. The Future of Biopesticides. Norfolk, Conn.: Business Communications Corporation.

Everitt, J. H., D. E. Escobar, R. Villarreal, M. A. Alaniz, and M. R. Davis. 1993. Integration of airborne video, global positioning system, and geographic information system technologies for detecting and mapping two woody legumes on rangelands. Weed Tech. 7:981–987.

Feldheim, K., and Conner, J. K. 1996. The effects of increased UV-B radiation on growth, pollination success, and lifetime female fitness in two *Brassica* species. Oecologia 106:284–297.

Fellowes, M. D. E., A. R. Kraaijeveld, and H. C. J. Godfray. 1998. Trade-off associated with selection for increased ability to resist parasitoid attack in *Drosophila melanogaster*. P Roy Soc Lond B Bio 265(1405):1553–1558.

Ferro, D. N. 1993. Integrated pest management in vegetables in Massachsetts. Pp. 95-107 in Successful Implementation Of Integrated Pest Management For Agricultural Crops, A. R. Leslie, and G. W. Cuperus, eds.. Boca Raton, Fla: Lewis Publications.

Ferro, D. N. and S. M. Lyon. 1991. Colorado potato beetle (Coleoptera: Chrysomelidae) larval mortality: operative effects of *Bacillus thuringiensis* subsp. *san diego*. J. Econ. Entomol. 84:806–809.

Ferro, D.N. and Gelernter, W.D. 1989. Toxicity of a new strain of *Bacillus thuringiensis* to Colorado potato beetle (Coleoptera: Chrysomelidae). J. Econ. Ent.. 82:750–755.

Fisher, J. R., J. J. Jackson and A. C. Lew. 1994. Temperature and diapause development in the egg of *Diabrotica barberi* (Coleoptera: Chrysomelidae). Env. Ent. 23(2):464–471.

Forgash, A. J. 1984. History, evolution, and consequences of insecticide resistance. Pestic. Biochem. Physio. 22:178–186.

Foster, G. G., W. G. Vogt, T. L. Woodburn, and P. H. Smith. 1988. Computer simulation of genetic control. Comparison of sterile males and field-female killing systems. Theor. Appl. Genet. 76:870–879.

Foyer, C. H. 1993. Interactions between electron transport and carbon assimilation in leaves: coordination of activities and control. Pp. 199-224 in Photosynthesis: Photoreactions to Plant Productivity, Y. P. Abrol, P. Mohanty, and Govindjee, eds. Dordrecht, The Netherlands: Kluwer Academic Publishers.

Franco, C. M. M., and L. E. L. Coutinho. 1991. Detection of novel secondary metabolites. Crit. Rev. in Bio. 11:193–276.

Frankenhuyzen, K. U. 1993. The challenge of *Bacillus thuringiensis*. Pp. 193-220 in *Bacillus thuringiensis*, an Experimental Biopesticide: Theory and Practice, P. F. Entwistle, J. S. Cory, M. J. Bailey, and S. Higgs, eds. Chichester, New York: John Wiley & Sons.

fFrench-Constant, R. H. 1994. The molecular and population geneti cs of cyclodience insecticide resistance. Insect Biochemistry and Molecular Biology. 24(4):335–345.

Frei, B., and P. Schmid. 1997. Development trends in pesticide formulation and packaging. Pp. 34-43 in. Pesticide Formulation and Adjuvant Technology, C.L. Foy and D. W. Pritchard, eds. Boca Raton, Fla.: CRC Press.

Gelernter, W. D., N. C. Toscano, K. Kido, and B. A. Federici. 1986. Comparison of a nuclear polyhedrosis virus and chemical insecticides for control of the beet armyworm (Lepidoptera: Noctuidae) on head lettuce. J. Ec.. Ent.. 79:714–717.

Georghiou, G. P. 1986. The magnitude of the resistance problem. Pp. 14-43 in Pesticide Resistance: Strategies and Tactics for Management, National Research Council, ed. Washington, D.C.: National Academy Press:

Georghiou, G. P., and A. Lagunes. 1988. The Occurrence of Resistance to Pesticides: Cases of Resistance Reported Worldwide Through 1988. Rome: Food and Agriculture Organization of the United Nations.

Ghate, S. R., and C. D. Perry. 1994. Ground speed control of pesticide application rates in a compressed air direct injection sprayer. Trans. of ASAE. 37:33–38.

Gold, W. G., and M. M. Caldwell. 1983. The effects of ultraviolet-B radiation on plant competition in terrestrial ecosystems. Physiological Plantarum. 58:435–444.

Good, A. G., G. A. Meister, H. W. Brock, T. A. Grigliatti, and D. A. Hickey. 1989. Rapid spread of transposable P elements in experimental populations of *Drosophila melanogaster*. Genetics 122:387–396.

Gould, F. 1988. Evolutionary biology and genetically engineered crops. BioScience 38:26–31.

Gould, F. 1991. Evolutionary potential of crop pests. Amer. Sci. 79:496–507.

Gould, F. 1995. Comparisons between resistance management strategies for insects and weeds. Weed Tech. 9:830–839.

Gould, F. and P. Schliekelman. In press. Reassess autocidal control. In New technologies for Integrated Pest Management, G. G. Kennedy and T. Sutton, eds. St. Paul, Minn: APS Press

Grace, S. C., and B. A. Logan. 1996. Acclimation of foliar antioxidant systems to growth irradiance in three broad-leaved evergreen species. Plant Phys. 112:1631–1640.

Grinstein, A., and G. A. Matthews. 1997. Preface. Phytoparasitics. 25:Suppl:9S–10S.

Gueldner, R., C. Reilly, L. Pusey, C. Costello, R. Arrendale, R. Cox, R., D. Himmelsbach, G. Crumley, H. Cutler. 1988. Isolation and identification of iturins as antifungal peptides in biological control of peach brown rot with *Bacillus subtilis*. J. Ag. Food Chem., 36:366–370.

Hall, F.R., and R. D. Fox. 1997. The reduction of pesticide drift. Pp. 209-240 in Pesticide Formulation and Adjuvant Technology, C. L. Foy and D. W. Pritchard eds. Boca Raton, Fla.: CRC Press.

Hall, F. R., A. C. Chapple, R. A. Downer, L. M. Kirchner, and J. R. M. Thacker. 1993. Pesticide application as affected by spray modifiers. Pesticide Science: 38(2/3):123–133.

Hanks, J. E., and J. L. Beck. 1998. Sensor-controlled hooded sprayers for row crops. Weed Tech. 12:308–314.

Harris, M. K., ed. 1980. Biology and breeding for resistance to arthropods and pathogens in agricultural plants. Texas Agric. Exp. Stn. Pub. MP-1451.

Hastings, I. M. 1994. Selfish DNA as a method of pest control. Proc. Trans. Roy. Soc. London B 344:313–324.

Hatcher, P. E., and N. D. Paul. 1994. The effect of elevated UV-B radiation on herbivory of pea by *Autographa gamma*. Ent. Exp. Et App. 71:227–233.

He, H., Silo-Suh, L.A., Handelsman, J., Clardy, J. 1994. Zwittermicin A, an antifungal and plant protection agent from *Bacillus cereus*. Tet. Letters 35:2499–2502.

Heap, I. 2000. International Survey of Herbicide Resistant Weeds. March 23, 2000. [Online]. Available: www.weedscience.com.

Hedin, P. A., ed. 1983. Plant resistance to insects. ACS Symposium series 208. Washington, D.C.: American Chemical Society.

Henter, H. J., and S. Via. 1995. The potential for coevolution in a host-parasitoid system 1.1. Genetic variation within an aphid population in susceptibility to a parasitic wasp. Evolution. 49(3):427–438.

Hill, M. J., G. E. Donald, P. J. Vickery, and E. P. Furnival. 1996. Integration of satellite remote sensing, simple bioclimatic models and GIS for assessment of pasture suitability for a commercial grazing enterprise. Australian J. of Exp. Ag. 36:309–321.

Hinkle, M. K. 1982. Impact of the imported fire ant control programs on wildlife and quality of the environment. Pp. 41-43 in Proceedings of the Symposium on the Imported Fire Ant, June 7-10, 1981, Atlanta, Georgia, S. L. Battenfield, ed. Washington, D.C.: US Department of Agriculture.

Holt, J. S., S. B. Powles and J. A. M. Holtum. 1993. Mechanisms and agronomic aspects of herbicide resistance. Ann. Rev. Plant Physiol. Plant Molec. Biol. 44:203–229.

Hoy, M. A., R. D. Gaskalla, J. L. Capinera, and C. N. Keierleber. 1997. Laboratory containment of transgenic arthropods. Am. Entomol. 43:206-209, 255–256.

IANR (University of Nebraska Institute of Agriculture and Natural Resources). 1994. From Lab to Label. [Online]. Available: HtmlResAnchor http://ianrwww.unl.edu/ianr/pat/hortpara/sld031.htm (9/22/98).

Ignoffo, C. M., and C. Garcia. 1992. Combinations of environmental factors and simulated sunlight affecting activity of inclusion bodies of the *Heliothis* (Lepidoptera: Noctuidae) nucleopolyhedrosis virus. Env. Ent. 21:210–213.

Jansson, R. K., W. R. Halliday, and J. A. Argentine. 1997. Evaluation of miniature and high-volume bioassays for screening insecticides. J. Ec. Ent. 90:1500–1507.

Jaques, R .P. and Laing, D. R. 1989. Effectiveness of microbial and chemical insecticides in control of the Colorado potato beetle (Coleoptera: Chrysomelidae) on potatoes and tomatoes. Can. Ent. 121:1123–1131.

Jordan, N. R. 1993. Prospects for weed control through crop interference. Ec. App. 3:84–91.

Kaiser, K. 1978. The rise and fall of Mirex. Env. Sci. Tech. 12:520–528.

Kareiva, P. 1993. Agriculture – transgenic plants on trial Nature. 363:580–581.

Kareiva, P. 1996. Contributions of ecology to biological control. Ecology. 77:1963–1964.

Kerremans, P. and G. Franz. 1995. Isolation and cytogenetic analyses of genetic sexing strains for the medfly, *Ceratitis capitata.* Theor. Appl. Genet. 9:255–261.

Kidwell, M. G. and A. R. Wattam. 1998. An important step forward in the genetic manipulation of mosquito vectors of human disease. Proc. Natl. Acad. Sci. USA 95:3349–3350.

Kidwell, M. G., and J. M. C. Ribeiro. 1992. Can transposable elements be used to drive disease refractoriness genes into vector populations? Parasitol. Today 8:325–329.

Kline, W., K. Holmstrom, and S. Walker. 1996. ECB and CEW Mapping for Vegetable Crop IPM. Rutgers Cooperative Extension/Grant F. Walton Center for Remote Sensing & Spatial Analysis is now Utilizing GIS/GPS/RS Technologies for Precision Agriculture. [Online]. Available: http://deathstar.rutgers.edu/projects/gps/gps.html#IPM

Klinger, T., P. E. Arriola and N. C. Ellstrand. 1992. Crop-weed hybridization in radish (*Raphanus sativus*): effects of distance and population size. Am. J. Botany 79(12):1431–1435.

Kogan, M. 1982. Plant resistance in pest management. Pp. 93-134 in Introduction to Insect Pest Management, R. L. Metcalf and W. H. Luckman, eds. New York: John Wiley and Sons.

Kommedahl, T., and Mew, I. 1975. Biocontrol of corn root infection in the field by seed treatment with antagonists. Phytopathol. 65:296–300.

Kooman P.L., M. Fahem, P. Tegera, and A. J. Haverkort. 1996. Effects of climate on different potato genotypes to dry matter allocation and duration of the growth cycle. Eur. J. of Agr. 5(3-4):207–217.

Korzybski, T., Z. Kowszyk-Gindifer, and W. Kurylowicz. 1978. Section C: Antibiotics isolated from the genus *Bacillus* (Bacillaceae). In Antibiotics - Origin, Nature and Properties, Volume III. Washington, D.C.: American Society for Microbiology. .

Krafsur, E. S., C. J. Whitten, and J. E. Novy. 1987. Screwworm eradication in North and Central-America. Parasitology Today 35:131–137.

Krimsky, S.,and R. Wrubel. 1996 Agricultural Biotechnology and the Environment. Chicago, Ill: University of Chicago Press.

Krysan, J. L., D. E. Foster, T. F. Branson, K. R. Ostlie, and W. W. Crenshaw. 1986. Two years before the hatch: rootworms adapt to crop rotation. Bull. Entomol. Soc. Am. 32:250–253.

Lamoreaux, R. J. 1990. Herbicide discovery and development: Emphasis on groundwater protection. Crop Protection 13:483–487.

Larson, R. A. 1988. The antioxidants of higher plants. Phytochemistry. 27:969–978.

Lass W. L., and R. H. Callihan. 1997. The effect of phenological stage on detectability of yellow hawkweed (*Hieracium pratense*) and oxeye daisy (*Chysanthemum leucanthemum*) with remote multispectral digital imagery. Weed Technol. 11:248–256.

Lass, L. W., H. W. Carson, and R. H. Callihan. 1996. Detection of yellow starthistle with high resolution multispectral digital images. Weed Technol. 10:466–474.

Leason, M., Cunfliffe, D., Parkin, D., Lea, P.J., and Miflin, B.J. 1982. Inhibition of pea leaf glutamine synthetase by methionine sulphoximine, phosphinothricin and other glutamate analogues. Phytochemistry 21:855–857.

Liedtke, J. 1997. High Resolution Imagery Will Enhance the Effectiveness of Precision Farming. Ag Innovator. [Online]. Available HtmlResAnchor http://www.agriculture.com/01/11/97.

Lyttle, M. H. 1995. Combinatorial chemistry: A conservative perspective. Drug Development Research. 35:230–236.

Lyttle, M. H., M. D. Hocker, U. C. Hui, C. G. Caldwell, D. T. Aaron, A. Engquist-Goldstein, J. E. Flatgaard, and K. E. Bauer. 1994 Isozyme specific glutathione-S-transferase inhibitors: Design and synthesis. J. Med. Chem. 37:189–194.

MacIntosh, S. C., G. M. Kishore, F. J. Perlak, P. G. Marrone, T. B. Stone, S. R. Sims, and R. L. Fuchs. 1990. Potentiation of *Bacillus thuringiensis* insecticidal activity by serine protease inhibitors. J. Agric. and Food Chem. 38:1145–1152.

Madronich, S. 1993. The atmosphere and UVB radiation at ground level. Pp. 1-39 in Environmental UVB Photobiology, A. R. Young. New York: Plenum Press.

Manker, D. C., R. L. Starnes, M. G. Beccio, W. D. Lidster, S. C. MacIntosh, J. K. Swank, D. R. Jiménez, and D. R. Edwards. 1995. A synergistic metabolite from *Bacillus thuringiensis* with potential for use in agricultural applications. Invited Symposia Pacifichem '95 Honolulu, Hawaii, December 1995. Washington, D.C.: American Chemical Society.

Manker, D.C., B. D. Lidster, S. C. MacIntosh, and R. L. Starnes. 1994. Potentiator of *Bacillus* pesticidal activity, US Patent Application Serial # 08/146,852, PCT # WO 9409630. (May 11)

Marrone, P. G. 1999. Microbial pesticides and natural products as alternatives. Outlook on Ag. 28(3):149–154.\par Maxwell, B. D., M. L. Roush, and S. R. Radosevich. 1990. Predicting the evolution and dynamics of herbicide resistance in weed populations. Weed Tech: 4(1):2–13.

McCloud, E. S., and M. R. Berenbaum. 1994. Stratospheric ozone depletion and plant-insect interactions: effects of UV-B radiation on foliage quality of *Citrus jambhiri* for *Trichoplusia ni*. J. Chem. Ecol. 20:525–539.

McCoy, C. W. 1990. Ventomogenous fungi as microbial pesticides. Pp.139-159 in New Directions In Biological Control: Alternatives For Suppressing Agricultural Pests And Diseases , R. R. Basker and P. E. Dunn, eds. Proceedings of a UCLA Symposia on Molecular and Cellular Biology, Series v. 112.

McGaughey, W., F. Gould, and W. Gelernter. 1998. Bt resistance management. Nature Biotech. 16:144–146.

McKenzie, J. A. and A. Y. Game. 1987. Diazinon resistance in *Lucilia cuprina*: Mapping of a fitness modifier. Heredity 59:371–381.

Meister, G. A. and T. A. Grigliatti. 1994. Rapid spread of a P element/Adh gene construct through experimental populations of *Drosophila melaogaster*. Genome 36:1169–1175.

Metcalf, R. L. 1989. Insect resistance to insecticides. Pest.. Sci. 26:333–358.

Milner, J. L., L. Silo-Suh, J. C. Lee, H. He, J. Clardy, J. Handelsman. 1996. Production of kanosamine by *Bacillus cereus* UW85. Appl. Env.. Microb. 62:3061–3065.

Moar, W. J., and Trumble, J. T. 1987. Biologically derived insecticides for use against beet armyworm. Calif. Agr.. Nov-Dec: 13-15.

Mochida, O. 1992. Management of insect pests of rice in 2000. Pp. 67-81 in Pest Management and the Environment in 2000, A. Aziz, S.A. Kadir, and H. S. Barlow, eds. Seminar Kuala Lampur, Malaysia.

Mooney, H. A., and W. Koch. 1994. The impact of rising CO2 concentrations on the terrestrial biosphere. Ambio 23:74–76.

Moore, J. 1991. Insecticide program stresses B.t. products. Am. Veg. Grower. Jan:32–34.

Moulin, A. P. 1996. Decision support systems and computer models, new tools for contemporary agriculture. Can. J. Pl. Sci. 76:1.

Mulrooney, J. E., K. D. Howard, J. E. Hanks, and R. G. Jones. 1997. Application of ultra-low-volume malathion by air-assisted ground sprayer for boll weevil (Coleoptera: Curculionidae) control. J. Econ. Ent.. 90:639–645.

Murdoch, W. W., J.D. Reeve, and C.B.Huffaker. 1984. Biological control of olive scale and its relevance to ecological theory. Am. Nat. 123:371–392.

Murdoch, W.W., R.F. Luck, and S.L. Swarbrick. 1995. Regulation of an insect population under biological control. Ecology 76:206–217.

Murdoch, W.W. and C.J. Briggs. 1996. Theory for biological control: recent developments. Ecology 77:2001–2013.

Myers, J. H., A. Savoie, and E. Vanreden. 1998. Eradication and pest management. Ann Rev. Ent. 43:471–491.

NASA, 1998. Data And Information Services For Global Change Research. Sioux Falls, S.Dak.: Earth Observing System (EOS), EROS Data Center.

Niemela, P. and W. J. Mattson. 1996. Invasion of North American forests by European phytophagous insects—Legacy of the European crucible. Bioscience. 46:741–753.

NRC (National Research Council). 1986. Pp. 100-110 in Pesticide Resistance: Strategies and Tactics for Management. Washington, DC: National Academy Press

NRC (National Research Council). 1991. Neem: A Tree for Solving Global Problems. Washington, DC: National Academy Press.

NRC (National Research Council). 1996. Ecologically Based Pest Management: New Solutions for a New Century. Washington, DC: National Academy Press

NRC (National Research Council). 1997. Precision Agriculture in the 21st Century: Geospatial and Information Technologies in Crop Management. Washington, DC: National Academy Press.

Olson, K. E., S. Higgs, P. J. Gaines, A. M. Powers, B. S. Davis, K. I. Kamrud, J. O. Carlson, C. D. Blair, and B. J. Beaty. 1996. Genetically engineered resistance to dengue-2 virus transmission in mosquitoes. Science 272:884–886.

Orth, A. B., H. Teramura, and H. D. Sisler. 1990. Effects of ultraviolet-B radiation of fungal disease development in *Cucumis sativus*. Am. J. of Botany. 77:1188–1192.

Osburn, R.M., Milner, J.L., Oplinger, E.S., Smith, R.S., Handelsman, J. 1995. Effect of *Bacillus cereus* UW85 on the yield of soybean at two sites in Wisconsin. Plant Dis. 79:551–556.

Otvos, I.S., J. C. Cunningham, and W. J. Knapp. 1989. Aerial application of two baculoviruses against the western spruce budworm, *Choristoneura occidentalis* Freeman (Lepidoptera: Tortricidae) in British Columbia. Can. Ent.. 121:209–217.

Ozkan, E, ed. 1997. Equipment Developments. Chemical Application Technology. Newsletter 1(1): 6p.[Online]. Available: http://www.ag.ohio-state.edu/~catnews/catequi.htm.

Paice, M. E. R, P .C. H. Miller, and W. Day. 1996. Control requirements for spatially selective herbicide sprayers. Comp. and Elect. in Ag. 14:163–177.

Panagopoulus, I., J. F. Borman, and L. O. Bjorn. 1991. Response of sugar beet plants to ultraviolet-B (280-320nm) radiation and *Cercospora* leaf spot disease. Physiologia Plantarum 84:140–145.

PANNA (Pesticide Action Network North America). 1998. PANUPS: Pesticide Updates – Asia, October 16, 1998. Adapted from Agrow: World Crop Protection News, 6/26/98 and 9/18/98.

Parlow, J. J., D. A. Mischke, and S. S. Woodard. 1997. Utility of complementary molecular reactivity and molecular recognition (CMR/R) technology and polymer-supported reagents in the solution-phase synthesis of heterocyclic carboxamides. J. Org. Chem. 62:5908–5919.

Penuelas, J., and M. Estiarte. 1998. Can elevated CO_2 affect secondary metabolism and ecosystem function? Trends Eco. Evo. 13:20–24.

Peterson, J. J. 1978. Development of resistance by the southern house mosquito to the parasitic nematode *Romanomermis culicivorax*. Env. Ent. 7:518–520.

Pettigrew, M. M., and S. L. O'Neill. 1997. Control of vector-borne disease by genetic manipulation of insect populations: technological requirements and research priorities. Aust. J. of Ent.. 36:309–317.

Peypoux, F., F. Besson, G. Michel, L. Delcambe, and B. Das, B. 1978. Structure de l'iturine C de *Bacillus subtilis*. Tetrahedron 34:1147–1152.

Pfadt, R. 1971. Fundamentals of Applied Entomology. New York: The Macmillan Company.

Pfeifer, T. A. and T. A. Grigliatti. 1996. Future perspectives on inspect pest management: engineering the pest. J. Invert. Pathol. 67:109–119.

Pimentel, D. 1995. Amounts of pesticides reaching target pests: environmental impacts and ethics. J. Ag. Env. Ethics. 8:17–29.

Pimentel, D., L. Lach, R. Zuniga, and D. Morrison. 2000. Environmental and economic costs of nonindignous species in the United States. Bioscience: 50(1):53–65.

Pimm, S. L., G. J. Russell, J.L. Gittleman, T. M. Brooks. 1995. The future of biodiversity. Science. 269(5222):347–350.

Plant, R.E. and N.D. Stone. 1991. Knowledge-based systems in agriculture. Toronto, Ontario: McGraw-Hill, Inc.

Podgwaite, J. D., R. C. Reardon, D. M. Kolodny-Hirsch, and G. S. Walton. 1991. Efficacy of ground application of the gypsy moth (Lepidoptera: Lymantriidae) nucleopolyhedrosis virus product, Gypchek. J. Econ. Ent. 84:440–444.

Porter, J. H., M. L. Parry, and T. R. Carter. 1991. The potential effects of climatic change on agricultural insect pests. Agric. Forest Meterol. 57:221–240.

Prout, T. 1978. The joint effects of the release of sterile males and immigration of fertilized females on a density regulated population. Theor. Popul. Biol. 13:40–71.

Purvis, A., P.M. Agapow, J.L. Gittleman, and G. M. Mace. 2000. Nonrandom extinction and the loss of evolutionary history. Science. 288(5464): 328-330.

Quarles, William. 1996. New Microbial Pesticides for IPM. IPM Pract.18(8), August: 5-10.

Raffa, K. F. 1989. Genetic engineering of trees to enhance resistance to insects. BioScience 39:524–534.

Raymond, M., A. Callaghan, P. Fort and N. Pasteur. 1991. Worldwide migration of amplified insecticide resistance genes in mosquitoes. Nature 350:151–153.

Ribeiro, J. M. C. and M. G. Kidwell. 1994. Transposable elements as population drive mechanisms: specification of critical parameter values. J. Med. Entomol. 31:10–16.

Robertson, H. M, and Lampe, D. J., 1995. Distribution of transposable elements in arthropods. Annu. Rev. Entomol. 40:333–357.

Robertson, S. T. 1980. History of gas odorization. Pp. 1-5(a). in: Odorization, F. H. Suchomel and J. W. Weatherly III , eds.. Chicago, Ill:. Institute of Gas Technologies.

Rodgers, P.B. 1993. Potential of biopesticides in agriculture. Pestic. Sci. 39:117-129.

Rogers, H. H. and R. C. Dahlman. 1993. Crop responses to CO_2 enrichment. Vegetatio 104/105:117–131.

Ros, J.,and M. Tevini. 1995. Interaction of UV-radiation and IAA during growth of seedlings and hypocotyl segments of sunflower. J. Plant Physiol. 146:295–302.

Rosenheim, J. A. 1998. Higher-order predators and the regulation of insect herbivore populations. Ann. Rev. Entomol. 43:421–447.

Roush, R. T. and W. M. Tingey. 1991. Evolution and management of resistance in the Colorado potato beetle, *Leptinotarsa decemlineata*. Resistance '91—Achievements and Developments in Combating Pesticide Resistance, Rothamsted Experimental Station, 15–17 July.

Roush, M. L., S. R. Radosevich, and B. D. Maxwell. 1990. Future outlook for herbicide-resistance research. Weed Tech.: 4(1):208–214.

Sammons, A. E., C. R. Edwards, L. W. Bledsoe, P. J. Boeve, and J. J. Stuart. 1997. Behavioral and feeding assays reveal a western corn rootworm (coleoptera, chrysomelidae) variant that is attracted to soybean. Env. Ent. 25(6):1336–1342.

Sasaki, A., and H. C. J. Godfray. 1999. A model for coevolution of resistance and virulence in coupled host-parasitoid interactions. P Roy Soc Lond B Bio 266(1418): 455–463.

Schoper, J.B., R. J. Lambert, and B. L. Vasilas. 1987. Pollen viability, pollen shedding, and combining ability for tassel heat tolerance in maize. Crop Sci. 27:27–31.

Schwinn, F. J. and H. V. Morton. 1990. Antiresistance strategies: design and implementation in practice. Pp. 170-183 in: Managing Resistance to Agrochemicals From Fundamental Research to Practical Strategies, M. B. Green, H. M. LeBaron and W. K. Moberg, eds. Washington, DC: American Chemical Society.

Sears, M. K., Jaques, R. P. and Laing, J. E. 1983. Utilization of action thresholds for microbial and chemical control of lepidopterous pests (Lepidoptera: Noctuidae, Pieridae) on cabbage. J. Econ. Ent.. 76:368–374.

Serebovsky, A. S. 1940. On the possibility of a new method for the control of insect pests. Zool. Zh. 19:618–630.

Sharga, B. M. 1997. *Bacillus isolates* as potential biocontrol agents against chocolate spot on Faba beans. Can. J. Microbio. 43:915–924.

Shelton, A. M. and J. A. Wyman. 1991. Insecticide Resistance of Diamondback Moth (Lepidoptera: Plutellidae) in North America. Proc. Second Internatl. Diamondback Moth Workshop. Taiwan.

Shemshedini, L. and T. G. Wilson. 1990. Resistance to juvenile hormone and an insect growth regulator in *Drosophila* is associated with an altered cytosolic juvenile hormone-binding protein. Proc. Natl. Acad. Sci. USA. 87:2072–2076.

Sholberg, P.L., Marchi, A., Berchard, J. 1995 Biocontrol of postharvest diseases of apple using *Bacillus spp.* isolated from stored apples. Can. J. Microbiol. 41:247–252.

Silo-Suh, L.A., B. Lethbridge, S. J. Raffel, H. He, J. Clardy, and J. Handelsman. 1994 Biological activities of two fungistatic antibiotics produced by *Bacillus cereus* UW85. Appl. Envi.. Microb. 60:2023–2030.

Sinkins, S. P., C. F. Curtis, and S. L. O Neill. 1997. The potential use of symbionts to manipulate arthropod populations. In Influential Passengers: Inherited Microorganisms And Arthropod Reproduction, S. L. O Neill, A. A. Hoffman, and J. H. Werren, eds. New York: Oxford Univ. Press.

Sinkins, S. P., H. R. Braig, and S. L. O'Neill. 1995. *Wolbachia* superinfections and the expression of cytoplasmic incompatibility. Proc. Roy. Soc. Lond B Bio. 261(1362):325–330.

Slansky, F. S., Jr. 1992. Allelochemical-nutrient interactions in herbivore nutritional ecology. Pp. 135-174 in Herbivores: Their Interactions with Secondary Plant Metabolites, Volume 2, Ecological and Evolutionary Processes, G. Rosenthal and M. Berenbaum, eds. New York: Academic Press.

Smith, K. P., M. J. Havey, and J. Handelsman. 1993. Suppression of cottony leak of cucumber with *Bacillus cereus* strain UW85. Plant Dis. 77:139–142.

Smits, P. H., I. P. Rietstra, and J. M. Vlak. 1988. Influence of application techniques on the control of beet armyworm larvae (Lepidoptera: Noctuidae) with nuclear polyhedrosis virus. J. Econ. Ent. 81:470–475.

Smits, P. H., M. C. van Velden, M. van de Vrie, and J. M. Vlak. 1987b. Feeding and dispersion of *Spodoptera exigua* larvae and its relevance for control with a nuclear polyhedrosis virus. Ent. Exp. Appl. 43:67–72.

Smits, P. H., M. van de Vrie, and J. M. Vlak. 1987a. Nuclear polyhedrosis virus for control of *Spodoptera exigua* larvae on glasshouse crops. Ent.. Exp. Appl., 43:73-80.

Spielman, A. 1994. Why entomological antimalaria research should not focus on transgenic mosquitoes. Parasitol. Today 10:374–376.

Stabb, E. V., L. M. Jacobson, and J. Handelsman. 1994. Zwittermicin A producing strains of Bacillus cereus from diverse soils. Appl. Env. Microb. 60:4404–4412.

Starnes, R. L., C. L. Liu, and P. G. Marrone. 1993. History, use and future of microbial insecticides. Am. Ent. 39:83–91.

Stetter, J. 1998. Pesticide innovation: trends in research and development. Med. Fac. Landbouwww. Univ. Gent.: 63(2):135–164.

Stewart, P A., and A. Sorensen. 2000. Federal uncertainty or inconsistency? Releasing the new agricultural-environmental biotechnology into the fields. Politics and Life Sciences 19(1): in press.

Sun, C. N., Y. C. Tsai and F. M. Chiang. 1992. Resistance in the diamondback moth to pyrethroids and benzoylphenylureas. Pp. 149-167 in Molecular Mechanisms of Insecticide Resistance: Diversity Among Insects, C. A. Mullin and J. G. Scott, eds., Washington, DC: American Chemical Society

Swanson, M. B., G. A. Davis, L. E. Kincaid, T. Schultz, J. E. Bartimess, S. Jones, and E. L. George. 1997. A screening method for ranking and scoring chemicals by potential human health and environmental impacts. Env. Tox and Chem. 16:372–383.

Taylor, A. G., and G. E. Harman. . 1990. Concepts and technologies of selected seed treatments. Ann. Rev. Phyto. 28:321–339.

Teramura, A. H. 1983. Effects of ultraviolet-B radiation on the growth and yield of crop plants. Phys. Plant. 58:415–427.

Teramura, A.H., and J. H. Sullivan. 1991. Potential impacts of increased solar UV-B on global plant productivity. Pp. 625-634 in Photobiology, E. Riklis, ed. New York: Plenum Press.

Tette J. P. and B. J. Jacobsen. 1992. Biologically intensive pest management in the tree fruit system. Pp. 57-83 in Food, Crop Pests and the Environment, the Need and Potential for Biologically Intensive Integrated Pest Management, Zalom F.G. and Fry, W. E., eds., St. Paul, Minn.: APS Press.

Tevini, M. 1993. Effects of enhanced UV-B radiation on terrestrial plants. Pp. 125-153 in UV-B Radiation and Ozone Depletion: Effects on Humans, Animals, Plants, Microorganisms, and Materials, M. Tevini, ed. Boca Raton, Fla: Lewis Publications.

Tevini, M., and A. H. Teramura. 1989. UV-B effects on terrestrial plants. Photochem. Photobiol. 50:479–487.

Thirumalachar, M., and M O'Brien. 1977. Suppression of charcoal rot in potato with a bacterial antagonist. Plant Dis. Reptr. 61(7):543–546.

Thomas, D.A., C. A. Donnelly, R. J. Wood, and L. S. Alphey. 2000. Insect population control using a dominant, repressible, lethal genetic system. Science 287(5462):2474–2476.

Thompson, G.D., J. D. Busacca, O. K. Jantz, and H. A. Kirst. 1994. Spinosyns: an overview of new natural insect management systems. Proceedings of the ESA Annual Meeting, Dallas, 13-17 December 1994.

Trumble, J .T., and B. Alvarado-Rodriguez. 1993. Development and economic evaluation of an IPM program for fresh market tomato production in Mexico. Agric. Ecosys.and Env. 43:267–284.

Trumble, J. T. 1985. Integrated pest management of *Liriomyza trifolii*: influence of avermectin, cyromazine, and methomyl on leafminer ecology in celery. Ag. Ecosys. and Env. 12:181–188.

Trumble, J. T. 1989. Leafminer and beet armyworm control on celery. Final Report for 1988-9, Project #9, University of California at Riverside.

Trumble, J. T. 1990. Vegetable insect control with minimal use of insecticides. Hort. Sci. 25:159–164.

Trumble, J. T. 1991. California Tomato Board Layman's Summary 13-O.

Tumlinson, J. H., C. M. Demoraes, W. J. Lewis, P. W. Pare, and H. T. Alborn. 1998. Herbivore-infested plants selectively attract parasitioids. Nature 393(6685):570–573.

Turchin, P. 1999. Population regulation: a synthetic view. Oikos. 84:153–159.

UC Davis (University of California, Davis). 1997. Proceedings of the Workshop on Remote Sensing for Agriculture in the 21st century, October 1996. Available. [Online]: http://cstars.ucdavis.edu/proj/ag-21/ag21cen.html.

Ustin, S. L., D. A. Roberts, and Q. J. Hart. 1997. Seasonal vegetation patterns in a California coastal savanna derived from advanced visible/infrared imaging spectrometer (aviris) data. In Remote Sensing Change Detection: Environmental Monitoring Applications and Methods, Elvidge, C.D., and Lunetta, R., eds. Ann Arbor, Mich: Ann Arbor Press.

Utkhede, R., and J. Rahe. 1983. Interactions of antagonist and pathogen in biological control of onion white rot. Phytopath. 73:890–893.

Vanderplank, F. L. 1944. Hybridization between *Glossina* species and suggested new method for control of certain species of tsetse. Nature 154:607–608.

Vargas-Teran, M. , B. S. Hursey, and E. P. Cunningham. 1994. Eradication of the screwworm from Libya using the sterile insect technique. Parasitol. Today 10:119–122.

Vitousek, P. M., C. M. D'Antonio, L. L. Loope, and R. Westbrooks. 1996. Biological invasions as global environmental change. Am. Sci. 84:468–478.

Wagner, P. 1993. Techniques of representing knowledge in knowledge-based systems. Ag. Sys. 41:53–76.

Warr, W. A. 1997a. Combinatorial chemistry and molecular diversity: An overview. J. Chem. Inf. Comput. Sci. 37:134–140.

Warr, W. A. 1997b. Commercial software systems for diversity analysis. Persp. Drug Disc. Design..7/8:115–130.

Warwick, S. I. 1991. Herbicide resistance in weedy plants: physiology and population biology. Ann. Rev. Ecol. Syst. 22:95-114.

Watson, R. T., M. C. Zinyowera, R. H. Moss, eds. 1996. Climate Change 1995. Contribution of Working Group II to the Second Assessment Report of the Intergovernmental Panel on Climate Change. Cambridge University Press.

Weller, M. 1988. Biological control of soilborne plant pathogens in the rhizosphere with bacteria. Ann. Rev. Phyto. 26:379–407.

Werren, J. H. 1997. Biology of *Wolbachia*. Ann. Rev. Ent. 42:587–609.

Westbrooks, R. G. 1998. Invasive Plants. Changing the Landscape of American: Fact Book. Washington, DC. Federal Interagency Committee for the Management of Noxious and Exotic Weeds.

White, A. D., and H. D. Coble. 1997. Validation of HERB for use in peanut (*Arachis hypogaea*). Weed Technol. 11:573–579.

Whitten, M. J., and G. G. Foster. 1975. Genetical methods of pest control. Ann. Rev. Ent. 20:461–476.

Williams, C. M. 1967. Third-generation pesticides. Sci. Am. 217:13–17.

Wilson, E. O. 1993. Is humanity suicidal. Biosystems. 31(2-3):235–242.

Wood, M. 1992. Microbes blow those hornworms away. Ag. Res. June: 4–7.

Yamamoto, H. 1985. Development of validamycin, its controlling effect on rice sheath blight. Jpn. Pest. Inf. 17:17.

Yamanaka, S., M. Hashimoto, M. Tobe, K. Kobayasii, J. Sekizawa, and M. Nishimura. 1990. A simple method for screening assessment of acute toxicity of chemicals. Archives of Tox. 64:262–268.

Zalom, F. G., and Fry, W. E. 1992. Biologically intensive IPM for vegetables. Pp. 107-165 in Food, Crop Pests and the Environment, the Need and Potential for Biologically Intensive Integrated Pest Management, F. G. Zalom, and W. E. Fry, eds., St. Paul, Minn.: APS Press.

5

Evaluation of Pest-Control Strategies

In an effort to identify the circumstances under which chemical pesticides might be required in future pest management, the committee received input from experts during the information-gathering phase of its study. Perspectives were received in the form of invited presentations, written input, and informal responses from university and industrial scientists, pest-management practitioners, policy analysts, and other people with expertise in current practices and impacts of pesticide use. On the basis of input from workshops and other information sources, the committee concluded that the diversity of the US agricultural enterprise and other sectors of pesticide use makes generalizing virtually impossible.

Pesticides are used in a multiplicity of settings—agricultural crop and livestock production, silviculture, homes and lawns, schools, golf courses, rights of way, wildlands, and others. Pest managers use an array of chemical pesticides, cultural practices, biological control, and genetically modified organisms to control a broad spectrum of pest species. Moreover, even in a single production system, the utility of chemical pesticides can vary. Although generalizing is difficult, experts who provided input to the committee agreed that pest-management practices can improve in all managed ecosystems. The intent here is to provide some insights on circumstances in which pesticides are in use and to illuminate the variation in pest-management practices in some managed and natural ecosystems.

PESTICIDE USE IN MANAGED AND NATURAL ECOSYSTEMS

For purposes of classification, the committee used both biological and cultural criteria to recognize six major classes of agroecosystems. In the context of agronomic crop production, biological constraints differ between perennial systems—which include silviculture, orchards and vineyards, forages and turf— and annual systems—which include row crops, vegetables, and cereals. Stored-products systems have unique attributes; climate and temperature are factors for all of the systems, but manifest themselves in unique ways in that stored-products systems are often indoors and spatially constrained. Animal-production systems (including those for swine, ruminants, poultry, such nonfood animals as horses and llamas, and aquaculture) present a different set of biological constraints. Urban pest-management systems—indoors for vermin, structural pests, and companion animals; and outdoors for lawn, garden, golf courses, ornamentals, rights of way, and nuisance insects—present cultural and biological constraints that differentiate the process of management from that in other systems. Finally, wildland systems (including rangelands, forests, conservation holdings, and aquatic systems) often present species conservation priorities that make nontarget effects important in pest-management strategies.

Perennial Cropping Systems

The longevity of perennial crop plants (particularly trees) creates a distinctive challenge in that both time and vegetational structure promote biological diversity (Lawton and Gaston 1989). Thus, management decisions in these systems targeted at particular pests often have community-level implications. For example, use of conventional pesticides for control of major pests can preclude adoption of nonchemical alternative methods of controlling other pests (Brunner 1994); use of conventional pesticides remains heavy in these agroecosystems. In 1995, over 90% of acres on which the five most widely grown fruit crops (grapes, oranges, apples, grapefruits, and peaches) were grown were treated with at least one pesticide, and most of the acres received herbicide, fungicide, and insecticide treatment (Economic Research Service 1995).

Explanations for the heavy reliance on conventional pesticides are numerous and include shortages of trained consultants, institutional limits on information transfer, and unavailability of pesticides with appropriate specificity (Brunner 1994). Cultural factors enter in as well; because of consumer aesthetic concerns, crops grown for fresh market receive more intensive pesticide use to ensure quality. Among perennial crops, regional and seasonal variation affects the intensity of pesticide use. Disease, weed, and arthropod pest complexes vary with locality, climate, and cultivars grown. For example, in the 1994–1995 growing season, while insecticides were applied to 96% of grape acreage grown in Michigan, they were applied to only 17% of grape acreage in Washington (Economic Research Service 1997). Many of the pesticides traditionally used in tree crop systems raise concerns regarding human and environmental exposure and have been either canceled or substantially restricted (for example, phosalone, ethyl parathion, daminozide, ethion, EBDC, and cyhexatin); positive effects of regulatory action are evident in reduced residue detection (chapter 3). The cancellations, however, have increased reliance on fewer products and raised concerns about the availability and competitiveness of alternatives (CAST 1999) and about increased rates of resistance acquisition. A subject of lingering concern, arising at least in part out of failure to enforce laws, is continuing exposure of applicators and farmworkers to remaining traditional chemical pesticides. Institutional and regulatory barriers constrain adoption of nonchemical alternatives under many circumstances (Brunner 1994).

Annual Cropping Systems

Because of the vast acreage dedicated to annual crops, variability—in regional, seasonal, climatic conditions and in cultivar availability—characterizes production in these agroecosystems. Consequently, pest-management practices for these crops reflect variability, as seen in Tables 5-1 and 5-2.

Corn is the most widely planted crop in the United States and production is overall chemically intensive; in 1995, herbicide was applied to 98% of corn acreage in 10 states surveyed, amounting to 55,850,000 acres (Economic Research Service 1997). Insecticide use in that year, however, was restricted to 26% of the acreage. Sweet corn, however, which is grown for fresh market, received insecticides in 41% of the acreage in Washington and 82% in Illinois. Soybean crops also are widely planted, in diverse soil types, climate regimes, and biotic communities. Herbicide use after planting is high in soybeans, in general, and rose from 52% of acreage in 1990 to 74% in 1995. The diversity of weeds in the weed seed bank, particularly across the diverse acreages planted in soybean (over 45 million acres in major producing states), presents opportunities for weed species shifts and a challenge to single-strategy management plans (Gunsolus et al., 2000).

In contrast with corn and soybean crops, wheat, although one of the largest field crops, currently demands considerably less pesticide use. In 1994, although wheat constituted 29% of all surveyed acreage, it represented only 4% of pesticide use. Season profoundly affects herbicide use in wheat; winter wheat can establish itself over the fall and winter and compete well with spring weeds, but spring wheat is often treated with herbicide to provide seedlings with a competitive edge against weeds. Cotton, another major crop, remains pesticide-intensive at least in part because of geographic and climatic factors. Winter survival of pests, particularly insects, tends to be higher in southern states, where cotton is extensively grown.

By virtue of volume alone, pest-management practices on field crops have a high potential for adverse environmental effects. Resistance is a major concern, at least in part because of the intensity of the selection pressure applied to pests. Nontarget effects can be considerable simply by virtue of volume. The fact that corn cultivars producing transgenic *Bacillus thuringiensis* endotoxin constitute over 20% of planted acres in 1999 has raised concerns that pollen shed by these plants and expressing the toxic protein can have unintended effects on nontarget butterflies on plants growing in the vicinity of cornfields (Losey et al. 1999). Roundup-Ready® soybeans were planted on more than 25 million acres in the United States in 1998 (K. Marshall, Industry Affairs Director, Monsanto, personal communication, August 8, 2000); there is concern that such herbicide-resistant varieties will not reduce weed control to one herbicide application because of a lack of residual weed control.

Although vegetable crops, like field crops, are annuals, characteristics of vegetable production systems resemble tree fruit crops in their diversity and in the cultural constraints on production. More than 60 types of vegetables are grown in the United States (Zehnder 1994); this biological diversity is accompanied by a diversity of crop-specific pest species. Regional, seasonal, and cultivar variation also contributes to the composition of the pest community. Potatoes, for example, differ in their response to defoliation by Colorado potato beetle, depending on the region in which they are grown (Zehnder and Evanylo 1989); composition of the potato pest fauna depends on region (Zalom and Fry 1992). Cultivar differences can also influence the efficacy of different pest-management approaches. For example, low-growing spinach varieties require greater amounts of pesticide for aphid control than do upright varieties because of inadequate coverage by the pesticide.

Thus, one barrier to adopting more nonchemical alternative management strategies is the idiosyncratic nature of each crop; integrated pest management (IPM) programs must be both crop-specific and region-specific. Box 5-1 provides an example of a weed management approach that includes the role of timing in effective weed control.

Broad-spectrum pesticides thus have considerable appeal to many vegetable growers (Gianessi 1993, Gooch et al. 1998). Yet another factor

TABLE 5-1 Pest Management Practices for Major Field Crops in Major Producing States, 1990-1997[a]

Crop	1990	1991	1992	1993	1994	1995	1996	1997
Wheat								
Planted area, 1,000 acres	33,600	26,950	30,400	31,550	29,750	28,840	26,810	29,900
Land receiving herbicides, %	38.2	30	35.2	44.5	49.9	59.3	54.6	46.9
Treatments, average no.	1.1	1.1	1.1	1.1	1.1	1.1	1.2	1.4
Ingredients, average no.	1.5	1.5	1.6	1.8	1.8	1.8	2	2
Acre-treatments, average no.	1.5	1.6	1.6	1.8	1.8	1.9	2	2.1
Amt. of herbicide applied, lb./acre	0.28	0.29	0.29	0.32	0.34	0.26	0.45	0.43
Before or at plant only, %	3.5	2.8	2.5	3.6	5	4.4	5.8	5.2
After plant only, %	33.7	25.8	31.9	39.6	42.9	53	45.2	36.9
Both, %	1	1.4	0.8	1.3	2	1.9	3.6	4.8
Land receiving insecticides, %	4.7	8.1	6.4	2.5	12.9	6.6	13.4	6.5
Treatments, average no.	1.1	1	1	1	1.1	1	1	1.2
Ingredients, average no.	1.1	1	1.1	1	1	1	1	1.2
Acre-treatments, average no.	1.1	1	1.1	1	1.1	1.1	1	1.3
Amt. of insecticide applied, lb./acre	0.46	0.23	0.35	0.27	0.4	0.37	0.38	0.49
Before or at plant only, %	0.2	0.4	na	0.1	0.5	0.2	1.4	0.1
After planting only, %	4.5	7.7	6.4	2.4	12.4	3.4	12	3.4
Spring and durum wheat								
Planted area, 1,000 acres	16,700	14,700	16,850	16,800	17,600	17,450	19,350	18,300
Land receiving herbicide, %s	92.3	93.5	90.7	95.2	96.1	94.6	82.9	81.7
Treatments, average no.	1.3	1.3	1.2	1.3	1.2	1.3	1.5	1.5
Ingredients, average no.	1.9	2.1	2.1	2.2	2.2	2.4	2.7	2.9
Acre-treatments, average no.	1.9	2.1	2.1	2.3	2.3	2.5	2.9	3.1

Amt. of herbicide applied, lbs./acre	0.59	0.53	0.53	0.53	0.55	0.57	0.74	0.79
Before or at plant only, %	1.1	3.1	5.8	4.3	4.2	1.9	0.7	4.4
After plant only, %	78.4	77.7	76.5	80.2	81	82.9	63.6	55
Both, %	12.8	12.7	8.4	10.7	10.9	9.8	19	22.3
Soybeans								
Planted area, 1,000 acres	39,500	42,050	41,350	42,500	43,750	45,150	45,950	49,250
Land receiving herbicide, %	95.8	96.8	98.2	97.5	98.4	97.6	97.4	97.7
Treatments, average no.	1.5	1.5	1.6	1.6	1.7	1.7	1.8	1.8
Ingredients, average no.	2.3	2.3	2.4	2.5	2.7	2.7	2.8	2.7
Acre-treatments, average no.	2.3	2.3	2.4	2.5	2.8	2.8	2.9	2.9
Amt. of herbicide applied, lbs./acre	1.42	1.32	1.14	1.12	1.17	1.1	1.26	1.26
Before or at plant only, %	44.2	39.1	35.9	27.7	28.1	23.4	20.2	16.4
After plant only, %	20.1	26.1	27.9	29.7	28.5	32.1	27.8	33.4
Both, %	31.5	31.5	34.4	35.1	41.6	42.2	49.4	47.9
Amount banded, %	13.2	11.8	11.5	9.5	8.4	8.5	6.7	8
Cotton[b]								
Planted area, 1,000 acre	9,730	10,860	10,200	10,360	10,023	11,650	10,025	9,265
Land receiving herbicides	94.7	91.5	90.6	91.9	93.6	97.1	92.8	96.1
Treatments, average no.	2.1	2.3	2.4	2.5	2.6	2.7	2.6	2.8
Ingredients, average no.	2.3	2.4	2.7	2.7	2.7	2.8	2.8	2.9
Acre-treatments, average no.	2.7	2.9	3.2	3.2	3.4	3.3	3.2	3.4
Amt. of herbicide applied, lbs./acre	1.81	2.07	2.08	2.04	2.33	2.07	1.91	1.98
Before or at plant only, %	57.7	52	49.1	45	41.2	46.3	36.4	34.2
After plant only, %	5.7	5.1	9.2	9.5	6.2	6.7	6.2	4
Both, %	31.3	34.5	32.5	37.5	46.1	44.1	50.2	57.9
Amount banded, %	42.1	42.5	43	44.6	42.9	47.4	48.7	48.6

TABLE 5-1 Continued

Crop	1990	1991	1992	1993	1994	1995	1996	1997
Land receiving insecticides, %	na	66.4	64.8	64.9	71	75.3	78.6	73.7
Treatments, average no.	na	3	4.5	4.9	5.7	6.1	4.4	4.9
Ingredients, average no.	na	2.3	3.2	3.4	3.5	3.8	3	2.8
Acre-treatments, average no.	na	3.7	6	6.6	7.6	7 .9	5.3	5.4
Amt. of insecticide applied, lbs./acre	na	1.49	3.7	3.96	2.47	2.35	1.8	2.33
Land receiving other pesticides, %	na	56.5	47.3	63.5	66.8	55.8	60	67.9
Treatments, average no.	na	2	1.8	1.7	2	2.1	1.9	1.8
Ingredients, average no.	na	2	2	1.9	2	2.1	2.4	2.2
Acre-treatments, average no.	na	2.4	2.3	2.2	2.6	2.7	2.8	2.5
Amt. of other pesticides applied, lbs./acre	na	1.59	2.29	1.79	1.71	2.44	2.15	1.59
Corn								
Planted area, 1,000 acre	58,800	60,350	62,850	57,350	62,500	55,850	61,500	62,150
Land receiving herbicides, %	94.8	95.5	96.9	97.5	97.9	97.5	93.3	96.7
Treatments, average no.	1.4	1.4	1.4	1.4	1.5	1.5	1.5	1.6
Ingredients, average no.	2.2	2.1	2.3	2.3	2.5	2.4	2.6	2.7
Acre-treatments, average no.	2.2	2.2	2.3	2.4	2.5	2.5	2.7	2.8
Amt. of herbicide applied, lbs./acre	3.25	2.97	2.91	2.94	2.8	2.77	2.85	2.77
Before or at plant only, %	39.3	38.4	33	34.7	29.4	30.4	23.4	22.4
After plant only, %	29.1	34.1	36.4	36.8	38.1	37.9	35.4	38.7
Both, %	26.4	23	27.2	25.6	30.2	29	34.5	35.6
Amount banded, %	12.8	13.6	15.2	13.7	13.7	11.6	11.3	10.2

Land receiving insecticides, %	32.3	30.5	28.6	28.2	26.6	26.1	29.1	30.4
Treatments, average no.	1.1	1.1	1.1	1.1	1.1	1.1	1.1	1.2
Ingredients, average no.	1.1	1.1	1.1	1	1.1	1.1	1.2	1.2
Acre-treatments, average no.	1.1	1.1	1.1	1.1	1.1	1.1	1.2	1.3
Amt. of insecticide applied, lbs./acre	1.18	1.14	0.97	0.94	0.82	0.75	0.67	0.7
Before or at plant only, %	25.8	22.7	22.5	22.3	18.5	18	18.5	19.1
After plant only, %	4.4	5.6	4.9	5.2	6.5	6.8	7.9	8.2
Both, %	2.1	2.1	1.3	0.7	1.5	1.3	2.7	3.1
Fall potatoes								
Planted area, 1,000 acre	605	620	586	616	640	625	641	609
Land receiving herbicides, %	93.8	90.6	93.1	91.3	91.5	93.6	91.4	88.2
Treatments, average no.	1.3	1.4	1.3	1.4	1.4	1.4	1.2	1.4
Ingredients, average no.	1.6	1.7	1.6	1.7	1.8	1.8	1.8	1.9
Acre-treatments, average no.	1.6	1.7	1.7	1.8	1.9	1.9	1.8	2
Amt. of herbicide applied, lbs./acre	2.18	2.31	1.88	2.13	2.49	2.53	2.58	2.31
Before or at plant only, %	22.5	16.3	18.1	18.3	16.6	11.9	25	15.9
After plant only, %	64.9	67.1	70	64.9	62.3	74.4	62.2	65.9
Both, %	6.4	7.2	5	8.1	12.6	5.3	4.2	6.4
Amount banded, %	4	5.2	1.7	1.9	2	1.4	0.2	0.3
Land receiving insecticides, %	86	92.9	87.6	86.3	82.7	84.6	91.5	90.9
Treatments, average no.	1.9	1.9	2	1.9	2.2	2.1	1.7	2.2
Ingredients, average no.	1.7	1.7	1.7	1.7	1.9	1.8	1.6	1.9
Acre-treatments, average no.	2	2	2.2	2	2.4	2.2	1.9	2.3
Amt. of insecticide applied, lbs./acre	3.62	2.89	3.13	2.89	3.7	2.91	2.15	3.08
Before or at plant only, %	24.6	20	23.2	22.6	25.3	19.7	17	20.2
After plant only, %	45	52.1	46.3	47.3	42	44.7	61.8	32.5
Both, %	16.4	20.8	18.1	16.4	15.4	20.2	12.7	38.2

TABLE 5-1 Continued

Crop	1990	1991	1992	1993	1994	1995	1996	1997
Land receiving fungicides, %	54.5	50.6	57.1	62.3	68.2	74.6	86.3	95.5
Treatments, average no.	2.6	2.5	3.1	3.1	3.5	4.8	4.2	5.6
.Ingredients, average no.	1.4	1.5	1.9	1.9	2.1	2.6	2.1	3.1
Acre-treatments, average no.	3.5	3.5	4.2	4	5	6.6	5.7	7.8
Amt. of fungicide applied, lbs./acre	3.19	3.21	3.69	3.65	4.95	5.92	5.03	6.6
Land receiving other pesticides, %	31.4	41	39.1	47.5	57.5	55.8	64.6	64.3
Treatments, average no.	1.5	1.5	1.6	1.4	1.5	1.4	1.4	1.4
Ingredients, average no.	1.2	1.3	1.4	1.2	1.3	1.3	1.3	1.2
Acre-treatments, average no.	1.5	1.5	1.7	1.4	1.5	1.6	1.5	1.4
Amt. of other pesticides applied, lbs./acre	99.8	88.6	120.6	131.5	152.1	133.4	178.5	132.1

[a]Represents planted area of corn (IL, IN, IA, MI, MN, MO, NE, OH, SD, & WI), soybeans (AR, IL, IN, IA, MN, MO, NE, & OH), cotton (AZ, CA, LA, MS, &TX), winter wheat (CO, KS, MT, NE, OK, SD, TX, & WA), spring wheat (MN, MT, & ND), durum wheat (ND), and fall potatoes (ID, ME, & WA), which are the surveyed states included in all years. For these crops, the area represented in 1997 was about 167 million acres, 75% of total planted acres of these crops.

[b]1990 survey for cotton collected only herbicide treatments.

Source: USDA, ERS, Cropping Practices Surveys, 1990-95 and ARMS, 1996-97.

promoting dependence on chemical pesticides is the nature of consumer expectations for vegetables for the fresh market; there is a considerable demand for damage-free produce. In the vegetable-processing industry, federal regulations permit only very low levels of contaminating material, such as insect parts. The US Food and Drug Administration (FDA) identifies acceptable levels for these contaminants—Defect Action Levels—and they are typically measured in either the number of foreign parts per weight of food item or by percentage of foreign contamination by weight of food product (FDA 1998). At the same time, consumer concern about pesticide residues in vegetables and fruits (NRC 1993) can restrict the utility of these chemical pesticides in the future, as might concern about worker exposures, given the nature of nonmechanized harvesting procedures in many cropping systems. Registrations of pesticides for use on vegetable crops are decreasing, in part because of the Federal Insecticide, Fungicide, and Rodenticide Act reregistration process; availability of either chemical or nonchemical alternatives is reduced for many crop systems.

Stored-Products Systems

Pest problems in stored products are influenced by a number of factors, including time in storage, grain temperature and moisture, and type of management (Kenkel et al. 1994). Moisture and temperature in turn are affected by locality; moisture at the time of harvest of wheat can vary by a factor of almost 2 (Hagstrum and Heid 1988). Problems are more likely to arise in the Southwest (Oklahoma and Texas) than in the relatively cool, dry Great Plains states (North and South Dakota and Montana) (Kenkel et al. 1994). Because chemical costs are low and alternative management practices few in number, use of chemicals predominates. Low cost also promotes scheduled fumigations irrespective of whether pest populations merit treatment. Accordingly, resistance has been a persistent problem, increasing in grain elevators, in flour mills and on farms (Beeman and Wright 1990). Many of the broad-spectrum chemicals on which stored product pest-management practices are based, such as phosphine, dichlorvos, methyl bromide, and ethyl dibromide are under review and, because of adverse health and environmental effects, are unlikely to be registered for many of their current uses (EPA 1999). The lack of ready alternatives, either chemical or nonchemical, is likely to present problems to the industry in the future.

TABLE 5-2 Fruit and Vegetable Acreage Treated with Pesticides, Major Produc

	1993		
	Herbicide	Insecticide	Fungicide
Fruit:			
Grapes, all types	64	66	93
Oranges	94	90	57
Apples, bearing	43	99	88
Grapefruit	93	93	85
Peaches, bearing	49	99	98
Prunes	40	93	84
Avocados	50	12	10
Pears	44	98	92
Cherries, sweet	45	94	87
Lemons	71	88	14
Cherries, tart	49	98	99
Plums	70	89	79
Olives	67	27	33
Nectarines	84	98	95
Blueberries	75	91	81

	Total Application, 1,000 lbs. (1997)			1,000s of Planted Acres	No. of States Surveyed
	Herbicide	Insecticide	Fungicide	(1997)	
Fruit:					
Grapes, all types	1306.4	3552.7	39875.6	893.6	6
Oranges	3399.3	47361.1	2088.6	832.9	2
Apples, bearing	710.6	9459.5	5170	350.8	10
Grapefruit	518.7	10604.5	1056.7	159	2
Peaches, bearing	160.2	2014.6	4376.5	135.9	9
Prunes	90.2	1220	476.2	100.5	1
Avocados	68.8	84.3	95.8	62.5	2
Pears	132.3	4655.9	1335.3	67.9	5
Cherries, sweet	73.5	1108.4	633.7	48	4
Lemons	142.5	6813.2	147	49	1
Cherries, tart	48.6	157.2	872.3	32.4	4
Plums	64.3	1141.8	360.3	44	1
Olives	66.2	162.1	95	37.4	1
Nectarines	75.5	1153.9	273.9	38	1
Blueberries	61.7	127.2	234.5	34.2	5

1992 - 1997

1995			1997		
Herbicide	Insecticide	Fungicide	Herbicide	Insecticide	Fungicide
74	67	90	75	60	87
97	94	69	91	88	65
63	98	93	60	96	90
92	89	86	58	62	71
66	97	97	54	82	84
	46	73	84	48	71
24	9	1	44	33	12
65	96	90	57	90	85
61	92	93	61	84	80
83	73	64	78	73	66
67	94	98	78	98	99
48	75	71	74	85	69
54	14	30	53	16	30
82	97	96	73	82	79
73	86	87	67	83	88

Animal-Production Systems

Alternatives to chemical pesticides are few for pest management in animal- production systems. Among the constraints on development of alternatives are mobility of both host animals and pest arthropods, general unavailability of economic-injury threshold data, and lack of competitiveness (in both cost and efficiency) of pesticide alternatives (Campbell 1994). Concerns about nontarget effects through food residues are of little concern, because of the existence of strict Environmental Protection Agency (EPA) and FDA regulations. The strict regulatory environment, however, has created reluctance among pesticide producers to invest in new products or to reregister old products (Campbell 1994). Their reluctance translates into a heavy reliance on a relatively small number of control chemicals and concomitantly high potential for resistance acquisition in target species (Campbell 1994). Estimates of the magnitude of environmental impact of many of the control chemicals of choice (including avermectins and their metabolites) are under debate (Wratten and Forbes 1996, Spratt 1997).

In the context of companion-animal pest management, the ready availability of many products and the absence of strict requirements for applicator training contribute to increased health risks to both private consumers and professionals. The Department of Pesticide Regulation of

TABLE 5-2 Continued

	1992		
	Herbicide	Insecticide	Fungicide
Vegetables:			
Sweet corn, proc.	92	75	19
Tomatoes, proc.	90	81	92
Greenpeas, proc.	91	49	1
Lettuce, head	68	97	76
Snap beans, proc.	95	68	55
Watermelon	37	53	71
Sweet corn, fresh	75	84	41
Onion	86	79	83
Broccoli	58	95	31
Tomatoes, fresh	75	95	86
Carrots	67	37	79
Cantaloupe	44	78	73
Cucumbers, proc.	74	34	32
Asparagus	86	64	28
Snapbeans, fresh	52	77	62

	Total Application, 1996			1,000s of Planted Acres	No. of States Surveyed
	Herbicide	Insecticide	Fungicide	(1996)	
Vegetables:					
Sweet corn, proc.	1177.5	214.6	12.2	416.6	5
Tomatoes, proc.	582.9	297	11436.3	318	1
Greenpeas, proc.	194.1	32.8	3.8	221.7	5
Lettuce, head	160.6	504.3	430.7	194.9	2
Snap beans, proc.	429	149.7	63.7	134.2	4
Watermelon	90.1	88.1	670.9	163.8	6
Sweet corn, fresh					
Onion	728.7	180.9	921.7	127.4	8
Broccoli	201.4	260.2	44.3	106	1
Tomatoes, fresh	765.9	400	1694.6	88.7	6
Carrots	153.2	55.9	469.1	108	6
Cantaloupe	na	na	na	na	na
Cucumbers, proc.	67.8	24.1	76.7	71.5	6
Asparagus	177.1	81.3	70.2	72	3
Snapbeans, fresh	63.4	75.6	258.2	66.5	7

Source: ERS, 1997; USDA, 1997; USDA, 1998.

1994			1996		
Herbicide	Insecticide	Fungicide	Herbicide	Insecticide	Fungicide
94	66	9	90	74	11
76	71	86	78	71	90
93	50		89	35	2
60	100	77	52	98	76
91	58	41	90	72	49
41	45	64	43	41	65
79	81	36	79	89	42
88	76	89	88	83	89
67	96	36	64	96	37
52	94	91	54	93	90
72	34	71	89	40	78
41	82	41	na	na	na
77	48	30	76	36	34
91	70	23	88	56	33
60	79	63	49	75	73

the California Environmental Protection Agency, for example, identified pet grooming facilities as a potential source of problems after several cases of pesticide poisoning were reported in 1995 (California EPA 1997). Investigations revealed that pet groomers in many establishments received virtually no training and regularly immersed their hands in pesticide solutions. Label changes might reduce such exposures.

Urban Pest-Management Systems

Residential pest control is performed or coordinated by consumers to manage nonstructural pests and enhance the value of properties for aesthetic or recreational purposes. Real expenditures for pesticides applied by homeowners were roughly constant from 1979 to 1995 (Templeton et al. 1998). About 12% of households in the United States hired lawn-care companies in 1995; fertilizer or pesticide application was the main service provided by the companies (Templeton et al. 1998). Lawn-care experts indicate that homeowners tend to worry less about costs than about having weed-free lawns. At the same time, some owners of lawn-care companies worry about applicator exposure in residential environments. Although applicator exposures have not generally been well characterized, at least in part because of the unstructured nature of the industry, there are indications that exposures and accompanying health effects are fre-

BOX 5-1
Assessing Integrated Weed Management from Biological Time Constraints and Their Impact on Weed Control and Crop Yield

In recent years, herbicide technology—combined with mechanical, cultural, biological, and other nonchemical methods of weed control—has increased the scope of available techniques for control of weeds in many crops. Herbicides decrease the manual labor required to control weeds, thus freeing producers to do other tasks. Herbicides can also cause unintentional problems such as off-target movement to sensitive crops (drift) and small amounts of residual herbicides can persist in the soil and disrupt future crop rotations. In addition, weed populations can shift in response to herbicides and other production practices, and establishment of the new weed pests can lead to crop losses. Because producers expect that clean fields free of weeds will increase agricultural productivity, they will probably continue to use chemical herbicides as a primary method of weed control. The continued efficacy of any weed control, however, will depend on its proper timing. For example, if herbicides are used too often, selection pressure for weed growth increases and the herbicide becomes a less effective control. Integration of chemical and nonchemical techniques will increase the complexity of weed management.

Introduction of genetically engineered row crops resistant to nonresidual broad-spectrum herbicides will continue to influence how many row-crop producers manage weed populations. Postemergence herbicide sprays, such as glufosinate and glyphosate, are sprayed onto fields to protect crops against a broad spectrum of weeds. These chemical sprays are nonselective; they do not distinguish between weeds and crop plants. To avoid crop injury, industrial scientists developed genetically enhanced crop plants, such as corn and soybeans that could tolerate over-the-top applications of these chemical herbicides. Producers of large-acreage commodity crops now can plant these herbicide-resistant varieties and apply the herbicide over the entire field without fear of injury to the crop. Because the herbicides have little soil residual activity, must be timed carefully to control weeds.

The producers need to be aware of biological time constraints that can affect weed control and crop yield. In the context of a single application of a nonresidual broad-spectrum postemergence herbicide, weeds must be considered that emerge before the treatment and are controlled by it and, emerge before the treatment but are not controlled by it, and that emerge after treatment. The relative impact of these weeds on crop yield depends on the time of weed emergence, the rate of weed growth, and the time of weed removal. Those factors are referred to as biological time constraints. Understanding these biological time constraints can help producers to evaluate whether an integrated weed-management system can be integrated into the crop-production system.

Weed-emergence Periods

Diversity of weed species and emergence patterns will influence the time of herbicide application. For example, research on annual weed-emergence patterns in the midwestern United States indicates that peak annual weed-emergence flushes can begin as early as mid-April (such as wild mustard) and as late as early July (such as common waterhemp). That is most important with nonresidual postemergence herbicides. Order and duration of weed emergence will influence the timing of postemergence application. Diversity of weed species often makes two-pass weed control necessary.

Rate of Weed Growth

The rate of weed growth depends upon environmental conditions. Slow early-season weed growth can lull one into an warranted sense of security. For example, under most Minnesota spring and early cropping conditions, foxtail reach 10 cm in height in about 4 weeks and 20 cm in about 6 weeks. If the label indicates that maximal efficacy is at a height of 10–20 cm, the producers have about 2 weeks to complete a glyphosate weed-management program in soybean. Disruption of the framework of application by wet fields, windy spray conditions, or time or labor constraints could result in poor weed control or yield loss because of weed-crop interference.

Critical Period of Weed Control

The greatest yield loss is associated with full-season competition. Crops can coexist with weeds for some period without yield loss (critical period). There are two critical periods: early-season competition and late-emerging weeds. For example, weeds usually do not interfere with corn growth until 2-5 weeks after their emergence; soybeans are more competitive with weeds for 4-6 weeks after emergence. The time over which weed-control efforts must be maintained before a crop can effectively compete with late-emerging weeds and prevent crop-yield loss is about 4–5 weeks after crop emergence for corn and soybean. Therefore, the critical period of weed interference indicates that postemergence weed-control programs in corn are associated with more risk of yield loss, because of untimely control of early-emerging weeds than are postemergence weed-control programs in soybean.

Implementation of a pest-management strategy based on biological timing should be considered in relation to its economic feasibility. A pest-management strategy is more likely to be adopted if it can reduce an important pest population to a level that no longer limits profitability. Because farmers generally are risk-averse, pest-management tools that minimize variability of crop yield and net returns across the field and over growing seasons are more likely to be implemented. In addition, producers have limited time and labor to complete operations in the field. Herbicides often meet this need, and in row crops, such as corn and soybean, at least two weed-control tactics per field per growing season are necessary. Communication between weed scientists and producers is essential to align time and labor-management issues with site-specific biological time constraints. It is the weed scientist's role to address biological time constraints and risk from the crop producer's perspective of time and labor constraints and to demonstrate how weed biology affects the economics of crop production. An understanding of biological time constraints and risk-management analysis can help producers to carry out integrated weed-management programs better.

Sources: Adapted from Gunsolus et al., 2000.

quent (Gadon 1996). Health and safety training is neither required nor uniformly available in the industry; certification requirements are highly variable. Health effects of homeowner exposure are even more difficult to measure; despite the potential risk, particularly for homeowner applicators, virtually no statistics are available to allow a thorough evaluation of the problem.

Wildland Systems

For the purposes of the report, natural ecosystems include rangelands, forests, conservation holdings, rights of way, and aquatic systems. Preservation of target species to conserve biodiversity is the main goal of management efforts; as a consequence, nontarget impacts have greater importance in dictating management practices than in many other systems. Biological control, for example, can have more nontarget effects if the biocontrol agent is insufficiently host-specific and is capable of shifting onto native hosts (Louda et al 1997, Onstad and McManus 1996). Researchers and practitioners indicated that improved weed control is necessary for all these systems. Land managers can use an array of tactics—including hand-pulling weeds, biocontrol, and chemical pesticides—to protect native flora and fauna of natural parks, wildlands, and habitats preserved for conservation. In some cases, herbicides might be selected in preference to other tools, but this decision depends on site-specific conditions. Much herbicide use involves spot treatments with backpack sprayers, and overall quantities of pesticides used are generally low—an entire national park might require less than 1,000 gallons/year (John Randall, The Nature Conservancy, August 10, 1999, personal communication). With spot treatments, low-quantity use, and selection of lower-hazard and less-mobile herbicides, human health and environmental impacts are considered low. Nonetheless, public concerns about herbicide use in private and public forests remain high because of potential effects of these chemicals on water quality, species biodiversity and habitats, and other environmental characteristics.

DECIDING AMONG ALTERNATIVE PEST-MANAGEMENT STRATEGIES IN DETERMINING THE UTILITY OF CHEMICAL PRODUCTS

Even a perfunctory examination of the diverse agroecosystems contributing to the US agricultural enterprise leads inevitably to the conclusion that such diversity precludes a simple formulaic approach to specifying which chemical products, if any, will play a role in the future. Evaluating alternative management approaches or alternative management products necessitates taking into account a wide array of variables. This section presents two approaches for comparing pest-management strategies in different agroecosystems.

METHODS FOR ASSESSING PEST-MANAGEMENT STRATEGIES

Assessment of pest-management strategies will improve if it is conducted through transparent processes that use quantitative analysis to take into account all the implications of alternative policy options. Here, we introduce the key elements of such a framework. It is presented as a cost-benefit analysis, a method that is increasingly used to assess resource management and environmental policies; for example, it has been used to assess water projects. This approach monetizes all costs and benefits so that they are measured in dollars, and its full implementation might be constrained by data limitations and difficulties in monetizing human and environmental health risks. But, we introduce it here as an integrative conceptual framework that will provide directions for future data collection and identify some key considerations that have to be quantitatively balanced in making strategic choices.

Regulatory agencies have used components of this framework in their analyses, but it has not been integrated, and decisions in many cases are based on partial quantitative information. EPA regulatory decisions about pesticide use have used separate analysis of economic benefits and environmental and health risks. For years, benefit analyses relied on partial budgeting (the yield and cost effects of banning a pesticide were considered without taking into the account the effect on prices), but that was recently changed. Present studies on the benefits of pesticide use recognize the market effect, the uncertainty of key parameters, and distributional effects on consumers and producers (Lichtenberg et al. 1988; Zilberman et al. 1991; White and Wetzstein 1995; Sunding 1996; Deepak et al. 1999; and Carpenter et al. 1999). Even these studies do not consider the effects of pesticide policies on the chemical industry. Furthermore, their analysis tends to be static, whereas effects vary over time. The framework presented below attempts to correct those flaws of the current model.

The policy process might not assign explicit values to life and limb, but actual choices imply economic evaluations of lives saved. Cropper et al. (1992) argue that the value of statistical life implied by pesticide regulations varies widely across regulations, and this is consistent with the findings on the impact of other regulations that affect environmental health (Cropper and Freeman 1990). Policy-making can improve if choices become more consistent. For example, decisions that imply a very low value of life are obviously too lenient, and decisions that imply a value of hundreds of millions of dollars might be too strict. Several studies (Harper and Zilberman 1989; Lichtenberg et al. 1989; and Sunding and Zivin 2000) estimated the economic and health effects of various pesticide regulations, derived implied values of life, and determined consistent optimal policies under various assumptions regarding value of life. The approach presented here can provide regulatory choices under alternative assumptions regarding the value of life and limb; alternatively, it can provide the quantitative tradeoffs between economic benefits and the risks that have to be explicitly considered in decision-making. The previous studies provide partial estimates of tradeoffs but suggest that, in principle, quantitative assessments of the tradeoff between health cost and environmental benefits are feasible. More data and policy-relevant research are required for tradeoffs to be useable in actual decision-making. Our framework provides basic formulations and identifies some of the data needed to implement them.

Benefit-cost figures must be adjusted for time differences through discounting. When there are uncertainties, the net benefits are weighted by their probabilities. This approach of expected discounted net benefits is used to evaluate a strategy that lasts several years and has uncertain outcomes.

What matters for our purposes is not the final presentation of results, but the exact information and assessments that we need to compare different pest-management strategies. Therefore, we first present market-outcome categories (impacts on quantities and prices and goods traded in markets) and then nonmarket outcomes of pesticide strategies (impacts on government, health, and assets not traded or valued directly by markets). We then discuss factors that will enable us to analyze them quantitatively. We emphasize some of the links between the biological considerations and economic outcomes, and we identify some of the gaps in our knowledge and suggest where new and better information will be needed. Although we present a general method for assessing the impact of a new strategy on the economy, at the end we provide a more detailed discussion that assesses some of the impacts of pest-management strategies at the farm level.

A pest-management strategy will affect several markets, including markets of final products (such as wheat) and markets of inputs in the agricultural production process (such as water, pesticides, and labor) and in the production of pest-control devices (such as research and development, and chemicals). It can also have a monetary effect on the environment in terms of drift damage to structures or animals that can be monetized. The welfare-economics discipline (Just et al. 1982) developed methods to quantify effects on different markets. Benefit-cost assessment studies the effects of various policies on buyers and sellers in each market.

Several important markets are affected by pest-control strategies, and we define below some of the most important categories within these markets.

Producers' Surplus

Producers' surplus (*PS*) is the difference between the income that an output generates and the cost of its production. New pesticide technologies affect several markets, including the markets for agricultural products and pest-control inputs. In evaluating a new pest-control strategy, we should establish a benchmark for comparison and consider two types of effects on producers' surplus.

Pest-control Suppliers' Surplus (SS)

SS is the difference between the income of pest-control suppliers and their production costs. Pest-control suppliers can consist of several integrated organizations. For example, in the case of chemical pesticides, the supply chain includes the chemical company and its dealers and representatives. The same strategy can be used on several crops or in several regions, so let *i* be an indicator of a distinct region and let it assume values from 1 to *I*. The acreage in region *i* used in the strategy is A_i, and the income per acre from providing input for the strategy is m_i. Thus, the total income of the suppliers is

$$SI = \sum_{i=1}^{1} A_i m_i ,$$

Pest-control suppliers have three primary cost categories:

Acreage related costs *(VSA)*

$$VSA = \sum_{i=1}^{1} A_i sca_i ,$$

where sca_i is the variable cost per acre at region *i*. These costs include scouting costs (if they are provided by suppliers) and material and application costs that are in proportion to acreage.

Variable costs of aggregate production *(VSP)*

$$VSP = vs\left(\sum_{i=1}^{1} A_i x_i\right),$$

where x_i is per-acre quantity in region *i*. These costs include the cost of producing chemical, biological, or other types of materials for pest control.

Fixed costs *(FS)*

These involve the costs of R&D and registration and other costs that do not depend on the acreage or volume of materials used.

The pest-control suppliers' surplus is

$$SS = SI - VSA - VSP - FS.$$

In analyzing the feasibility of new pest-control strategies, we should first evaluate whether they generate surplus for pest-control suppliers (whether $SS > 0$). Several issues can arise in quantifying SS.

1. Dynamic Considerations. Introduction of new strategies is a dynamic process that takes time, and each element of the surplus has a time dimension. For a more detailed analysis, let Ss_t denote SS at year t. Assume that we evaluate strategies being developed at year $t = 0$, where r is the discount rate and the planning horizon is of T years. In this case

$$SS = \sum_{t=0}^{T} \frac{Ss_t}{(1+r)^t}.$$

Each of the SS elements can be decomposed similarly to the discounted sum of annual items. The various elements of SS evolve differently over time. For example, most fixed costs are initial investments in R&D, production, and marketing capacities that occur during the product introduction stages. But, pest-control supply incomes occur much later. It can be 6–10 years from the initial investment in a new strategy to the marketing of products. Furthermore, the use acreage at various locations changes after a diffusion process, which is generally an S-shaped time function. The variable cost of production may decrease over time, reflecting the process of learning by using.

The choice of discounting factors may significantly affect the value of SS due to the relatively heavy load of expenditures at the earlier stages of a strategy and the concentration of earning at the later stages. Higher discount factors will reduce the weight of benefits in SS and increase the likelihood of negative outcome, while lower r will likely increase the SS[1]. Thus, availability of low interest-funds for investment in a new strategy is an important condition in facilitating such investments.

2. Uncertainty Considerations. The parameters of the SS elements

[1]There is a large literature on the selection of discount rate for public project assessment. When the pest-control suppliers are part of the private sector, an appropriate market interest rate will be used in deriving SS.

are not known with certainty. Technology assessment exercises must use different kinds of estimates. Quantitative policy assessments will obtain estimates of reliability (variance) and provide a range of values for *SS* and each component (for example, the estimated mean of *SS* and a confidence interval where *SS* will likely exceed a 95 % probability). The confidence interval is defined by lower and upper limits that represent the outcome of low- and high-case scenarios.

3. Regulatory Costs. Pest-control supplier's expenditures are affected by regulations. The fixed cost of R&D includes costs needed to comply with registration requirements. Environmental and safety regulations and the legal framework make suppliers liable for some of the environmental costs caused by pest-control activities, and these costs (insurance costs) are incorporated into both the fixed and variable costs of operation. Stricter regulations can increase the cost of operation and investment, reduce *SS*, and make some strategies economically less feasible. However, imposing payment of environmental costs on the parties responsible will prevent introduction of strategies that are undesirable from an overall societal perspective and will reduce excess environmental and health costs. Thus, legislators face the challenge of designing the optimal legislative framework that will enable innovation and technological change to take into account production, environmental, and health costs.

4. Price Determination. The suppliers' income per acre reflects prices received for materials and services. The prices vary throughout the life of the pest-control strategy, reflecting supply, demand, and market-power considerations. Suppliers have monopoly power if a patent protects their strategy for several years immediately after it is introduced. Because a strategy can be used in several separate markets, the supplier will behave like a discriminatory monopolist (Carlton and Perloff 1990) and set different prices for different markets. Technically, the economic rule for setting prices in each market is that the increase in supplier revenue resulting from incremental sales increases is equal to the incremental increase in production costs. For example, in the case of a biological pesticide, the supplier will set the price for a particular region; thus, for the quantity sold, the incremental increase in revenues is equal to the cost of producing the extra quantity.

The supplier's revenue in a region is the product of price and quantity and reflects the acreage on which a strategy is being used and its value to the user. The latter value (discussed in more detail later) depends on the benefits to the user, which depend on price of the output, the impact of the technology on yield and costs, and the availability, impact, and costs of alternative technologies. As technology adoption increases and the technology is applied on more acres, the total revenue potential increases. However, the value per acre from adoption by later users of the technology is likely to be smaller; thus, the price might actually decrease over time. For example, if a new technology to address soilborne disease in tomatoes is introduced, the earlier adopters might be producers of fresh tomatoes who have a higher ability to pay, and adoption by producers of lower-value processed tomatoes will lag behind.

The factor that will probably maintain the price of a technology and even reduce it over time in spite of increasing sales volumes is a reduction in the cost of supply. This reflects both learning by doing and increased productivity that capitalizes on increasing returns to scale in production and distribution of strategy components.

To estimate the price dynamics of a pest-control strategy in different markets, we need a good understanding of the demand for the technology and the evolution of cost over time. Those elements have to be incorporated into the calculus of the monopolistic supplier to generate estimated prices during the period when the patent is in effect (see Stoneman and Ireland 1983, for a formula for pricing products and the resulting adoption behavior when the supplier is a monopolist).

When patent rights expire there can be several suppliers of the strategy, the market for strategy components will be more competitive, and the price will be closer to the average cost of production. The suppliers' surplus in the later years of a strategy's life will be substantially reduced. That suggests that suppliers' surplus will increase as the time it takes to develop and introduce a technology becomes shorter and as the length of the patent life increases. Thus, policy regulations that affect development time (through registration requirements) and patent life can significantly affect *SS* and the introduction of new technologies.

Users' Surplus (US)

US is the difference between the income of pest-control users and the cost of their products. The analysis here is geared largely to assessing the impact of a pest-control strategy on farmers' surplus. *US* can be divided into several main categories:

Users' Revenues *(UR)*

UR, is the sum of the products' per acre revenues in each region, , and the acreage of the region. Thus,

$$UR = \sum_{i=1}^{I} A_i rv_i$$

The per-acre revenues are the product of output price, p_i, in the region and yield per acre, y_i; thus, $rv_i = p_i y_i$. A pest-control strategy can affect revenues by reducing crop-damage, increasing yield (relative to the initial situation, which serves as a benchmark), or increasing product quality and its price. Assessment of yield and quality effects is essential for quantitatively estimating the impacts of a pest-control strategy.

Users' Nonpest Costs *(UNC)*

UNC is the sum of nonpest costs in all regions.

$$UNC = \sum_{i=1}^{I} A_i nc_i$$

where nc_i is the nonpest-control cost per acre in region i. *UNC* includes land preparation, fertilizer, harvesting, and nonpest-control postharvesting costs.

Pest-control Costs *(UPC)*

UPC is the sum of pest control in all regions.

$$UPC = \sum_{i=1}^{I} A_i pc_i$$

where pc_i is pest control per acre in region i. *UPC* includes application, material, and other costs per acre associated with pesticide use. We separate pest control and nonpest control for simplicity, but this is not always feasible. Separation is difficult, for example, when the cultivation practices (such as crop rotation) are a major element of the pest-control effort. Pest-control costs can be further divided by the pest problems they address and can also include the cost of pest damage to the environment and the cost of worker safety that are borne by users.

Using those definitions,

$$US = UR - UNC - UPC.$$

A pest-management strategy is feasible only if it generates a surplus for its users; therefore, when assessing strategies, we should investigate first whether *US* is greater than zero.

Several issues can arise in quantifying *US*.

1. Dynamic considerations. New technologies have a long life, and key elements evolve. For a more detailed analysis, let denote users' surplus at year t. With this notation,

$$US = \sum_{t=1}^{T} \frac{Us_{it}}{(1+r)^t} = \sum_{t=1}^{T} \frac{A_{it}(rv_{it} - nc_{it} - pc_{it})}{(1+r)^t}$$

where A_{it} is acreage in region i used under the strategy in year t and rv_{it}, nc_{it}, and pc_{it} are per-acre revenues, nonpest costs, and pest-control costs, respectively, at year t in region i.

A_{it} varies over time, reflecting an adoption process. As mentioned earlier, adoption decisions depend on investment costs and on future and present benefits. Users recognize dynamic processes that affect the costs and benefits associated with investment in new technologies, and they might wait to adopt a new strategy (McWilliams and Zilberman 1996). A formal quantitative assessment of the impact of technologies requires estimates of the key parameters that determine the diffusion process. The estimates are used to trace an estimated path of acreage used with a new strategy over time.

Revenues per acre with strategy i, rv_{it}, can vary over time. This is a product of price (p_{it}) and yield per acre (y_{it}) in region i at year t. In many cases, yields increase over time as farmers' use of a new strategy increases because of learning. But, prices can decline in response to increased supply. In some cases, a new strategy's yield declines over time because of pest resistance or infestation of secondary pests, and this must be taken into account in the modeling of yield dynamics and the resulting pest-management strategy (see models by Hueth and Regev 1974, Regev et al. 1976). Similarly, nonpest costs can change over time because of input-use efficiency (which can reduce cost) and increased input price (which may increase cost). Pest-control costs can decline after an initial period that requires investment in new pest-control equipment and training; however, pest resistance can increase costs over time. Thus, assessment of a new strategy's profitability requires quantitative modeling of cost and yield dynamics derived from a quantitative understanding of adoption dynamics, learning processes, and pest resistance.

2. Uncertainty and Risk Considerations. Estimations of US are affected substantially by risk and uncertainty considerations for two reasons. First, there are significant variances in pest-control users' behavior and estimates of price productivity and cost parameters. The high uncertainty of the estimated parameters reflects lack of knowledge and a high degree of heterogeneity even within regions. Therefore, rather than obtaining a single estimate of US, it is useful to obtain a confidence interval for a range of values in which US might be within, say, a 95 % probability.

Second, many farmers are not neutral to risk, and their choices are affected strongly by the risks (related to yields, prices, and so on) that they face. Risk aversion can lead to lower supply (Sandmo 1971) or induce further adoption of risk-reducing techniques. Carlson and Wetzstein (1993) have surveyed results that demonstrate the important role of risk and risk aversion in pesticide use and pest-control technology choices. Pest-control strategy assessments should estimate the impact of risks facing users and analyze how risk considerations influence the extent of the adoption of the strategy on *US*.

3. Regulatory and Policy Considerations. Pest-control users are affected by government policies and regulations. The policies can be in the form of price supports and subsidies that affect revenues. Lichtenberg and Zilberman (1986) showed that when supports play an important role in agriculture, models of revenue generation and market equilibrium have to be modified to reflect it. Price-support policies tend to reduce risks faced by farmers and to increase average prices, and this can lead to a tendency to adopt new technologies including new pest-control strategies. Agriculture is now going through a period where price supports are reduced, and analyses of the impact in some sectors (peanuts, tobacco, and dairy) should not ignore the impact of support policies. Those policies might also be introduced in the future. Thus, computation of *US* should be adjusted to take these support payments into account.

Derivation of pesticide costs should take into account the costs associated with pesticide regulation. These can include purchase costs of extra protective equipment and insurance, compensation, and penalty costs associated with pest control use. The costs affect the profitability of farm operations but also influence the tendency of farmers to adopt particular pest-control strategies. Although financial incentives do not play a major role in efforts to reduce pesticide use, they could play a bigger role in the future, and regulatory costs might have a bigger impact on *US*. One set of incentives involves granting entitlements to participate in price support programs based on conduct of specific practices. An obvious example is the past use of a land diversion requirement as a condition for receiving deficiency payments. Such incentives might induce adoption of agricultural practices, and they should be considered in policy evaluation and design.

4. Price determination. When a pest-control strategy has a significant effect on user output, and these users have substantial market shares in their products, then the strategy can affect the output price. Agricultural industries tend to be competitive, and producers are price takers. Acreage and output with a given technology are functions of the output price, and we can denote as the supply curve of the industry, where

$$S(p) = \sum_{i=1}^{1} A_i(p) y_i(p).$$

The supply curve is an increasing function of price. The product also has a demand curve, *D(p)*, that represents the quantities that consumers are willing to buy at a given price. Industry equilibrium, which is defined by output price and aggregate output, is determined at the point where supply and demand intersect.

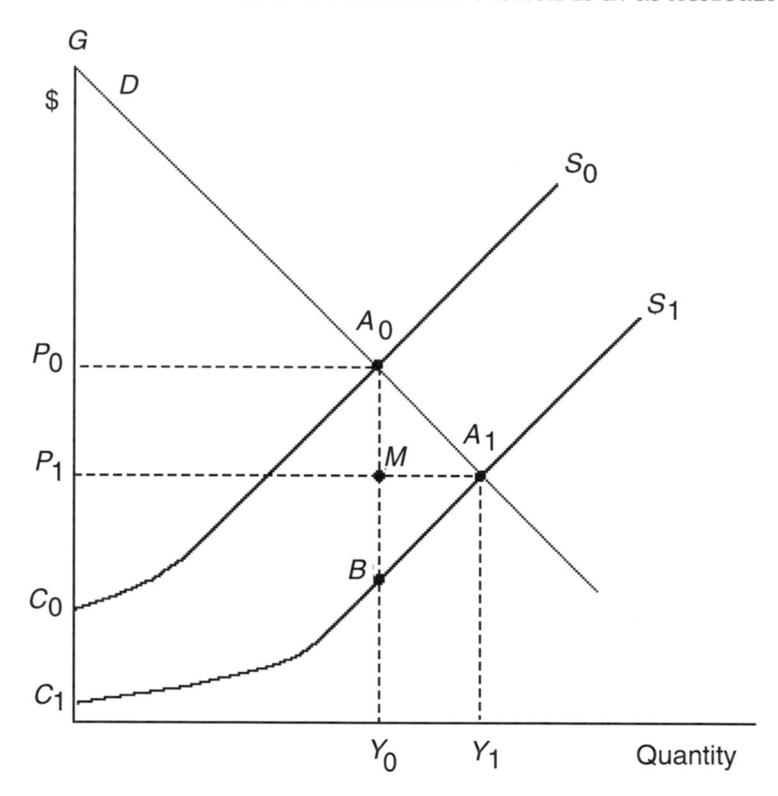

FIGURE 5-1 Equilibrium in output market when supply shifts as a result of technological change.

Price determination of the output market is demonstrated in Figure 5-1. Curve *D* denotes the demand for output, and curve S_0 is the supply curve for an initial pest-control technology 0. The intersection of the two curves, at A_0, and it results in output price P_0 and output quantity Y_0. If a new pest-control technology, technology 1, will increase supply, it will result in a shift of the supply curve to the right, from S_0 to S_1. As supply shifts from S_0 to S_1, the intersection of supply and demand moves from A_0 to A_1, at which total output increases from Y_0 to Y_1 but price decreases from P_0 to P_1.

The analysis in Figure 5-1 suggests that assessment of impacts of pest-control strategies should not always take output prices as given. When the introduction of a pest-control strategy is likely to increase output supply significantly, and the demand for this output is inelastic (in the sense that prices change significantly with relatively small changes in quantity consumed), introducing the strategy will reduce output price.

The negative price feedback that can be associated with yield-increasing pest-control strategies is important because it will lead to reduction in adoption relative to what would have happened with the initial price. It might also lead to reduction in producers' profits.

Figure 5-1 provides a graphic presentation of *US;* this is the area between the supply curve and the output price. For technology 0, this is $P_0A_0C_0$, and for technology 1, it is $P_1A_1C_1$. The increase in output associated with the transition from technology 0 to technology 1 is a source of increase in *US*, but the reduction in price is a source of *US*, and the net effect depends on the properties of the specific demand and supply curve. When the demand for the final product is very inelastic, *US* can actually decline from the introduction of a supply-increasing pest-control strategy because of the significant drop in price and the relatively small increase in output that the technology entails.

Consumers' Surplus (CS)

This category of market impact can intuitively be defined as the difference between the value of a volume of output to consumers (the maximal amount that consumers will be willing to pay for a quantity of product) and the amount that they actually do pay. The magnitude of *CS* depends on the responsiveness of demand to prices. If the price does not change as quantity changes (for example, in the case of agricultural products that have close substitutes), *CS* will not be very substantial. But in products with inelastic demands, where changes in quantity can substantially affect price, the impact of supply enhancement on *CS* can be substantial because of the significant price reduction. Therefore, *CS* considerations might not be very important when one analyzes the impact of pest control that affects commodities that are internationally traded and where the United States is playing a small role or that are produced in regions with relatively small impact on the market. They also might not be important for specialty crops with close substitutes. However, *CS* effect could be substantial when pest-control strategies are used by producers of agricultural commodities that have a significant market share and products with inelastic demand. For example, Knutson et al. (1993), found *CS* effects to be substantial in their analysis of the impact of banning pesticides in field crops. Zilberman et al. (1991) found similar results in their analysis of impacts of pesticide bans in major fruit and vegetable crops in California.

In terms of Figure 5-1, when pest-management strategy 0 is being used, *CS* is equal to GP_0A_0. When strategy 1 is introduced and the price goes down to P_1, *CS* will increase and consumers will gain because they will consume more and pay less. That gain is measured by the area

$P_0A_0A_1P_1$. It represents the reduced expenditures associated with prior production quantities and some of the added value associated with increased consumption of pesticides. The sum of the consumers' and users' surpluses from the use of strategy 0 is measured by the area GA_0C_0, and the output market surplus with strategy 1 is GC_1A_1. The effect of the introduction of the supply-increasing strategy is to increase *CS* and the overall surplus in the output market (the net market effect on producers and consumers is given by the area A_0A_1B in Figure 5-1), but the impact on *US*, as mentioned earlier, is unclear and might be negative.

Some special *CS* considerations associated with pest-control strategies are the following:

1. Product Quality. Pest-control strategies can affect consumer welfare by increasing product quality. Babcock et al. (1992) found that the impact on apple quality accounted for at least 33 % of the benefit of fungicides used on apples in North Carolina. Consumers will pay much more for produce that is available earlier in the season, has higher sugar content, and is without blemishes. The method of hedonic pricing by Rosen has been introduced to estimate the price effect of improved product characteristics (see Cropper et al. 1988, Parker and Zilberman 1993).

The impact of higher product quality can be represented by an outward shift of the demand curve. For example, if an increase in sugar content will increase the value of consumption by 5 % per unit, the demand curve will shift upward appropriately because consumers will be willing to pay a higher price for every level of quantity demanded. As shown in Figure 5-2, a shift from the initial strategy (strategy 0) to a new strategy that increases quality but has the same supply costs (strategy 1) will shift demand from D_0 to D_1. The output price will increase from P_0 to P_1, and output will increase from Y_0 to Y_1. The overall increase in the output market surplus in this case will be equal to the area $C_0A_0A_1C_1$ in Figure 5-2. The analysis suggests that assessment of the impact of pest-control strategies should quantify the product-quality effect and assess its influence on production, prices, and overall welfare.

2. Dynamic Considerations. An assessment of impact of pest-control strategy on consumer welfare must recognize that the impact changes over time. If Cs_t is consumer surplus at year *t*, *CS* associated with a given pest-control strategy is a discounted sum of surpluses over the years and can be written as

$$CS = \sum_{t=0}^{T} \frac{Cs_t}{(1+r)^t}.$$

The price effect of the pest-control strategy can change over time. For example, if a new supply-increasing strategy is introduced, adoptions in

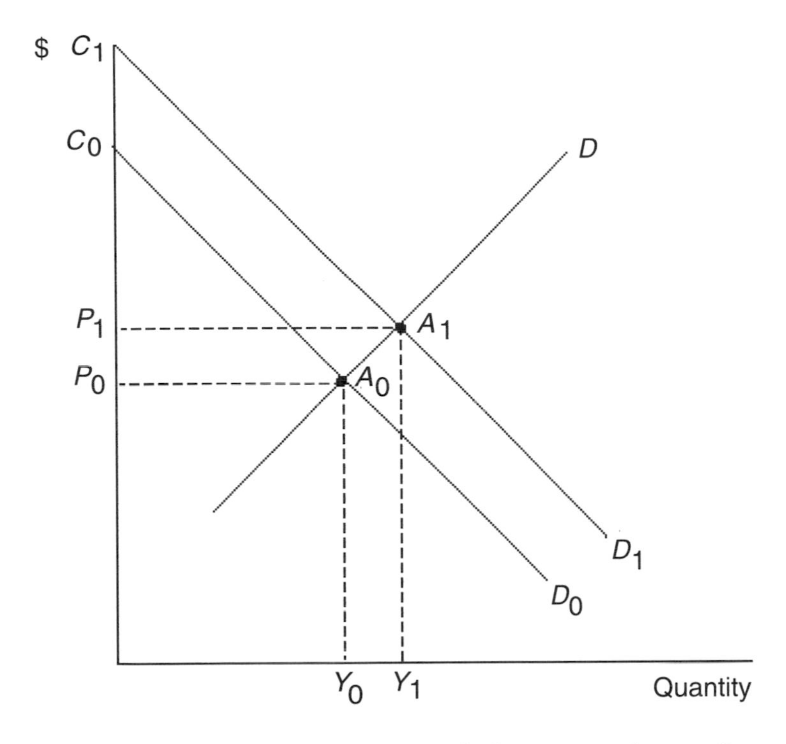

FIGURE 5-2 Equilibrium in output market with change in product quality.

early years might be low and the price effect not very significant. Over time, increased adoptions would lead to outward price reduction. The *CS* can change over time because of exogenous factors, such as population growth. Increased population might tend to increase the demand for a product and the annual *CS* associated with the pest-control strategy.

3. Final versus Intermediate Good. Many agricultural commodities are intermediate goods for the production of final goods. For example, alfalfa, corn, and soybeans are used as feed products in the production of meats. Therefore, analysis of *CS* effects have to incorporate factors that affect both the grain market and the final-product market. Knutson et al. (1990) show that a pesticide ban in grain crops will reduce consumer welfare substantially through its impact on the price of meats. They separated the impacts of the ban into impacts on feed-crop growers, livestock operators, and final consumers. When agricultural products go through several stages of processing, intermediate surpluses have to be derived.

Environmental and Health Costs (EHC)

Pest-control strategies have significant effects on human health and various dimensions of environmental quality. Some of the effects can be quantified and monetized, and the quantification should be incorporated into the cost-benefit analysis.

The health costs associated with pest-control strategies are denoted by *HC* and are in several categories shown below.

- **Worker safety costs *(WSC)*.** *WSC* is the sum of several cost categories. These consist of sick days and their attributed costs (e.g., the costs of medical treatment, earnings lost, and pain and suffering (Viscusi and Magat 1987). An even more sensitive category is statistical mortality, which is multiplied by an attributed value. Zeckhauser (1975) and Thaler and Rosen (1976) provided methodological guidelines to value statistical life, and Cropper et al. (1996) provide an overview of the implied valuation of life associated with existing pesticide policies. (The notion of statistical death reflects the estimate that there is a small probability of accidental death due to worker use of a pest-control strategy. This probability, multiplied by the size of the population involved, provides an estimate of statistical death.)

Risk-assessment method have been developed to quantify the various types of accidents associated with different production activities, including pest-control activities. If we consider the number of worker sick-days per period (year) associated with a particular pesticide strategy, *SD*,

$$SD = \sum_{i=1}^{1} A_i \alpha_i$$

where A_i is the acreage in region$_i$ and *i* is the sick-day-per acre coefficient, *SD* depends on the material used in the pest-control strategy and the exposure level (which depends on quantity of material used, how it is applied, type of protection gear worn, weather conditions during application, and vulnerability of the exposed population).

- **Food-safety costs *(FSC)*.** *FSC* includes the effects on human health of exposure to materials used in pest-control strategies during and after production. The effects vary with age, location, and individual and are a source of major concern. *FSC* is also a sum of subcategories, including sick-day costs, mortality costs, and costs of disposal and treatment of contaminated food. For each subcategory of cost, the expected number of accidents is proportional to the quantity of output produced in each region multiplied by coefficients that reflect the risk per unit of output. The coefficients are affected by residue levels, toxicity of materials, and vulnerability to the materials.

When it comes to *FSC*, costs are associated with perceptions and uncertainty. Consumers might view some treatment of food products as less desirable than others because they feel that it makes products less safe or wholesome. Some studies have estimated willingness to pay for pesticide-free food (van Ravenswaay and Hoehn 1991). Assessment of pest-control strategies should take into account that some perceptional cost or benefit is associated with the use of strategies in production of foods. When everything else is equal, these considerations could make some strategies more desirable.

- **Other exposure costs (*OEC*).** *OEC* include exposure to residues of material used in pest-control strategies. People might be exposed to toxic materials when they enter a sprayed field or fumigated storage area. *OEC* also includes accidental poisoning associated with mistaken consumption of chemicals. Humans might be exposed to chemical residues in the water or materials carried through the air. Again, assessment of strategies has to recognize the magnitude of accidental effects on third parties. When the effects are deemed substantial, the estimation of mortality and morbidity cases will provide the foundation for computing the *OEC*.

HC can be estimated:

$$HC = WSC + FSC + OHC.$$

Several important issues can arise in quantifying *HC*.

1. Dynamic Considerations. The health effects of different pesticide strategies might occur at different times, and they require adjustment in computation procedures. Therefore, it could be useful to break each category of *HC* into annual components. *HC* is then the discounted sum of the annual costs. Some materials might cause acute problems, others may cause chronic problems that will be discovered years later. There is also the issue of cumulative exposure. It is useful to break the annual health costs into these subcategories.

2. Uncertainty Considerations. Quantification of health risks is an immense challenge, and variances of risk assessment are important. A major problem is that different risk assessment coefficients are estimated at various levels of reliability. For example, some studies provide average risks posed by a particular pest-control strategy, and others provide risk estimates that may not be exceeded at 99 % probability. For consistent quantification, all risk estimates should be converted to the same degree of reliability, say, the mean of a level that is not exceeded at 95 % probability. Harper and Zilberman (1989) demonstrate that the quantitative health risk estimates associated with alternative pesticide use regulations in California's Imperial Valley vary by several orders of magnitude when derived with different degrees of reliability. They and others also demonstrate how tradeoffs between health risks and economic benefits can be meaningful only when all health risk estimates are consistent.

Environmental Costs and Benefits (ECB)

Several categories of environmental costs can be associated with pesticide strategies. The strategies might also have some benefits relative to an initial situation. The categories of environmental costs and benefits are described briefly below.

Nontarget Species Costs (NTC)

Environmental nontarget costs are borne by society and distinguished from production nontarget costs (damage to beneficial pests that reduces productivity), which are borne by users. Pest-control activities can harm species that provide benefits through various use categories (for example, hunting, fishing, birdwatching, and pollination), through biodiversity, and through existence (see Randall and Stoll (1983) for discussion of such benefits and how to assess them). *NTC* can be presented as

$$\sum_{i=1}^{1} \beta_i A_i$$

where β_i is nontarget species cost per acre in region i, and it is multiplied by the acreage to provide an estimate of the regional *NTC*.

Damage to Property and Resources (DRC)

Various pesticide strategies can lead to residues that contaminate nearby property. For example, aerial spraying of pesticides might cause damage to nearby land and structures. A more severe problem is groundwater and surface-water contamination by chemicals, which can cost hundred of millions of dollars to clean (see Lichtenberg et al. (1988) for studies on the clean-up costs of 1,2-dibromo-3-chloropropane, DBCP). Environmental damage depends on the manner of application. When applying materials like chemicals, one can separate applied input and effective input, and the application technology will determine the input use efficiency (percentage of material that is actually used in production). The residue coefficient is 1 minus the input use efficiency. Higher penalties for environmental contamination would lead to adoption of more precise application technologies that would reduce residues and environmental damage (Khanna and Zilberman 1997).

Environmental Costs and Benefits of Resource Use (*ECBR*)

Various pest-control strategies affect directly and indirectly the amounts of resources used in agricultural activities. For example, a pesticide strategy might increase yield per acre substantially and thus reduce agricultural land and water use relative to an initial benchmark. The released land might provide substantial environmental benefits as habitat for wildlife- preservation activities. Another pest-control strategy that might be perceived as environmentally friendly but decreases productivity per acre and lead to expansion of the land base and water resources used. If γ_i is the environmental cost of an acre added to production in region i because of an environmental strategy,

$$ECBR = \sum_{i=1}^{1} \gamma_i A_i$$

As with other cost categories, dynamic and uncertainty considerations affect environmental costs and benefits. When it comes to the nontarget costs, quantification and monetization present serious problems. It is extremely important in an assessment to at least be aware of the different types of environmental costs and, when possible, quantify them in physical units.

Government Net Costs (*GNC*)

Implementation of different pest-control strategies might require substantial government involvement. Government might need to finance some of the basic and applied research that leads to establishment of such strategies and need to spend resources on education, extension, monitoring, and enforcement of regulations. Or, government might receive income through taxes and penalty payments. All those costs have to be taken into account when benefits and costs of various categories are considered.

Evaluation

All the different benefit and cost items that are associated with a pesticide strategy are summed to generate the net benefit (*NB*) of the strategy:

$$NB = CS + PS - EHC - GNC.$$

NB equals the sum of consumer surplus and producer surplus, minus environmental health costs and government net costs. Detailed calculations require quantification of each of the subcategories, recognizing time variation and discounting net benefits of different years, and finally adjusting for uncertainty. Such an adjustment can be provided by a range of net benefit values rather than one number.

The analysis thus far might have ignored international trade considerations, which can be important in assessing pesticide strategies, especially because more than 50 % of the output of some products is exported. Knutson et al. (1990) developed applications that specifically consider the trade effect of pesticide bans. Our analysis has emphasized the importance of economy wide changes and regional heterogeneity in obtaining realistic assessments of various strategies. For a more detailed understanding of how some of the biological considerations should enter into impact assessment, we present below a farm-level framework of analysis.

MODEL FOR DESIGN AND EVALUATION OF PEST-MANAGEMENT STRATEGIES ON THE FARM SCALE

A generalized model for design and evaluation of pest-management strategies on the farm scale is based on modification of the model presented by Higley and Wintersteen (1992). The model evaluates the benefits and costs associated with a set of alternative strategies. In addition, it identifies the pest-management strategies that fit some minimal criteria and those which maximize "net good". Net good could include net economic returns, low environmental impact, maximal durability, or an optimization over a set of these and other criteria. Optimization over all the factors requires a common currency for net good and can be used to determine the pest densities required to use a pesticide on the basis of accumulated costs. For example, the conceptual model can be expressed as follows for a crop on a farm, with the common currency units (in this example, dollars) listed below each parameter:

net return = [Agricultural Production × (price received – dockage from pest)] – costs

$/acre/year = unit production/acre $/unit production
$/unit production $/acre
year

where agricultural production (yield) is a function of pest abundance (here assumed to be additive among different pests),

$$\text{Pest Abundance} = (N_w + N_i + N_p + N_j)$$

where N_w is weed abundance, N_i is insect pest abundance, N_p is pathogen abundance and N_j is any other pest abundance that can have an impact on agricultural production.

Costs include all the management costs associated with growing a crop or producing an agricultural product on the farm. The pest-management costs can be summed as follows:

$$M_w + M_i + M_p + M_j$$

where M_w is cost of weed management, M_i is cost of insect management and M_p is cost of pathogen management and M_j is cost of other pest management. The cost associated with the evolution of resistance to a pesticide is referred to as "durability cost" and is the cost difference between the current pesticide and an alternative management method (M_d) required after the evolution of resistance in the pest. The cost associated with pest impact on human health (H_p) and pest management on human health (H_{pm}) can also be included. The cost of environmental degradation associated with pest management (E_{pm}) and associated with the pest itself (E_p) on the farm are also variables that can be included. Higley and Wintersteen (1992) demonstrated the use of contingent valuation surveys to estimate monetary costs associated with environmental degradation resulting from agricultural pest management. The cost of special technology and information for pest management (T_{pm}), and all other costs associated with growing the crop (O) are additional variables that can be included in the estimation of net return.

Net return = [yield x (price received – dockage from pest) – $(H_p + E_p)$] – $[(M_w + M_i + M_p + M_j + M_d + H_{pm} + E_{pm} + T_{pm} + O)]$

Assessment with Model

If net return (NR) is less than zero across the possible range of pest abundance, the logical conclusion is to reject the pest-management strategy. NR must be greater than zero to conclude that a pesticide is worth using in the system. If NR is greater than zero, then its value relative to other practices assessed in the same manner to manage the same pest(s) determines the relative importance of the pesticide in the management system. This example uses NR as the dependent variable, but other measures of response to pesticides—such as crop yield, environmental quality, or human effects—could be used.

Identifying Research Needs

The model presented above could be used to identify which parameters and their interactions are most important in determining the net return from a particular pest- management practice. Research would focus on the interrelationships between parameters in the model. Various portions of the model require research:

- The relationship between cost of management and its impact on the pest needs study. For example, herbicide evaluation is usually based on the proportion of the weed population killed, whereas crop yield might be adequately increased by weed injury that could be achieved with rates of much lower herbicide use.
- The model provides the ability to quantify interactions among pests (insects, pathogens, and weeds) and their management. Different pests in the same agroecosystem are often managed independently, and this can lead to wasted pesticide applications if there are interactions among the pests or the methods used to manage them. For example, two major pests of dry-land spring wheat are wild oats (weed) and wheat-stem-sawfly (insect). The wheat-stem-sawfly life cycle is broken in the presence of wild oats because the larvae do not survive in the stems of wild oats and there is no selective preference for laying eggs in the wheat or wild oats (Sing et al. 1999). Thus, the weed has positive value for reducing the impact of the insect pest, which would increase the economic injury level for the weed.
- The model does not include future impacts of pests that are left behind when the current pest density is below the economic injury level (EIL). Studies that determine parameters for pest population temporal and spatial dynamics

REFERENCES

Babcock, B.A., E. Lichtenberg, and D. Zilberman. 1992. Impact of damage control and quality of output: estimating pest control effectiveness. Am. J. Ag. Econ. 74:163–172.

Beeman, R. W., and V. F. Wright. 1990. Monitoring for resistance to chlorpyrifos-methyl, pirimiphos-methyl and malathion in Kansas populations for stored-product insects. J. Kansas Ent. Soc. 63(3):385–392.

Brunner, J. F. 1994. Integrated pest management in tree fruit crops. Food Rev. Int. 10:135–157.

California EPA. 1997. Overview of the California Pesticide Illness Surveillance Program – 1995. Sacramento: California Environmental Protection Agency, Dept. of Pesticide Regulation.

Campbell, J. B. 1994. Integrated pest management in livestock production. Food Rev. Int. 10:195–205.

Carlson, G. A. and M. E. Wetzstein. 1993. Pesticides and pest management. Pp. 268–318 in Agricultural and Environmental Resource Economics, G. A. Carlson, D. Zilberman, and J. A. Miranowski, eds. New York and Oxford: Oxford University Press.

Carlton, D. W., and J. M. Perloff. 1990. Modern Industrial Organization. Glennview, Ill. and London: Scott, Foresman/Little, Brown Higher Education.

Carpenter, J., L. Gianessi, and L. Lynch. 1999. The Economic Impact of the Scheduled US Phaseout of Methyl Bromide. National Center for Food and Agricultural Policy, Report supported by USDA's Economic Research Service.

CAST (Council for Agricultural Science and Technology). 1999. The FQPA: A Challenge for Science Policy and Pesticide Regulation. [Online]. Available: HtmlResAnchor http://www.cast-science.org/fqpa/fqpa-06.htm

Cropper, M. and A. M. Freeman III. 1990. Valuing Environmental Health Effects. Discussion Paper QE90-14. Washington, D.C.: Resources for the Future, Quality of the Environment Division.

Cropper, M. L. 1996. The determinants of pesticide regulation: a statistical analysis of EPA decision making. In The Political Economy of Environmental Protection: Analysis and Evidence, R. D. Congleton, ed. Ann Arbor: University of Michigan Press.

Cropper, M. L.,L. B. Deck, and K. E. McConnell. 1988. On the choice of functional form for hedonic price functions. Review of Economics and Statistics LXX/4:668–675.

Cropper, M., W. N. Evans, S. J. Berardi, M. M. Ducla-Soares, and P. R. Portney. 1992. The determinants of pesticide regulation: a statistical analysis of EPA decision making. Journal of Political Economy 100:(1):175–197.

Deepak, M. S., T. J. Spreen, and J. J. Van Sickle. 1999. Environmental Externalities and International Trade: The Case of Methyl Bromide. In Flexible Incentives for the Adoption of Environmental Technologies in Agriculture, C. F. Casey, A. Schmitz, S. M. Swinton, and D. Zilberman, eds. Boston: Kluwer.

Economic Research Service. 1995. Agricultural Resources and Environmental Indicators. Agricultural Handbook No. 712, Washington, DC: Economic Research Service (US Department of Agriculture).

Economic Research Service. 1997. Agricultural Resources and Environmental Indicators, 1996–1997. Agricultural Handbook No. 712. Washington DC: US Government Printing Office.

EPA (US Environmental Protection Agency). 1999a. Implementing the Food Quality Protection Act. EPA 735-R-99001. Washington, DC: EPA, Office of Pesticide Programs.

FDA (US Food and Drug Administration). 1998. The Food Defect Action Level Handbook. . [Online]. Available: http://vm.cfsan.fda.gov/~dms/dalbook.html

Gadon, M., 1996. Pesticide poisonings in the lawn care and tree service industries. J. of Eco. 38:794–799.

Gianessi, L. 1993 . The quixotic quest for chemical-free farming. Iss. in Sci. and Tech. 1993:29–36.

Gooch, J. J., A. Sray, and C. Greenleaf. 1998. The fight to save OPs, carbamates. Farm Chemicals 161:18–22.

Gunsolus, J. L., T. R. Hoverstad, B. D. Potter, and G. A. Johnson. 2000. Assessing integrated weed management in terms of risk management and biological time constraints. Pp. 373–383 in Emerging Technologies for Integrated Pest Management: Concepts, Research, and Implementation, G. G. Kennedy and T. B. Sutton, eds. St. Paul, Minn: APS Press.

Hagstrum, D. W., and W. G. Heid, Jr. 1988. US S. wheat-marketing system: an insect ecosystem. Bull. Ent. Soc. Am. 34(1):33–36.

Harper, C. R. and D. Zilberman. 1989. Pest externalities from agricultural inputs. Am. J. Ag. Ec. 71(3):692–702.

Higley, L. G.,. and W. K. Wintersteen. 1992. A novel approach to environmental risk assessment of pesticides as a basis for incorporating environmental costs into economic injury levels. Am. Ent.38:34–39.

Hueth, D. And U. Regev. 1974. Optimal agricultural pest management with increasing pest resistance. Am. J. Ag. Ec. 56:543–552.

Just, R. E., D. L. Hueth, and A. Schmitz. 1982. Applied Welfare Economics and Public Policy. Englewood Cliffs, NJ: Prentice Hall.

Kenkel, P., J. T. Criswell, G. W. Cuperus, R. T. Noyes, K. Anderson, and W. S. Fargo. 1994. Stored product integrated pest management. Food Rev. Int. 10:177–193.

Khanna, M., and D. Zilberman. 1997. Incentives, precision technology and environmental protection. Ecol. Ec. 23(1):25–43.

Knutson, R. D., C. R. Hall, E. G. Smith, S. D. Cotner, and J. W. Miller. 1993. Economic Impacts of Reduced Pesticide Use on Fruits and Vegetables. Washington, DC American Farm Bureau Federation.

Knutson, R. D., C. R. Taylor, J. B. Penson, and E. G. Smith. 1990. Economic Impacts of Reduced Chemical Use. College Station, Tex.: Knutson and Associates

Lawton, J. H., and K. J. Gaston. 1989. Temporal patterns in the herbivorous insects of bracken: A test of community predictability. J. An. Ec. 58:102–1034.

Lichtenberg, E., and D. Zilberman. 1986. The welfare economics of regulation in revenue-supported industries: The case of price supports in US S. agriculture. Am. Ec. Rev. 76(5):1135–1141.

Lichtenberg, E., D. D. Parker, and D. Zilberman. 1988. Marginal Analysis of Welfare Costs of Environmental Policies: The Case of Pesticide Regulation. Am. J.Ag. Ec. 70(4):867–874.

Lichtenberg, E., D. D. Parker, and D. Zilberman. 1988. Marginal analysis of welfare effects of environmental policies: The case of pesticide regulation. Am. J. Ag. Ec. 70:867–874.

Lichtenberg, E., D. Zilberman, and K. T. Bogen. 1989. Regulating environmental health risks under uncertainty: groundwater contamination in California. J. Env. Ec. and Mgmt. 17:22–34.

Losey, J. E., L. S. Rayor, and M. E. Carter. 1999. Transgenic pollen harms monarch larvae. Nature. 399:214.

Louda, S. M., D. Kendall, J. Connor, and D. Simberloff. 1997. Ecological effect of an insect introduced for biological control of weeds. Science. 277:1088–1090.

Maxwell, B. D. 1992. Weed thresholds: the space components and considerations for herbicide resistance evolution. Weed Tech. 6:205–212.

McWilliams, B., and D. Zilberman. 1996. Time of technology adoption and learning by using. Econ. Innov. New. Tech. 4(2):139–154.

NRC (National Research Council). 1993. Pesticides in the Diets of Infants and Children. Washington, DC: National Academy Press.

Onstad, D. W., and M. L. McManus. 1996. Risks of host range expansion by parasites of insects. Bioscience 46(6):430–435.

Parker, D. D., and D. Zilberman. 1993. Hedonic estimation of quality factors affecting the farm-retail margin. Am. J. of Ag. Ec. 75:458–466

Randall, A. and J. Stoll. 1983. Existence Value in a Total Valuation Framework. In Managing Air Quality and Scenic Resources at National Parks, R. Rowe and L. Chestnut, eds. Boulder, Col.: Westview Press.

Regev, U., A. P. Guitierrez, and G. Feder. 1976. Pests as a common property resource: a case study of alfalfa weevil control. Am. J. Ag. Ec. 58(2):1861–197.

Sandmo, A. 1971. On the theory of the competitive firm under price uncertainty. Am. Ec. Rev. 61:65–73.

Sing, S., B. Maxwell, and G. D. Johnson. 1999. Wheat stem sawfly-wild oat interactions in Montana dryland spring wheat. Bull. Ecol. Soc. Am.

Spratt, D. M. 1997. Endoparasite control strategies: implications for biodiversity of native fauna. Int. J. Parasitol. 27:173–180.

Stoneman, P., and N. J. Ireland. 1983. The role of supply factors in the diffusion of new process technology. Ec. J. Supp. Mar. 83:66–78.

Sunding, D. 1996. Measuring the Marginal Cost of Non-Uniform Environmental Regulations. Am. J. Ag. Ec. 78:1098–1107.

Sunding, D., and J. Zivin. 2000. Insect Population Dynamics, Pesticide Use and Farmworker Health. Am. J. Ag. Ec. forthcoming.

Templeton, S., D. Zilberman, and S. J. Yoo. 1998. An economic perspective on outdoor residential pesticide use. Env. Sci. Tech.. 21:416–423.

Thaler, R. and S. Rosen. 1976. The value of saving a life. In Household Production and Consumption, N.E. Tereckyj, ed. New York: National Bureau of Economic Research.

USDA (US Department of Agriculture). 1997. Agricultural Chemical Usage. Vegetables. 1996 Summary. US Department of Agriculture, National Agricultural Statistics Service (NASS), Agricultural Statistics Board. Washington, D.C.: US Government Printing Office.

USDA (US Department of Agriculture). 1998. Agricultural Chemical Usage. Fruits. 1997 Summary. US Department of Agriculture, National Agricultural Statistics Service (NASS), Agricultural Statistics Board. Washington, D.C: US Government Printing Office.

van Ravenswaay, E. O., and J. P. Hoehn. 1991. Willingness to pay for reducing pesticide residues in food: Results of a nationwide survey. Staff paper no. 91-18. East Lansing: Michigan State University, Department of Agricultural Economics.

Viscusi, W. K. and W. A. Magat. 1987. Learning About Risk: Consumer and Worker Responses to Hazard Information. Cambridge, Mass.: Harvard University Press.

White, F. C., and M. E. Wetzstein. 1995. Market effects of cotton integrated pest management. Am. J. of Ag. Ec. 77:602–612.

Wratten, S. D., and A. B. Forbes. 1996. Environmental assessment of veterinary avermectins in temperate pastoral ecosystems. Ann. Appl. Biol. 128:329–348.

Zalom, F.G., and W. E. Fry. 1992. Food, Crop Pests, and the Environment. St. Paul, Minn: American Phytopathological Society.

Zeckhauser, R. 1975. Procedures for valuing lives. Public Policy 23: 427-464.

Zehnder, G. W. 1994. Integrated pest management in vegetables. Food Rev. Int. 10: 119-134.

Zehnder, G. W., and G. K. Evanylo. 1989. Influence of extent and timing of Colorado potato beetle (Coleoptera: Chrysomelidae) defoliation on potato tuber production in eastern Virginia. J. Ec. Ent. 82(3): 948-953.

Zilberman, D., A. Schmitz, G. Casterline, E. Lichtenberg, and J.B. Siebert. 1991. The economics of pesticide use and regulation. Science 253: 518-522.

6

Conclusions and Recommendations

The National Research Council charged this committee with providing insight and information on the future of chemical pesticide use in United States agriculture. The committee was charged to:

• Identify the circumstances under which chemical pesticides may be required in future pest management.
• Determine what types of chemical products are the most appropriate tools for ecologically based pest management.
• Explore the most promising opportunities to increase benefits and reduce health and environmental risks of pesticide use.
• Recommend an appropriate role for the public sector in research, product development, product testing and registration, implementation of pesticide use strategies, and public education about pesticides.

The scope of the study was to encompass pesticide use in production systems—processing, storage, and transportation of field crops, fruits, vegetables, ornamentals, fiber (including forest products), livestock, and the products of aquaculture. Pests to be considered included weeds, pathogens, and vertebrate and invertebrate organisms that must normally be managed to protect crops, livestock, and urban ecosystems. All aspects of pesticide research were to be considered—identification of pest behavior in the ecosystem, pest biochemistry and physiology, resistance management, impacts of pesticides on economic systems, and so on.

Because its task was so broad and it had a relatively short period for its study, the committee met five times over 11 months in 1998 and held three workshops to seek input from the public. A critical early challenge was to refine the charge. The committee defined the future of agriculture to be the next 10–20 years. Beyond 20 years, it was felt that predicting technological innovations and their effects is extremely difficult. For example, it would have been extremely prescient for someone to predict in 1979 that within 20 years transgenic varieties would constitute upwards of 25% of all planted acreage of some crops. A 20-year span apparently is also sufficient for adverse unexpected effects of technology to manifest themselves. In the case, for example, of the chlorinated hydrocarbons, widespread use beginning in the early 1950s culminated in regulatory restrictions in the early 1970s (Chapter 2).

The committee also felt that the term *pesticide* required a more precise definition for the purpose of its report. The legal definition, set forth in the Federal Insecticide, Fungicide, and Rodenticide Act (FIFRA), is in part inconsistent with biological definitions of pesticides. The definition also has social aspects; public perceptions color related policy discussions and decisions (Chapter 2). Accordingly, the committee took a broad view of *pesticide*, including the strict legal definition, but also including microbial pesticides, plant metabolites, and agents used in veterinary medicine to control insect and nematode pests (Chapter 2).

With respect to the first charge—to identify circumstances in which chemical pesticides will continue to be needed in pest management—the committee decided early during its deliberations that an assessment of the full range of agricultural pests and of the composition and deployment of chemical pesticides to control pests in various environments would be an impossible task because of the large volume of data and the number of analyses required to generate a credible evaluation. The committee reviewed the literature and received expert testimony on the potential effects of pesticides on productivity, environment, and human health (Chapter 2) and on the potential to reduce overall risks by improving IPM approaches that use chemicals under diverse conditions—soils, crops, climates, and farm-management practices. The committee concluded that uses and potential effects of chemical pesticides and alternatives to improve pest management vary considerably among ecosystems. That conclusion was reinforced by expanded solicitation of expert opinion (Chapter 5). Overall, the committee concluded that chemical pesticides will continue to play a role in pest management for the foreseeable future, in part because environmental compatibility of products is increasing—particularly with the growing proportion of reduced-risk pesticides being registered with the Environmental Protection Agency (EPA) (Chapter 4), and in part because competitive alternatives are not universally available. In many situations, the benefits of pesticide use are high relative to risks or there are no practical alternatives.

With respect to the second charge—to determine the types of chemical products that are most appropriate for ecologically based pest management—the committee concluded that societal concerns, scientific advances, and regulatory pressures have driven and continue to drive some of the more hazardous products from the marketplace. Synthetic organic insecticides traditionally associated with broad nontarget effects, with potentially hazardous residues, and with exposure risks to applicators are expected to occupy a decreasing market share (Chapter 3). This trend has been promoted by regulatory changes that restricted use of older chemicals and by technological changes that lead to competitive alternative products (Chapters 3 and 4). Many products registered in the last decade have safer properties and smaller environmental impacts than older synthetic organic pesticides (Chapter 4). The novel chemical products that will dominate in the near future will most likely have a very different genesis from traditional synthetic organic insecticides; the number and diversity of biological sources will increase, and products that originate in chemistry laboratories will be designed with particular target sites or modes of action in mind (Chapter 4). Innovations in pesticide-delivery systems (notably, in plants) promise to reduce adverse environmental impacts even further but are not expected to eliminate them.

The committee recognized, however, that the new products share many of the problems that have been presented by traditional synthetic organic insecticides. For example, there is no evidence that any of the new chemical and biotechnology products are completely free of the classic problems of resistance acquisition, nontarget effects, and residue exposure. Genetically engineered organisms that reduce pest pressure constitute a "new generation" of pest-management tools, but genetically engineered crops that express a control chemical can exert strong selection for resistance in pests (Chapter 4). Similarly, genetically engineered crops that depend upon the concomitant use of a single chemical pesticide with a mode of action similar to that of the transgenically expressed trait could increase the development of pest resistance to the chemical. Moreover, adverse environmental impacts are still considerations (Losey et al., 1999). A recent study, for example, has revealed that a variety of transgenic *Bacillus thuringiensis* (*Bt*) corn has been found to release *Bt* toxins into the soils via root exudates (Saxena et al. 1999). Thus, the use of transgenic crops will probably maintain, or even increase, the need for effective resistance-management programs, novel genes that protect crops, chemicals with new modes of action and nonpesticide management techniques.

There remains a need for new chemicals that are compatible with ecologically based pest management and applicator and worker safety. Food residues have been addressed in previous National Research Council reports (for example, Pesticides in the Diets of Infants and Children,

NRC 1993), but applicator and worker safety remains a concern for this committee.

Recommendation 1. There is no justification for completely abandoning chemicals per se as components in the defensive toolbox used for managing pests. The committee recommends maintaining a diversity of tools for maximizing flexibility, precision, and stability of pest management.

No single pest-management strategy will work reliably in all managed or natural ecosystems. Indeed, such "magic bullet" fantasies have historically contributed to overuse and resistance problems (Chapter 2). Chemical pesticides should not automatically be given the highest priority. Whether they should be considered tools of last resort depends on features of the particular system in which pest management is being used (for example, agriculture, forest, or household) and on the degree of exposure of humans and nontarget organisms. Pesticides should be evaluated in conjunction with all other alternative management practices not only with respect to efficacy, cost, and ease of implementation but also with respect to long-term sustainability, environmental impact, and health.

With regard to the second charge—identifying what types of chemical products will be required in specific managed ecosystems or localities in which particular chemical products will continue to be required—the committee decided that there is too much variability within and among systems to provide coherent and consistent recommendations. Differences among managed and natural ecosystems in biological factors, such as pest pressure, and in economic factors, such as profitability, make generalizations about particular products of little value. Indeed, generalizing across systems as to the necessity of pesticides is responsible in part for many concerns and conflicts of opinion.

As for the third charge—to identify the most promising opportunities for increasing benefits of and reducing risks posed by pesticide use—the committee identified these:

- Make research investments and policy changes that emphasize development of pesticides and application technologies that pose reduced health risks and are compatible with ecologically based pest management.
- Promote scientific and social initiatives to make development and use of alternatives to pesticides more competitive in a wide variety of managed and natural ecosystems.
- Increase the ability and motivation of agricultural workers to lessen

their exposure to potentially harmful chemicals and refine worker-protection regulations and enforce compliance with them.

- Reduce adverse off-target effects by judicious choice of chemical agents, implementation of precision application technology and determination of economic- and environmental-impact thresholds for pesticide use in more agricultural systems.
- Reduce the overall environmental impact of the agricultural enterprise.

The most promising opportunity for increasing benefits and reducing risks is to invest time, money, and effort into developing a diverse toolbox of pest-management strategies that include safe products and practices that integrate chemical approaches into an overall, ecologically based framework to optimize sustainable production, environmental quality, and human health.

With respect to the fourth charge—recommending an appropriate role for the public sector in research, product development, product testing and registration, implementation of pesticide use strategies, and public education about pesticides—the committee agreed that specific policy actions in research, education, regulation, and management can enhance the likelihood that the opportunity for public-sector contributions is not missed.

Research topics that should be targeted by the public sector include

- Minor-use crops.
- Pest biology and ecology.
- Integration of several pest-management tools in managed and natural ecosystems.
- Targeted applications of pesticides.
- Risk perception and risk assessment of pesticides and their alternatives.
- Economic and social impacts of pesticide use.

TRADEOFFS IN PESTICIDE DECISION-MAKING

Members of a society differ in their evaluation of and preferences for pesticide use in agriculture. Because a society has limited resources and government cannot meet everyone's expectations, tradeoffs become central to policy decision-making. Making wise tradeoffs is one of the most difficult challenges for policy-makers. A role for government is to achieve a fair balance between the risks that a community bears and the benefits that it receives.

In general, cost-benefit analyses are important tools for informing public-policy decisions regarding use of chemical pesticides. The impacts of pesticides on the economy and environment are measured in monetary terms, and estimates are incorporated into comparisons of costs and benefits associated with alternative public actions. A considerable body of evidence exists on potential risks posed by and benefits of established categories of some pesticides; however, there are many uncertainties in measuring the full array of benefits and costs of pesticide use. Furthermore, the Food Quality Protection Act has fundamentally changed the use of cost-benefit analyses in that the law now largely precludes consideration of benefits.

Pesticides provide economic benefits to producers (Chapter 2) and by extension to consumers. One of the major benefits of pesticides is protection of crop quality and yield. Pesticides can under some circumstances prevent large crop losses, thus raising agricultural output and farm income (Zilberman 1997). The increased supply of crops can be passed on to consumers in the form of lower food prices (although low prices can adversely affect farmers regardless of production increases). Many farmers are responsible land stewards and are concerned with potential environmental impacts of pesticides, but it is unrealistic to expect most farmers to adopt alternative pest-management strategies that would decrease their profits without the use of some policy incentives and disincentives.

Recommendation 2. A concerted effort in research and policy should be made to increase the competitiveness of alternatives to chemical pesticides; this effort is a necessary prerequisite for diversifying the pest-management "toolbox" in an era of rapid economic and ecological change.

Many involved with agriculture recognize that pest-control practices can be improved. Lack of efficacy, risks to workers who apply pesticides, and threats to the environment often are cited as constraints of current pest-management strategies. Some producers are adopting alternatives to pesticides as part of a holistic management approach that integrates multiple farming objectives with agroenvironmental goals. A grower's decision to use pesticides depends on a broad array of biological and economic factors. However, a decision to use pesticides can alter the environment in ways that affect later pest-management choices.

Organic foods and ecolabeling markets are creating new opportunities for growers who are willing to reduce or exclude synthetic chemicals in their production practices. Environmentally friendly products appeal to consumers, too; organic food sales are growing at a rate of 20%/year in the United States. Yet policy analysts report that only 0.1% of agricultural research is devoted to organic farming practices. (Chapter 3). Availability of alternative pest-management tools will be critical to meet the production standards and stiff competition expected in these niche markets.

Globalization policies and practices are affecting pest management on and off the farm. Reduction in trade barriers increases competitive pressures and provides extra incentives for United States farmers to reduce costs and increase crop yields. In a global marketplace, United States farmers can compete with farmers from other countries where labor, land, and input costs are lower only by being more "productive," with higher yields per acre. Other forms of trade barriers create disincentives for adopting new technologies (such as the reluctance of the European Union to accept genetically modified organisms). It is not well understood how globalization will affect agriculture in developing countries, where 80% of the world's population lives (Schuh 1999). It is likely that trade will increase the spread of invasive pest species and pose risks to domestic plants and animals, as well as populations of native flora and fauna. Increased pressure to phase out the widely used ozone-depleting chemical methyl bromide has raised questions as to the availability of cost-effective substitutes. To meet those emerging global pest problems, researchers will need to develop effective, environmentally compatible, and efficient pest controls as a complement to a suite of prevention strategies.

Nontarget effects of exposure of humans and the environment to pesticide residues are a continuing concern. The side effects of pesticide use vary over time and space. In many cases, the environmental damage associated with the application of a chemical in a riparian zone, for example, is much larger than that associated with an application in other areas. When it comes to a local environmental-quality problem caused by pesticide use, application technologies and the location matter much more than the volume of pesticide applied. Numerous studies have shown that pesticides decrease crop loss, but the potential indirect environmental impacts of pesticides are not easily determined. The application of pesticides results in indirect effects on ecosystems by reducing local biodiversity and by changing the flow of energy and nutrients through the system as the biomass attributable to individual species is altered (Chapter 4). Pesticide policies should be based on sound science; where there is uncertainty, expert judgment will become more important in decision-making. Across-the-board pesticide policies that do not account for biological and ecological factors and for socioeconomic influences are likely to be less effective.

Pesticide resistance now is universal across taxa (Chapter 2). Pests will adapt to counter any control strategy that results in the death or reduced fitness of a substantial portion of their population. Cultural and biological controls are not immune to evolution of resistance. Pesticide resistance is conspicuous because of the intensity of selection by high-efficacy chemicals. Most cases of insecticide resistance are to chlorinated hydrocarbons because these compounds have been in widespread continuous use for many years. Herbicide resistance is becoming more of a problem, and severe impacts can occur when there are no alternative herbicides to control the resistant genotypes or when available alternatives are relatively expensive. By spreading the burden of crop protection over multiple tactics, rather than relying on a single tool, farmers will face less risk of crop loss and lower rates of pest adaptation to control measures (Bessin et al. 1990, Vangessel et al. 1996). Because pests will continue to evolve in response to pest controls, research needs to support development of pest-management tools that reduce selection pressure, delay selection for resistance, and thus increase the life of chemical and other products.

GOVERNMENT SUPPORT OF RESEARCH AND DEVELOPMENT

There are several justifications for public engagement in these lines of research. First, there is underinvestment from a social perspective in private-sector research because companies will aim to maximize only what we have called suppliers' surplus (difference between suppliers' income and their production costs) rather than the social surplus. Companies will compare their expected profits from their patented products resulting from research and will not consider the benefits to consumers and users. For example, the development and release of a self-sustaining biological control agent, such as a parasitoid or entomopathogen, or the development of an integrated crop-rotation program might benefit farmers and food consumers but not provide a marketable commodity to a company. Second, research activities are very risky, and even big corporations might have a higher aversion to risks than society does, and that will lead to underinvestment in risky research activities (Sandmo 1971). Recent developments in computers and medical biotechnology illustrate this point. Many of the breakthrough innovations in these fields were the result of publicly funded university research whose results were patented. The rights to use the patents were sold to new startup companies that in many cases were the result of partnerships between university professors and venture capitalists (Zilberman et al.1998, Parker et al. 1998). That is the case with important companies that include Sun, Cisco, Netscape, Genentech, Chiron, and, in agriculture, Calgene. Major corporations often take over such companies once they are established. Publicly supported university research, through the process of technology transfer, has become a source of economic growth in the United States, and this model is now being imitated elsewhere (Levy 1998). One reason that biotechnology research in agriculture is lagging in its development relative to medical biotechnology is the paucity of public support of discoveries in agriculture.

A third reason why public research might lead to innovations that elude the private sector is the different incentives that researchers in the private and public sectors face. For the most part, private sector researchers emphasize projects that improve existing product lines. The advancement of public researchers is affected by their publications in refereed journals, where novelty and originality have a premium. This argument suggests that public research grants should be allocated competitively to generate the highest-quality research.

A fourth argument for public support of research is that much of the funding is allocated to institutions of higher learning and used to train future scientists for the private sector. Availability of trained scientists will be a key to future innovation in pest-management technologies. Finally, the public sector should conduct research in areas that are pursued by the private sector to have the information and background for regulatory purposes.

Provision of information to policy-makers and the need to design and enforce policies to address problems of externalities is a justification of public research on benefit-risk assessment, producers' behavior, and environmental fate (externalities occur when activities of one economic agent indirectly affect the well-being of another—for example, by generating pollution). Because one avenue to reduce side effects of pesticides is improving application technologies, and the private sector might not invest in developing such innovations until policy incentives are enacted, the public sector can conduct some basic research in application technology to identify feasible avenues that will provide basic information in assessing new regulations.

The private sector is likely to invest in development and regulatory activities needed to introduce pest-control technologies for major markets. However, the private sector often cannot capture these costs in the case of specialty crops (such as, fruit, vegetable, and nursery crops). If one adds consumer and user surpluses to supplier surplus for these crops, the net social benefits might be positive. Therefore, one idea for government intervention is to support and provide extra incentives for the introduction of pest control in specialty crops where consumer surpluses could be important and supplier surpluses are typically small.

Pests will continue to thrive, and a strong science and technological base will be needed to support management decisions. We need to continue to use the best science to resolve these questions, and policy-makers need to use the best science in their decision-making. As new technologies develop, theoretical frameworks for resolving such questions continue to be developed.

Recommendation 3. Investments in research by the public sector should emphasize those areas of pest management that are not now being (and historically have never been) undertaken by private industry.

Federal funding of pesticide research has historically had a very narrow base, as seen with the Agricultural Research Service and National Science Foundation (NSF) funding (Chapter 1).

To diversify the range of tools available for managing pests, a diversity of approaches would be beneficial. The chief desirable policy changes to diversify the research enterprise are highlighted below.

Recommendation 3a. Investment in pest management research at USDA should be increased and restructured in particular to steadily increase the proportion and absolute amounts directed toward competitive grants in the National Research Initiative Competitive Grants Program (NRI), as opposed to earmarked projects.

A greater emphasis on research—not only on chemicals themselves, but also on the ecological consequences of pesticide use—can increase the probability that new products will be readily integrated into ecologically based pest-management systems. Pest biology and management studies represented some 15–17% of the funding allocations under NRI in 1998 and 1999; however, only 15 studies of biologically based pest management were funded in each of the last 2 years (USDA 2000).

Recommendation 3b. Total investment in pest management and the rate of new discoveries should be increased by broadening missions at funding agencies other than USDA—specifically, the National Institutes of Health (NIH), the NSF, EPA, Department of Energy (DOE), and the Food and Drug Administration (FDA) to address biological, biochemical, and chemical research that can be applied to ecologically based pest management.

The perspective on research and development and new products should be global and should take into account the collaboration and partnerships in research.

Investment in basic research applicable to ecologically based pest management is consistent with the missions of the funding agencies. Such initiatives could include

- Obtaining the ecological and evolutionary biological information necessary for design and implementation of specific pest-management systems.

• Identifying ways to enhance the competitiveness of alternatives or adjuncts by investing in studies of cultural and biological control.

• Elucidating fundamental pest biochemistry, physiology, ecology, genomics, and genetics to generate information that can lead to novel pest-control approaches.

• Examining residue management, environmental fate (biological, physical, and chemical), and application technology to monitor and reduce environmental damage and adverse health effects of both pesticides and pesticide alternatives.

The lack of basic information on pest population spatial and temporal dynamics is a major impediment to implementation of ecologically based pest management. NSF and EPA could make an important contribution by funding research associated with understanding of pest-population and community dynamics. This type of research is funded by these agencies, but it focuses mostly on natural, as opposed to managed, ecosystems. In addition, all agencies could improve the basic understanding of pests and their impacts by funding longer-term projects that would adequately capture the variability in pest dynamics, including pesticide-resistance evolution, under alternative management systems.

Recommendation 3c. On-farm studies, in addition to laboratory and test-plot studies, are a necessary component of the research enterprise. Investment in implementation research, which helps to resolve the practical difficulties that hinder progression from basic findings to operational utility, is needed.

The idiosyncratic nature of individual agroecosystems limits the utility of both laboratory and test-plot studies in predicting the efficacy of pest-management strategies. An increased emphasis on large-scale and long-term on-farm studies through the use of the global positioning system (GPS) and global information system (GIS) technologies could contribute substantially to diversifying management tools and approaches. The goal might be best achieved by investment of at least some earmarked funds to ensure stability of funding. Such research programs should remove the gap between "basic" and "demonstration" research for all managed and natural ecosystems. USDA needs to fund applied research because there are limits to models that serve basic science as well. Such models advance fundamental knowledge, but often the major economic problems involve organisms that are hardly ideal from a fundamental scientific viewpoint. These problems can be best addressed by research on the organisms in question. The information generated by applied on-farm research is crucial to extension scientists, crop consultants, and producers.

Recommendation 3d. Basic research on public perceptions and on risk assessment and analysis would be useful in promoting widespread acceptance and adoption of ecologically based management approaches.

The body of literature evaluating public responses to agrochemicals in particular and pest management in general is not extensive. Surveys that have been done (e.g., Potter and Bessin, 1998, Chapter 3) indicate that communication with the public about pesticides and their alternatives has been ineffectual. Media coverage of integrated pest management (IPM) in widely circulated urban newspapers is sketchy and tends to focus on urban issues, thus providing little useful information to readers relevant to integrated pest management in agricultural settings (Caldwell et al. 2000). Although there have been substantial advances in research on risk perception in recent years, risk communication is a relatively new discipline (Petersen, 2000). Research priorities include elucidating impacts of increasing benefit perceptions in risk communication, developing empirical methods for more accurate characterization of public perceptions, identifying reasons for differing qualitative and quantitative perceptions about pesticide technology and agrobiotechnology, and determining whether risk communication can reduce the gap that exists between public perceptions and scientific risk assessments (Petersen 2000).

Funding of New Ventures in Agricultural Biotechnology and Biopesticides

A broad body of research in economics and policy has investigated the role of the public sector in the economy (Laffont 1988). A major justification of government intervention is that it can improve the efficiency of the economy by instituting policies that expand the human well-being derived from the resource base. To improve efficiency, government intervention might be required to establish and enforce property rights, overcome deficiencies in information availability, regulate monopolistic behavior, provide the institutional foundation to settle externality problems, or augment the provision of public goods. Public goods are products and activities to which people have free access for which they do not need to compete; free air is a pure public good, as is national defense. Because of open access, individual consumers and producers might underinvest in public goods and have a tendency to take these resources for granted. The need for public-sector intervention in the economy is quantitatively determined by comparing socially optimal resource outcomes with market outcomes. For example, if monopolies underproduce and overcharge for their products, government regulation could lead to increased production and lower prices. Government investment is needed to provide optimal levels of public goods so that incremental social benefits are equal to incremental social costs.

The public sector consists of various layers of government (local, state, federal, and international). In theory, each level of government addresses problems that affect its constituency. The justifications of government intervention in the management of pest control include the need to address the externality problems associated with the human and environmental health effects of pesticides and the information uncertainties regarding pesticides and their impacts. The performance and value of pest-control technologies depend on their specific properties and the manner of their application. The regulatory process has been designed to screen out the riskier materials. However, few incentives exist for efficient and environmentally sound pest-control management strategies. Introduction of incentives that would reduce the reliance on riskier pest-control strategies and encourage the use of environmentally friendly strategies is likely to lead to increased efficiency in pesticide use. Such incentives as taxes and fees for the use of various categories of chemicals have been recommended, but because of user objections they might not always be politically feasible. Users might prefer subsidies to reduce pesticide loads, but this policy may strain the public budget.

Establishing regional pesticide targets and implementing them through tradable permits is a better solution that will achieve the same outcome. Because the environmental effects of pesticides can vary with location, one policy approach to reduce pesticide use in an environmentally sensitive area is to institute programs like the Conservation Reserve Program, which pays farmers to modify their behavior in a way that will promote improved environmental quality. Another practical policy that seems to have worked is to condition entitlement to government programs on meeting specified criteria of environmental stewardship. Because pest-control devices are often used as mechanisms to reduce uncertainty, experiments with subsidized insurance schemes against pest damage might constitute another avenue to induce adoption of improved pesticide practices.

Worker-safety concerns have emerged as a major problem associated with pesticide use. There have been some important improvements, owing, in part, to the mandates of the 1992 Worker Protection Standards (WPS) and the 1996 Food Quality Protection Act, but the search for more-efficient policies should continue. Development of these policies might entail investment in research to improve monitoring on the farm to allow more precise responses to changes in environmental conditions. Some of the presentations to the committee suggested that worker-safety problems are the result, not of the laws on the books, but of monitoring and enforcement (see Chapter 3, for discussion of WPS).

Sometimes objections to pesticides are an issue of subjective preference even when scientific evidence cannot support the objections. In this case, a government role in banning a pesticide might not be appropriate; an appropriate role might be to establish a legal framework that enables organic and pesticide-free markets to emerge and prosper so that consumers can be given an informed choice between lines of products that vary with pest management.

Although agricultural biotechnology is more successful now than it was 2–5 years ago, raising money for agricultural ventures is still difficult for various reasons. First and most important, few investors have expertise in agricultural biotechnology. Venture capitalists invest in industries that they know and have a history of successful investing in. Second, although investors who exited early from their investments in 1980s agricultural-biotechnology companies made good returns, there are no blockbuster successes such as agricultural versions of Genentech or Amgen. Third, agricultural biotechnology must compete with telecommunication, software, Internet, and health-care biotechnology for venture capital. Fourth, few new agricultural-biotechnology companies have continued to generate investor interest (there is no "critical mass"). For United States agriculture to stay in the forefront with safer, environmentally friendly pest-management tools, there needs to be a continuing cycle of innovative new companies that research risky cutting-edge technologies.

Recommendation 4. Government policies should be adapted to foster innovation and reward risk reduction in private industry and agriculture. The public sector has a unique role to play in supporting research on minor use cropping systems, where the inadequate availability of appropriate chemicals and the lack of environmentally and economically acceptable alternatives to synthetic chemicals contribute disproportionately to concerns about chemical impacts.

The public sector can foster innovation in product development and pest-management practices by continuing to reduce barriers to investment by the private sector and by increasing implementation of regulatory processes that encourage product and practice development. Pesticide use in minor-use crops, which include most fruits and vegetables, is of great concern to the public because of the potential for worker exposure and residues on foods (chapter 3). These crops are not the focus of industry research, because the small acreage available for treatment is insufficient to result in substantial net profits. To accomplish these ends, the base for public funding should be broadened so as to take advantage of multiple opportunities for innovation.

Recommendation 4a. The Department of Commerce Advanced Technology Program should be encouraged to fund high-risk R&D for IPM, EBPM and alternatives that have commercial potential for early stage companies.

The Advanced Technology Program (ATP) funded by the Department of Commerce awards grants that average $1-5 million, making it a valuable source for an early stage company. Typically ATP awards companies grants for risky, cutting edge R&D that has commercial potential. New tools for ecologically based pest management could get a boost if new companies could successfully compete for ATP funding for developing new IPM tools and alternatives.

Recommendation 4b. Incentives should be increased for private companies to develop products and pest-management practices in crops with small acreages, including access to compete for Interregional Research Project 4 (IR4) funds used to obtain product registrations for minor-use crops.

The Interregional Research Project 4 (IR4) program exists to assist in getting products registered on minor crops. IR4 awards grants to university researchers for biopesticide research. It has a long history of success in getting registrations for products for minor crops when there are no incentives for large companies to do so. We expect over the next few years to see as much success with biopesticides as IR4 has had with chemicals. Private companies are not allowed to obtain grants, but they are most capable of moving new products to market. The IR4 program should broaden its scope and allow private companies to obtain grants. IR4 should also better measure the outcomes (such as impact on farmers) of its current biopesticide grant program for academic researchers.

Recommendation 4c. Redundancy in registration requirements should be reduced to expedite adoption of safer alternative products (such as biopesticides and reduced-risk conventional pesticides)

Increased harmonization of review processes between EPA and the California Environmental Protection Agency (CAL-EPA), for example, would reduce the time requirements for registration. To reduce duplication, such agencies as EPA and CAL-EPA should divide the review tasks up front. Streamlining registration, however, must not come at the expense of public safety and local preferences.

Incentives can also be put into place to foster the development of products and pest-management practices in minor-use crops. Currently, the EPA's Biopesticide and Pollution Prevention Division (BPPD) has responsibility for registering new microbial and biochemical pesticides under subdivision M of the FIFRA. It takes 12–24 months to complete a biopesticide registration in BPPD. If that time could be reduced to less than 12 months for minor crops without compromising human and environmental-safety screening for minor-use crops, there would be an even greater favorable financial impact on small companies, and farmers would benefit by having earlier access to products. A recent problem for BPPD is that the division is responsible for transgenic crops in addition to biopesticides. BPPD has had to immediately address environmental groups' and Congress's issues on transgenic crops, which has slowed down reviews of new biopesticides.

Recommendation 4d. Evaluation of the effectiveness of biocontrol agents should involve consideration of long-term impacts rather than only short-term yield, as is typically done for conventional practices.

Many biocontrol agents are not considered acceptable by growers, because they are evaluated for their immediate impact on pests (that is, they are expected to perform like pesticides). Some biocontrol pathogens used against weeds might cause as little as a 10% reduction in fecundity, which might not be a visible result but has a major long-term effect causing population decline. Low-efficacy biocontrol agents alone might not be acceptable for pest management but, in combination with other low-efficacy tactics, they could be preferable because they avoid the selection for resistance for that is associated with high-efficacy tactics.

Recommendation 4e. At the farm level, incentives for adopting efficient and environmentally sound integrated pest-management and ecologically based pest-management systems can come from

- Expanding crop insurance for adoption of integrated pest-management and ecologically based pest-management systems practices.
- Implementing taxes and fees on environmentally higher-risk practices.
- Setting up tradable permit systems to reduce overall pollution emissions.
- Ensuring availability of funds in support of resource conservation (such as the Conservation Reserve Program).
- Conditioning entitlement to government payment on environmental stewardship.
- Assessing and more stringently enforcing regulations designed to protect worker health and safety.

Innovative crop-insurance policies can be developed to promote the adoption of pesticide alternatives and to increase their economic competitiveness. USDA is developing and piloting some innovative crop-insurance programs to increase incentives to farmers to use alternative products and IPM systems (James Cubie, USDA, and Tom Green, IPM Institute of America, October 10, 1998, personal communications) that reduce the number of pesticide sprays. They include crop insurance for use of IPM in potatoes designed to reduce chemical applications in late blight (so that growers wait until conditions exist for disease instead of using prophylactic sprays) and in corn to control corn rootworm (so that growers scout for corn rootworm adults and do not spray if numbers are below economic thresholds). Another potential approach includes U.S. Senator Richard Lugar's Farmers' Risk Management Act of 1999 (S. 1666), which would change the way crop insurance has traditionally been used to a risk-management approach that would involve landowners in helping them to financial viability. The Act, if it passes, would provide eligibility for crop insurance if IPM or crop advisers are used.

Recommendation 4f. Funds should be assigned to assess compliance with Worker Protection Standards and to improve worker health and safety in specialty crops.

Worker safety in specialty crops is a serious concern (Chapter 3). There are two interacting problems: a more intensive interface between worker and crop and inadequate effort in developing safer products and practices. Those problems are exacerbated by the collective importance of such crops in diversifying and enriching the United States diet (Chapter 3). The current Agriculture Health Study (Alavanja et al. 1996) emphasizes only two states (Iowa and North Carolina) and does not specifically address specialty (minor-use) crops. Funds should be assigned to study worker health and safety in specialty crops and to assess compliance with WPS. Without more detailed objective information on compliance, there is a reasonable doubt that the 1992 WPS is accomplishing its goal. Conducting an objective study of compliance with WPS will be difficult but important. It is imperative that the organization and individuals conducting such a study be unbiased and have no conflict of interest. There is a need for better data regarding exposure under different conditions in the field and a need for better models of this exposure for accurate worker risk assessment.

Furthermore, the diffuse nature of the data demands sophisticated sociological, epidemiological, and statistical expertise (as would be possessed by university scientists). Before such a study is conducted, however, the WPS needs to be studied to determine the degree of compliance that will be considered sufficient to protect worker health.

As mentioned earlier, efforts are needed to address safer practices for workers. Farm workers typically do not know when or what pesticides have been applied to fields, so they must rely on their employers to protect them from hazardous exposure. Because some employers might not follow WPS regulations, funds should be assigned to develop pesticide formulations that contain specific odors or dyes that would provide farm workers with direct information on the presence of hazardous pesticide residues.

Tools are available that allow inexpensive monitoring of worker exposure to pesticides and monitoring of the pharmacokinetics of compounds at concentrations commensurate with actual environmental exposures. These tools should be used.

EDUCATION AND INFORMATION

It is clear from the committee's study that the general public has a critical function in determining the future role of pesticides in US agriculture. Consumer interest in food and other goods perceived as safe and healthy fuels the rapid growth of the organic-food market; at the same time, consumer use of pesticides in the home and on the lawn continues to grow (Chapter 2). Many of the paradoxical decisions made by the voting and consuming public arise from a relatively poor grasp of the science behind crop protection (Chapter 3).

The public sector has responsibility for providing education and information. Publicly supported education at all levels is essential for facilitating equality of opportunity. The working of a democracy and the efficiency of the market are infeasible without an educated public. Availability of information is essential for efficient resource allocation. Because knowledge also has public-good properties, a major responsibility of the public sector is to provide basic knowledge and information for decisionmakers, in both the public and private sectors (Ruttan 1980).

Education in scientific and technical fields is designed to meet anticipated demands in the private and public sectors. As long as there is a demand for pesticide-based solutions to pest-control problems, the education system has to train people to work in this field and to provide independent pesticide expertise in the public sector. Because we agree that pest-control choices have to be determined in the context of a perspective that incorporates biophysical, ecological, and economic considerations, education should emphasis basic principles and knowledge that will lead to informed decisions. Schultz (1975) emphasized the value of a good understanding of basic principles in a modern economy. He argued that people with a solid general education in the principles of science have a better ability to deal with the disequilibrium associated with continuous changes in technology.

The broad set of considerations associated with pest-control decisions requires more interdisciplinary education in land-grant universities. People trained in life sciences and agriculture should also have a strong background in decision theory, risk evaluation, ethics, and economics to be able to handle pest-control problems in the commercial world. Many of the decisions associated with pest control are subject to public choice and public debate. To obtain rational and efficient outcomes, it is essential that scientists be able to communicate with the public in a clear and nontechnical manner about the tradeoffs associated with alternative pest-control issues.

All citizens should be familiar with the basic principles of applied biology and risk evaluation, which can be provided as part of basic education. The general public, including children from kindergarten to 12th grade, should be educated about basic principles of environmental risk and of pest and disease control.

Agriculture has become more science-based and requires much more specific expertise to enhance productivity. As the support and funding for extension increase, new types of institutions and private consultants emerge. Transmission of knowledge in the past was the responsibility mostly of the public sector, but it has become more privatized. That changes how pesticide-use decisions are made and has introduced into the process a value system that might not always have the public's interests at its core.

Recommendation 5. The public sector must act on its responsibility to provide quality education to ensure well-informed decision-making in both the private and public sectors.

This effort encompasses efforts in the agricultural sector, in the academic sector, and in the public sector at large.

Recommendation 5a. In the agricultural sector, a transition should be made toward principle-based (as opposed to product-based) decision-making. The transition should be encouraged throughout the continuum from basic to implementation research in universities, in extension, in the USDA Agricultural Research Service, and among producers.

Formulaic approaches to pest problems that are aimed at yield maximization rather than at sustainability approaches (product-based decision-making) have contributed to many of the problems plaguing agriculture. A sound grasp of fundamental principles should provide decision-makers with the flexibility needed to select from a menu of alternatives and to tailor practices to particular production systems.

A basic premise of an ecological approach to pest management is to manipulate biological processes to manage pests. For example, the impact of weeds on crop yield will depend on a number of biological factors, such as time of weed emergence and rate of weed growth (Box 6-1). Understanding these physiological factors can help a pest manager to determine optimal timing for intervention strategies, such as herbicide use to manage weed population to a level where benefits outweigh costs of control. University research and extension scientists, in collaboration with producers, should strive to develop biologically based decision-making that optimizes pest-management strategies for economic and environmental goals.

5b. Land-grant universities should emphasize systems-based interdisciplinary research and teaching and foster instruction in applied biology and risk evaluation for nonscientists.

There is a need to educate legislators and the general public about ecologically based pest management in research and in practice. Investment in increasing K-12 exposure to concepts of risk evaluation, food agriculture, and general biology can also have enormous benefits in creating a more knowledgeable and educated electorate.

5c. An effort should be made, in the government and in the land-grant system, to educate and train scientists about the value of public outreach.

The public sector should provide incentives and training for scientists to communicate effectively to the public about principles and practices of ecologically based pest management. Such incentives are almost nonexistent in many institutions, particularly outside the agriculture colleges. Outreach efforts, because of their overall value in providing popular support for the research enterprise, should be accorded some commensurate value in decisions related to career advancement and professional stature.

CODA

Our goal in agriculture should be the production of high-quality food and fiber at low cost and with minimal deleterious effects on humans or the environment. To make agriculture more productive and profitable in the face of rising costs and rising standards of human and environmental health, we will have to use the best combination of available technologies. These technologies should include chemical, as well as biological and recombinant, methods of pest control integrated into ecologically bal-

BOX 6-1
Working Model for Assessing Integrated Weed-Management Strategies

Highly efficacious pesticides often mask the complexity of a cropping system-pest relationship. Many crop producers and scientists are just beginning to understand the relationships between biological and agronomic practices and the impacts of a pest on a crop. Herbicides, in particular, have made the recognition and integration of more biologically based alternatives difficult. Commodity-grain producers now use highly efficacious and fast-acting herbicides as time- and labor-management tools, giving less regard to the biological and agronomic principles that can increase or decrease a herbicides effectiveness and durability (see box 5-1).

For more than 30 years, herbicide use has been the dominant tactic for weed management. Over this period, herbicide technology has been a primary focal point of the agrochemical industry-university relationship. Historically, university herbicide-evaluation trials were established to demonstrate the efficacy of individual products and to promote weed-management research. In its technology-transfer role, the University Extension Service also provided a public service by offering a third-party neutral perspective to this manufacturer-crop producer relationship.

During the 1990s major changes occurred in both institutions. As agrochemical markets became saturated, herbicides were differentiated more by marketing and product endorsements than by efficacy. Past agrochemical and university relationships were centered around basic and applied research and technology transfer. The current economic climate has resulted in agrochemical industries doing more of their own basic and applied herbicide research and has tended to push university herbicide evaluation and technology transfer to more of a marketing and product-endorsement focus. Although a marketing perspective might address attributes of a single herbicide, this approach to weed management does not address the complexities of integrating a pesticide into a cropping system of an individual producer.

The University of Minnesota is responding to public demand for alternatives to herbicide weed-control technology by adopting a biological and ecological research strategy with a goal of developing integrated weed-management systems, including herbicides as a component. University of Minnesota Extension and Branch Station personnel have formed a weed-management working group. The group focuses on evaluating weed management systems rather than individual herbicides. Its goals are as follows:

• Maintain critical information links between the agrochemical industry and the University of Minnesota.

• Develop research efforts that enhance extension activities required to address specific regional and site-specific producer concerns.

• Integrate herbicide-based weed control with other practices, such as cultivation, rotary hoeing, tillage, and weed-crop interference models.

• Evaluate integrated weed-management systems for efficacy, economic viability, and degree of risk to the producer.

• Develop and maintain a database of weed-control research for studying the interaction between weed control and environmental and agronomic variables.

These goals are integrated into the annual agrochemical industry-university protocols with a four-step process:

1. Meet with crop consultants, farmers, area agronomists, extension personnel, and seed representatives on a regional basis to identify grower concerns.
2. Have university personnel design and draft protocols to address the concerns.
3. Give agrochemical representatives the opportunity to review draft protocols for accuracy of label restrictions and requirements.
4. Conduct protocols at Branch Stations and cooperating on-farm sites, and maintain databases of research findings for analysis of long-term trends.

That approach to herbicide analysis has been in place since 1994.

Crop producers and many organizations involved in information dissemination have responded favorably to the model approach, primarily because the objectives of the research are well defined and issues of profitability and integrated weed management are addressed. Obstacles in developing the weed-management group have been related to communication of objectives to agrochemical-industry representatives and other university personnel and to finding funding sources that will sponsor this type of activity. The model system has yet to address more complex and time-intensive biological approaches to weed management, such as smother crops or biological control agents. Because the studies are not conducted on the same field year after year, natural weed population-regulating mechanisms cannot be assessed. The Minnesota model system, however, has stimulated crop producers to begin considering the benefits of integrated weed management and how integrated weed management is affected by environmental and agronomic variables. Studies of this nature need to be conducted on a long-term basis to assess the economic and biological stability of various integrated weed-management systems.

Sources: Box composed by Jeff Gunsolus and adapted from Gunsolus et al., 1995; Hoverstad et al., 1995; and Maxwell, 1999.

anced programs. The effort to reach the goal must be based on sound fundamental and applied research, and decisions must be based on science. Accomplishing the goal requires expansion of the research effort in government, industry, and university laboratories.

REFERENCES

Alavanja, M. C. R., D. P. Sandler, S. B. McMaster, S. H. Zahm, C. J. McDonnell, C. F. Lynch, M. Pennybacker, N. Rothman, M. Dosemeci, A. E. Bond, and A. Blair. 1996. The agricultural health study. Env. Health Pers. 104:362–69.

Bessin, R. T., E. B. Moser, and T. E. Reagan. 1990. Integration of control tactics for management of the sugarcane borer (Lepidoptera: Pyralidae) in Louisiana sugarcane. J. Eco. Ent. 83(4):1563–569.

Caldwell, A.M., K. Bantle-Stoner, G. W Cuperus, M. E. Payton, and R. C. Berberet. 2000. Media coverage of integrated pest management in major urban newspapers, Am. Ent. 46:56–0

Gunsolus, J. L., G. A. Johnson, T. R. Hoverstad, F. R. Breitenbach, and J. K. Getting. 1995. Reevaluation of university herbicide evaluation trials. Proceedings of the North Central Weed Science Society. 50:100–01.

Hoverstad, T. R., J. L. Gunsolus, G. A. Johnson, W. E. Lueschen, F. R. Breitenbach, and J. K. Getting 1995. Corn herbicide evaluation for southern Minnesota, efficacy and an economic analysis. Proceedings of the North Central Weed Science Society: 101–102.

Laffont, J.J. 1988. Fundamentals of Public Economics. Cambridge, Mass.: M.I.T. Press.

Levy, S. 1998. Beyond Silicon Valley. Newsweek. 132(19):44–50.

Losey, J. E., L. S. Rayor, and M. E. Carter. 1999. Transgenic pollen harms monarch larvae. Nature 399:214.

Maxwell, B. D. 1999. My View. Weed Science 47:129.

NRC (National Research Council). 1993. Pesticides in the Diets of Infants and Children. Washington, DC: National Academy Press.

Parker, D. D. Zilberman, and F. Castillo. 1998. Office of technology transfer, the privatization of university innovations, and agriculture. Choices 1998 (1st quarter):19–25.

Petersen, R. K. D. 2000. Public perceptions of agricultural biotechnology and pesticides: recent understandings and implications for risk communication. Am. Ent: 46:8–16.

Ruttan, V. W. 1980. Agricultural Research and the Future of American Agriculture. Staff Paper P80-17. University of Minnesota.

Sandmo, A. 1971. On the theory of competitive firm under price uncertainty. Am. Ec. Rev. 61:65–73.

Saxena, D., S. Flores, and G. Stotzky. 1999. Transgenic plants: insecticidal toxin in root exudataes from *Bt* corn. Nature 402:480

Schuh, G. E. 1999. Providing food security for a burgeoning global population. Talk presented at AAAS conference on Science and Technology for a Changing World. November 1, 1999, Washington, DC.

Schultz, T. W. 1975. The value of the ability to deal with disequilibrium. Jo. of Ec. Lit.:3:827–846.

USDA (US Department of Agriculture). 2000. NRI Annual Report: Fiscal Year 1998. Washington, DC: USDA.

Vangessel, M. J., E. E. Schweizer, D. W. Lybecker, and P. Westra. 1996. Integrated weed management systems for irrigated corn (*Zea mays*) production in Colorado – a case study. Weed Science. 44(2)423–428.

Zilberman, D. 1997. World food needs toward 2020: Discussion. Am. J. Ag. Ec. 79(5):1487–88.

Zilberman, D., C. Yarkin, and A. Heiman. 1998. Intellectual property rights, technology transfer and genetic resource utilization. In Farmers, Gene Banks and Crop Breeding: Economic Analyses of Diversity in Wheat, Maize, and Rice. Boston, Mass.: Kluwer Academic Publishers.

APPENDIX A
Selected National Research Council Publications About Pesticides

1951 *Conference on Insecticide Resistance and Insect Physiology.*

1956 *Safe Use of Pesticides in Food Production.*

1962 *Chemicals in Modern Food and Fiber Production.*

1962 *New Developments and Problems in the Use of Pesticides.*

1969 *Principles of Plant and Animal Pest Control Part 3: Insect-Pest Management and Control.*

1970 *Evaluating the Safety of Food Chemicals.*

1972 *Pest Control Strategies for the Future.*

1975 *Pest Control: An Assessment of Present and Alternative Technologies. Vol. 1. Contemporary Pest Control Practices and Prospects.*

1982 *An Assessment of the Health Risks of Seven Pesticides Used for Termite Control.*

1986 *Pesticide Resistance: Strategies and Tactics for Pest Management.*

1987 *Regulating Pesticides in Food: The Delaney Paradox.*

1991 *Frontiers in Assessing Human Exposure to Environmental Toxicants.*

1993 *Pesticides in the Diets of Infants and Children.*

1996 *Ecologically Based Pest Management: New Solutions for a New Century.*

APPENDIX
B
Workshop Agendas

COMMITTEE ON THE FUTURE ROLE OF PESTICIDES

March 23, 1998
National Research Council
2001 Wisconsin Avenue, Washington, DC

8:30 a.m. Introductory Remarks
May Berenbaum, Chair

8:40 Public Sector Research
Nancy Ragsdale, USDA, Agricultural Research Service
Ernie DelFosse and Ray Carruthers, USDA Agricultural Research Service
Mary Purcell and Kel Wieder, USDA CSREES, NRI Competitive Grants
Michael Alavanja, NIH NCI, Agricultural Health Study
Kate Aultman, NIH, National Institute of Allergy and Infectious Diseases
Margriet Caswell, USDA Economic Research Service
Sam Rives, USDA National Agricultural Statistics Service

10:10 Discussion
10:40 Break

11:10	Regulatory and Registration *Peter Caulkins, EPA OPP, Food Quality Protection Act* *Young Lee, FDA Pesticide Program: Residue Monitoring* *Robert Epstein, USDA, Agricultural Marketing Service Pesticide Data Program* *Dick Guest, USDA CSREES, InterRegional-4 Program*
12:10	Discussion and Morning Wrap-Up
12:40–1:20	Lunch (in NAS Cafeteria)
1:20	Pest Management Implementation *Harold Coble, USDA-CSREES National IPM Coordinator* *Helene Dillard, New York State Agric. Expt. Station* *Jim Cranny, US Apple Association* *Larry Elworth, Gettysburg, PA* *Kathleen Merrigan, Henry A. Wallace Institute*
2:35	Discussion
3:05	Break
3:35	Pesticide Development And Future Trends *Ray McAllister, American Crop Protection Association* *Forrest Chumley, Dupont Agricultural Products* *Bob Peterson, Dow AgroSciences* *Leonard Gianessi, National Center for Food and Agriculture Policy*
4:35	Discussion and Wrap-Up
5:05	Adjourn

May 16, 1998
Arnold and Mabel Beckman Center
Irvine, California

8: 15	Introductions and Announcements *May Berenbaum, Chair* *Mary Jane Letaw, NRC staff*

8:30	Roundtable Discussion: Pest Management Decision-Making (12–15 minute presentations) *Joe Boddiford, Georgia Peanut Commission, Sylvania, Georgia* *George Ponder, Curtice Burns Foods, Montezuma, Georgia* *Sam Lang, Fairway Green, Raleigh, North Carolina* *Charles Mellinger, Glades Crop Care, Jupiter, Florida* *Tony Thompson, Willow Lake Farm, Windom, Minnesota* *Bob Quinn, Millenial Farm and Ranch, Big Sandy, Montana* *Ron Cisney, Olocco Ag Services, Santa Maria, California* *Mike McKenry, UC Kearney Agricultural Experiment Station* *Jenny Broome, UC Davis, SAREP and BIFS project* *Barry Brennan, USDA/CSREES, Pesticide Applicator Training* *Steve Pavich, Pavich Family Farms, Porterville, California*
10:00	Break
10:30	Continue Roundtable Discussion
11:30	Question and Answer period (includes public comments)
12:15	Morning Wrap-Up
12:30	Lunch
1:30	Reconvene Workshop Roundtable Pesticide Safety: Mitigating Human Health and Environmental Risks (15-Minute Presentations) *Bob Gilliom, US Geological Survey, Sacramento* *Rupa Das, California Department of Health Services, Berkeley* *Michael O'Malley, Employee Health Services, University of California, Davis* *Tobi Jones, California Department of Pesticide Regulation, Sacramento* *David Pimentel, Department of Entomology, Cornell University*
2:45	Questions and Answers
3:15	Wrap-Up
3:30	Adjourn Open Session
4:00	Committee on Future Role of Pesticides Reconvenes In Closed Session

July 18, 1998
Rosemont Suites
5500 N River Road
Rosemont, Illinois

8:25	Introductory Remarks *May Berenbaum, Chair*
8:30	Breakfast Discussion Exchange
8:30	*Mark Whalon, Pesticide Research Center, Michigan State University, East Lansing*
8:45	*Wolfram Koeller, New York State Agricultural Experiment Station, Cornell University, Geneva, New York*
9:00	*Tom Sparks, Discovery Research, Dow AgroSciences, Indianapolis, Indiana*
9:15	*Ann Sorensen, Center for Agriculture in the Environment, American Farmland Trust, Dekalb, Illinois*
9:30	*Mike Owen, Iowa State University, Ames, Iowa*
9:45	*Dean Zuleger , Hartland Farms, Hancock, Wisconsin*
10:00	Break
10:15	Continue Discussion
10:15	*John G. Huftalin, Grower, Rochelle, Illinois*
10:30	*Eldon Ortman, Purdue University, West Lafayette, Illinois*
10:45	*Wayne Sanderson, NIOSH, Cincinnati, Ohio*
11:00	Question and Answer
11:30	Public Testimony (3 minute segments)
12:00	Adjourn

APPENDIX

C

About the Authors

MAY BERENBAUM, *Chair,* is professor of entomology and department head at the University of Illinois. Berenbaum has made major contributions to the understanding of the role of chemistry in the interactions between plants and herbivorous insects. She has identified key plant defensive chemicals and determined their modes of action. Her investigations have encompassed both proximate physiological mechanisms and their evolutionary consequences for both plants and insects. Berenbaum received her PhD in ecology and evolutionary biology at Cornell University. She was elected to the National Academy of Sciences in 1994 where she is a member of OPUS (Office for Public Understanding of Science). She serves on the National Research Council's Board on Agriculture and Natural Resources, and previously served on the Board on Environmental Studies and Toxicology.

MARK BRUSSEAU is professor of subsurface hydrology and environmental chemistry at the University of Arizona. His research is focused on developing a fundamental understanding of the factors and processes influencing the transport and fate of chemicals in the subsurface. Brusseau's approach integrates theoretically and experimentally based investigations with the development and use of mechanistically accurate mathematical models. He received his PhD from the University of Florida.

JOSEPH DIPIETRO is dean and professor of veterinary parasitology at the University of Florida. DiPietro studies the chemotherapy, epidemiology, and control of internal parasites in livestock. DiPietro's research includes clinical evaluations to measure anthelmintic efficacy of pyrantel salts, benzimidazoles, and avermectins in livestock animals. He has extensive background in research and use of ivermectins in horses, including ecological implications. DiPietro received his DVM and MS in veterinary parasitology from the University of Illinois.

ROBERT M. GOODMAN is a member of the Department of Plant Pathology, the interdepartmental program in Plant Genetics and Plant Breeding, the Institute for Environmental Studies, the graduate program in cellular and molecular biology, and the Biotechnology Training Program at the University of Wisconsin. His laboratory works on the molecular regulation of plant defense genes and the role of plant genotype in associations with noninvasive, beneficial microorganisms. Goodman is well known for his groundbreaking research at the University of Illinois, where he was the first to describe the molecular biology of a group of single-stranded DNA plant viruses, now called geminiviruses. Formerly, he was executive vice president for research and development at Calgene, Inc. Goodman has served on the National Research Council's Board on Agriculture and Natural Resources and numerous study committees. He received his PhD from Cornell University.

FRED GOULD is professor of entomology at North Carolina State University. Gould has researched ecological genetics of pest adaptation to chemical, biological, and cultural control tactics. His major emphasis in recent years has been focused on developing methods for delaying pest adaptation to transgenic crops that produce insecticidal proteins derived from *Bacillus thuringiensis*. Gould was an author of the National Research Council report *Ecologically Based Pest Management* (1996) and participated in the workshop on "Pesticide Resistance: Strategies and Tactics for Management" (1986). He received his PhD in ecology and evolution at State University of New York at Stony Brook.

JEFFREY GUNSOLUS is professor of research and extension at the University of Minnesota. He performs and publishes research on weed crop interactions, including evaluations of decision-making processes for weed management in corn and soybeans. In his extension role, Gunsolus provides pest-management expertise to growers. His extension bulletins include topics on herbicide mode of action, herbicide-resistant weeds, and chemical and cultural weed control of field crops. In a previous position, Gunsolus was an extension associate at Iowa State University. He received his PhD at North Carolina State University.

BRUCE HAMMOCK is professor of entomology and environmental toxicology at the University of California, Davis. Hammock is a leader in insect-toxicology research and development of immunochemical assays for environmental monitoring. He has extensive research in detection of metabolites produced by exposure to a wide range of agrochemicals and recombinant biocontrol products (such as *Bacillus thuringiensis*; baculoviruses). His group has isolated novel peptides from scorpion venom toxic only to insects and used the genes for these peptides in the development of recombinant viral insecticides. His group has carried out studies on mammalian safety of biological insecticides. He received his PhD in entomology-toxicology from the University of California, Berkeley. Hammock participated in a Research Council workshop Pesticide Resistance: Strategies and Tactics for Management (1986).

ROLF HARTUNG is professor emeritus of environmental toxicology at the University of Michigan. Hartung brings a wealth of knowledge of wildlife toxicology and public-health related to environmental pollutants. He has researched coactions between chlorinated hydrocarbon pesticides and aquatic pollutants; environmental dynamics of heavy metals; risk assessment; effects of polluting oils on waterfowl; and toxicity of aminoethanols. He has co-authored several Research Council reports including *Building a Foundation for Sound Environmental Decisions* (1997); *Review of the Department of the Interior's National Irrigation Water Quality Program* (1996); *Irrigation-Induced Water Quality Problems* (1989); and *Testing for Effects of Chemicals on Ecosystems* (1981). Hartung received his PhD in wildlife management from the University of Michigan.

PAMELA MARRONE is president and CEO of AgraQuest, a firm she founded that has a portfolio of proprietary natural-product pesticide discoveries and products. Marrone has substantial management expertise in startup and multi-international biotechnology firms. Before this endeavor, Marrone was president of Entotech, a subsidiary of Novo Nordisk, and senior group leader of insect control at Monsanto Agricultural Company. Marrone received her PhD degree in entomology from the North Carolina State University.

BRUCE MAXWELL is professor of agroecology at Montana State University. He focuses on research to predict evolution and dynamics of herbicide resistance in weed populations. He also has researched weed thresholds; effects of weed dispersion and competition on crop yield; development of successional weed-management strategies on rangeland; weed-management expert systems; and agricultural sustainability through ecosystem management and planning. Maxwell received a PhD degree in crop and forest science from the Oregon State University.

KENNETH RAFFA is professor of entomology in the department of entomology at the University of Wisconsin at Madison. Raffa conducts research on insects affecting forest-resource management, plant-insect interactions, insect ecology, population dynamics, and biological control. He is also adjunct professor in the department of forest ecology and management. He has researched chemical defenses of trees against insects and fungi, tri-trophic interactions, roles of microorganisms in mediating plant-insect interactions, insect-pheromone ecology, and deployment strategies for transgenic plants. Raffa formerly worked as section research biologist at E.I. Dupont and Company, where he investigated insecticide-resistance management, insecticide synergists, and behavioral manipulation of insects using natural products. Raffa received his PhD in entomology at Washington State University.

JOHN RYALS is the Chief Executive Officer and President of ParadigmGentics, Inc. in Research Triangle Park, North Carolina, a company that specializes in genomics and functional genomics for products in the areas of crop production, foods for human nutrition, industrial applications, and therapeutic foods. Dr. Ryals is also an adjunct professor of Crop Science at North Carolina State University. Prior to founding Paradigm Genetics in 1997, Dr. Ryals was the head of agricultural biotechnology research for Ciba-Geigy Corporation and Novartis, Inc. His research focus has been on the molecular biology of plant-pathogen interactions, and in particular, in the field of systemic acquired resistance in plants.

JAMES SEIBER is professor of environmental sciences at the University of Nevada. He has a broad background in analytical chemistry of pesticides, industrial byproducts and naturally occurring toxicants; ecological chemistry of plant-derived poisons; modeling for chemical environmental fate; trace organic analysis; and origin and fate of trace organics in the atmosphere, including pesticides. In previous positions, Seiber worked for Dow Chemical Company and chaired the department of environmental toxicology at the University of California. He is co-author of several Research Council reports, *Science and Judgement in Risk Assessment* (1994), and *Pesticides in the Diets of Infants and Children*(1993). Dr. Seiber received his PhD in chemistry at Utah State University. He resigned from the committee in December 1998, after appointment as Director, Western Regional Research Center, USDA-ARS, Albany, California.

DALE SHANER is the director of Ag Biotech at American Cyanamid in Princeton, New Jersey. He directs the agricultural-biotechnology program, which is aimed at finding new pesticide target sites and identifying genes associated with output traits in major crops. Before to becoming director, Shaner was a senior research fellow for herbicide discovery, and directed basic research in plant biochemistry, plant physiology, and plant-chemical-environment interactions. He has published on the mode of action of imidazolinone herbicides, biochemistry of resistance development, and selection and utility of herbicide-resistant crops. He chairs the Herbicide Resistance Action Committee, an international, intercompany committee. Before to his position at American Cyanamide, Shaner was an assistant professor at the University of California, Riverside. He received his PhD in plant physiology at the University of Illinois.

DAVID ZILBERMAN is professor of agricultural and resource economics at the University of California, Berkeley. He also serves as director of the Center for Sustainable Resource Development. Zilberman has diverse research interests: economics of technological change, economics of natural resources, microeconomic theory, and agricultural and nutrition policy. Zilberman has published numerous articles on topics of water quality in irrigated agriculture, pollution prevention, financial incentives and pesticide use, biotechnology and precision agriculture. Zilberman researched market incentives to control environmental problems for his PhD degree in agricultural economics at the University of California, Berkeley. Zilberman participated in a Research Council workshop on precision agriculture affiliated with the Committee on Assessing Crop Yield.

Index

G

H

I

N

O

P

R

S

T